Applied Mathematics for Personal Finance

Aaron Stevens
Boston University

DRAFT EDITION JANUARY 2015

Preface

*Applied Mathematics for Personal Finance*provides a general introduction to the ways that mathematics can be applied to personal financial decision-making. This book is suitable for college students with no previous background in economics or finance; only familiarity with high school algebra is assumed.

This book demonstrates how you can utilize math skills you already know in application areas that may be unfamiliar; it also introduces some new math skills that you can apply to familiar problems. The book emphasizes the development and application of the economic life-cycle model as the framework for evaluating all of your personal financial decisions. Economists, including six Nobel Laureates, have spent close to a century developing the concept of *life-cycle consumption smoothing.* "Smoothing" refers to the need to spread your economic resources over your lifetime, taking into account that your future is highly uncertain.

I wrote this book to remedy the challenge that students face when learning about personal finance. To put it politely, most personal financial advice – whether from trade books, textbooks, television talk show pundits, or professional personal financial advisors – is not substantiated by the body of economic science nor by its underlying mathematics.

About This Draft Edition

This book began as a textbook for an economics course about personal lifecycle economics. After running for 4 years as EC171 at Boston University, the material had become increasingly quantitative in nature. In 2014, the course has moved to the mathematics department as an experimental course under the umbrella of MA120. The content is evolving; I have produced this draft edition to give students access to the material while I continue to work on the details of several modules and copy-edit the entirety.

This draft is far from perfect, and I need your help. Please bring typographical and grammatical errors to my attention so I can improve this text! Email Aaron Stevens azs@bu.edu, subject: Correction in AMPF.

How this Book is Organized

*Module 1: Introduction*provides an introduction to the terminology and metrics of basic personal finance, and the fundamental mathematical equations used throughout this book. The main idea of the economic Life-Cycle model is introduced, and two case studies are presented and used to compare and contrast the financial fitness of neighboring families. This module also begins to introduce spreadsheets, which are used throughout the text for developing applied financial models.

*Module 2: Modeling Systems of Equations and the Economic Life-Cycle Model*develops the life-cycle model as the way to maximize your economic happiness over time. A simplified life-cycle model is developed to illustrate the key concepts of income, saving, and consumption through time in a world without taxes or uncertainty. Module 2 develops most of the conceptual framework (and equations) we will return to throughout the book, including non-smooth income, borrowing, and borrowing constraints.

Module 3: Borrowing Constraints and Dynamic Programming explains the problem of smoothing consumption under borrowing constraints. When you have most of your wealth tied up in human capital (or many other conditions), you might be borrowing constrained and unable to smooth your consumption. Dynamic programming is a technique to solve for the smoothest financial plan without borrowing.

Module 4: Thinking in Present Value addresses the tradeoffs you face when you make an investment or borrow money to be repaid at interest. Exponential growth is presented as a way to explain the future value of an investment or the present value of an amount to be received in the future. Changes in the price level over time are discussed along with the effects of inflation on future values.

*Module 5: Paying on Time*discusses streams of cash flow payments to be received or paid in the future. Tools for valuing annuities and calculating loan payments are studied in detail. Investments in bonds and borrowing via loans are discussed in detail. Risk-free inflation-indexed investments are presented as a baseline against which other investments must be compared.

Module 6: Modeling the Economic Life-Cycle with Interest revisits the life-cycle model while taking into consideration interest rates and inflation, and how these factors affect your decisions about lifetime saving and consumption. In particular, the module discusses the effect of interest rates on your standard of living now and in the future.

*Module 7: The Odds Are…*introduces basic concepts of random events, discrete probability and expected values. We approach probability from the perspective of games of chance, and use this background to think about financial outcomes that depend on random events. In particular, this module explores insurance as a mechanism for risk transfer and the role of expected values in insurance pricing.

Module 8: For Whom the Bell Tolls introduces concepts from actuarial science. We explore mortality and survival rates, and explore the Gompertz law of mortality as an analytical method to model survival rates within a cohort. Actuarial present value is introduced as the expected value of a future payment that depends on survival to a certain (old) age. Case studies include single-premium immediate annuities, life insurance, and Social Security benefits.

Module 9: Introduction to Investing in Stocks discusses the tradeoffs between investment risk and investment returns. Specifically, this module explains investments in the stock market. Many bad ideas about investing in stocks have become common knowledge, and this module identifies them and why they are wrong.

*Module 10: Descriptive Statistics for Stock Market Investors*introduces several concepts from the field of descriptive statistics that are important to analyzing the risk and return characteristics of investing in stocks. The standard deviation is presented as a way to measuring the variability of stock returns and confidence intervals and presented as a way to set reasonable expectations about future stock returns based on historical performance. Correlation between data series is introduced, and the stage is set for thinking about stock portfolios.

Module 11: Diversification and Portfolio Construction presents the benefits of diversification in a non-technical way. Fundamentals of portfolio construction are discussed with the goal of achieving the best possible expected return for a given amount of risk.

*Module 12: The Truth About The Risk in the Stock Market*debunks the conventional wisdom about investing in stocks for the long run. The numerical technique of Monte Carlo simulation is used to estimate the chance that investing in stocks will make you worse off in retirement, compared to investing in risk-free assets. This module explains why your living standard is at risk due to a bad sequence of investment returns. Finally, a strategy is presented to help think about how you should allocate your retirement investments between safe and risky assets so that your investment choices do not put your standard of living at risk.

*Module 13: Uncle Sam And You*addresses your life-long relationship with Uncle Sam, i.e., the United States federal government. The module begins with a general description of taxation and the methods of assessing taxes as well as explaining the various taxes paid by households in the United States. Federal income taxes are presented with numerical examples. Social Security benefits are presented as an inflation-indexed lifetime income. The effects of taxation and benefits on life-cycle consumption smoothing are discussed, and a framework is developed for a consumption-smoothing model to include the time value of money and taxes.

Acknowledgments

Many friends, colleagues, students, and family members contributed their time to reading, editing, or just providing ideas for this text.

Among my colleagues I am fortunate to include Professors Zvi Bodie and Laurence Kotlikoff, who provided countless ideas and subject matter expertise as we developed the EC171 class at Boston University. In addition to ideas and advice, Larry has graciously allowed me to use screenshots and outputs from the ESPlanner software program distributed by his company, Economic Security Planning, Inc. Stan Sclaroff and Mark Crovella, my former and current department chairs in the Department of Computer Science at Boston University have been extremely supportive of my efforts in teaching EC171 and MA120 and writing this book.

My colleague and friend John Magee provided ideas about organizing this work and engaging my students in the writing process. My friend and mentor Bert White provided important feedback on writing style along with constant encouragement. My Colleague Robert Puelz, of Southern Methodist University, has read many drafts and provided excellent feedback from his use of this book in his classes at SMU. Paula Hogan, a fee-only Certified Financial Planner and former chair of the Wisconsin chapter of the National Association of Personal Financial Advisors (NAPFA), read the manuscript and provided many important ideas and comments.

To the extent that this book is any good, it is due to the efforts of my students who have read, reviewed, corrected, copy-edited and commented at many phases of its development. While I can honestly claim that the content of the book stands on the shoulders of giants, the efforts of my students have really helped shape this book.

From my Fall 2011 EC171 class, I wish to thank the following students who read early drafts of the modules on the life-cycle model and Dynamic Programming and provided valuable feedback: Sam Howe, Isabella Jones, James Leong, and JavierMartinez.

From my Spring 2012 EC171 class, I wish to thank the following students who provided detailed mechanical and substantive copy-editing help: Naaji Adzimah, Steven Barros, Douglas Bolton, Kelly Burke, James Cicalo,Benjamin Cohen, Hannah Farber, Victoria Tavares-Finison,Brianna Galvin, Jehan Hamedi, Mike Latona, David Lee, Clarissa Keen, Sean Muller, Joshua Tang, Shana Richards,Chen-Hsu Wang,Michael Wexler, Rob Winter, Tracey Workman, and Michelle Zaniboni. With their help, the manuscript underwent tremendous improvements in a short period of time, and I thank them for their detailed review.

From my Summer Session 2012 EC171 class, I wish to thank the following students who provided detailed mechanical and substantive copy-editing help: Thomas Chen, Erik Herold, Amrita Kotak, Nikhil Manocha, and Amanda Robertson.

From my Fall 2012 EC171 class, I wish to thank the following students who provided detailed mechanical and substantive copy-editing help: Sarah Alfaiz, Rosanna Hok, Ashley Johnson, Richal Kaul, Shruti Patel, Amber Robinson, Dhruv Sharma, and Chris von Keitz.

From my Spring 2013 EC171 class, I wish to thank the following students who provided detailed mechanical and substantive copy-editing help: Jennifer Bernier, Paige Coles, Natalie Counts, Michelle Denorscia, Jerry Ho, Jacquelyn Ozburn, Rui Yi Woo, and Ziqing Zeng. In particular, I wish to thank Matthew Bourhis, who helped me to substantially revise modules 14, 15, 16, and 17 on his own time during the summer of 2013.

From my Summer 2013 EC171 class, I wish to thank Faisal Alhajj, who provided detailed mechanical and substantive copy-editing help.

From my Fall 2014 MA120 class, I wish to thank the following students who provided detailed mechanical and substantive copy-editing help: Christianna Gilbert, Julia Finke, Richard Jagolta, Kei Yatsu, Arianna Johnson, Jun Jang, Max Abranowicz, Kaitlyn Tran, Madison Litwin, Kathryn Lezynski, Charles Fidler, and Kian Mcgee.

My father Michael Stevens read multiple drafts and provided helpful comments and constant encouragement.My wife Jennifer read and discussed more material than she probably wanted to, and has provided encouragement and support throughout this project. Even though she says she hates personal finance, but she does not give herself credit for how much she *does* know!

Table of Contents

1 Introduction

Learning Objectives:

- Provide the context for this subject within the fields of mathematics and personal finance.
- Identify and explain the terminology and metrics of personal finance.
- Introduce spreadsheets as an interactive modeling tool, and begin to use spreadsheets to calculate a financial statement called the balance sheet.

1.1 Introductionto Applied Mathematics for Personal Finance

You have a great wealth of mathematical knowledge, acquired through numerous previous courses. You've learned about counting and numbers, arithmetic and algebra, geometry and trigonometry, and perhaps even calculus or statistics. Most mathematics courses are organized around a set of related mathematical topics, and convey skills about those topics that apply to many different disciplines or applications. This is a different kind of course.

This is a book about applying mathematical tools to a broad set of issues most generally described as personal finance. The mathematical tools required to model personal finance include systems of equations, exponential growth and compounding, probability, expected value, descriptive statistics, and numeric simulation.

To set the stage for the applications we will encounter, the book will include a great deal of background material about the underlying economic science that provides the basis for personal financial modeling. Further, most of the applications will be developed using spreadsheets software (e.g., Microsoft Excel or Google Spreadsheets), and the text will introduce many terms, concepts, and tools that are required to create interactive calculators using spreadsheets.

This module provides a broad overview of the ideas of personal financial literacy. Many students have had previous exposure to these ideas, especially if they have taken a personal finance or accounting class in the past. For some students, this will be completely new. The goal of this chapter is to bring everyone up to the same level of knowledge.

1.2 Personal Financial Literacy

Financial literacy requires an understanding of the vocabulary used to describe financial things, as well as the tools of analysis to make optimal financial decisions. This section begins with the vocabulary of personal finance, and the next section provides a detailed example using these ideas. Module 2 continues the introduction to financial literacy by introducing the economic life-cycle model, which is used to make optimal spending and saving decisions.

Money, Income, Consumption, and Saving

"I want to have a lot of money" is a common phrase we hear,but what exactly is money, and what do people mean when they say they want it? Often when people say "money," they actually mean "income" or "wealth."

Money

Money is a unit of account, a means of exchange, and a store of value. A *unit of account* is used to measure a quantity. In the United States, we measure the prices of goods and services in dollars, so dollars are a unit of account. A *means of exchange* is an accepted way of performing transactions between the buyers and sellers of goods and services. In the United States, dollars are the accepted means of exchange for sales. A *store of value* is the quality of being able to hold value through time. If you get paid in dollars on Friday, and plan to spend those dollars on Monday, the dollars are acting as a store of value between the time you receive your income and the time you use it to consume.

In different countries, and even throughout the history of the United States, there have been many things that meet the definition of money. Gold used to be money. To understand this, keep in mind that gold can be used as a unit of account, e.g., 1-ounce gold coins. Gold also acts as a store of value because it can be bought now and sold later. However, gold is not a common means of exchange, i.e., Campus Convenience does not accept payment in gold coins. Ice cream might be a means of exchange and a unit of account, but it hardly acts as a store of value because it melts too easily.

In the United States, our dollars (paper currency) and cents (coins) are money because they fulfill the definition of a unit of account, a means of exchange, and a store of value. Dollars are a form of *fiat currency*, which means that dollars are a promise of repayment by the government. Fiat currency has no intrinsic value because it cannot be consumed (for example, the way ice cream can be eaten). The dollars we use have value simply because the government that created them asserts that they do, and the citizenry have agreed to use dollars as money.

Bank deposits in checking, savings, and certificate of deposits accounts are also money. Bank deposits meet the definition of money because they are measured in dollars, can be used for exchanges, and act as stores of value. *Checks and debit cards* are mechanisms that instruct a bank to transfer money from one account to another, but in and of themselves are not money. *Credit cards* are not money, even though they are often accepted as means of exchange, because they do not provide a store of value or a unit of account. In fact, the balances owed on credit cards represent claims against the cardholder, and require the cardholder to pay out money in the future. Credit card balances are best thought of as *negative money*, since the amount owed on credit cards must be paid out of other money holdings.

An *emergency fund* or *reserve fund* is a store of wealth to be used only in case of a financial emergency such as a short-term job loss or medical emergency. These funds are not to be used for ordinary spending, so that they are available in the case of an actual financial emergency.Emergency funds are usually held in savings accounts, certificates of deposit at banks, or in U.S. Government savings bonds. These instruments of guaranteed value are highly liquid and easily converted to cash. If an emergency fund was subject to risk (for example, invested in the stock market), it might lose value and not provide the security it was intended to provide.

Income

When people say, "I want a job that pays more money," they probably mean that they want more income from their employment. *Income* is the payment one receives in exchange for producing goods or services. It is a flow (a quantity in movement, like water flowing in a river), so measuring it requires describing the quantity that flows in some period of time. For example, your income might be a wage of $10 per hour, or a salary of $40,000 per year.

Consumption or Spending

Consumption is the amount of spending (i.e., using financial resources) that you do in a given period of time. For example, if you buy food or pay the rent, buy clothing or go on vacation, you are consuming resources.
We can divide overall consumption into discretionary and non-discretionary spending. *Non-discretionary spending* is for contractually obligated expenses such as taxes, housing, and loan payments that must be paid "off-the-top," before you can spend freely on other wants and needs. *Discretionary spending* refers to consumption including all expenditures on food, clothing, transportation, entertainment, travel, etc.

Saving

Saving is the difference between income and consumption. Put another way: when you have income, you can choose to allocate it to consumption or saving. For example, if you have an income of $44,000 per year, and consumption of $35,000 per year, this would leave $44,000 – $35,000 = $9,000 of saving.

A *budget constraint* describes how your income must be allocated over time. We can express the relationship between income, consumption and savings using this budget constraint:

$$Income = Consumption + Savings$$

Saving can be positive or negative. Negative saving, or *dissaving*, is when you consume more than your income in a given period of time (e.g., per month or year). For example, if you have an income of $10,000, but spend $12,000, then you need to dissave $2,000.

It is important to note that dissaving is not always possible. For dissaving to occur, you must either have assets from which to dissave, or else you must borrow from some other person or economic institution. In many cases, borrowing is not possible, a subject to which we will return later. In cases where you have no assets and borrowing is not possible, the amount of savings is $0 and the budget constraint simplifies to:

$$Income = Consumption$$

But I'm a College Student!

College students are in a special situation of quasi-financial adulthood. For some students, many large expenses such as college tuition, housing, and health insurance might still be paid by their parents. Other students are financing these costs via borrowing, i.e., consuming more than their income.

Whether you are working part-time while in college or during the summers, or you receive an allowance from your parents, you have some income. Your income is probably small and does not cover all of your expenses. At some point in your future when you are working full-time, you will have a much higher income.

How does the budget constraint apply to you now, when you are a student with limited income? The same way it does for anyone else. Your level of consumption is must equal income minus savings. If you are spending more than your income, then you must have negative savings (i.e., borrowing) to make the budget constraint equal.

Borrowing when young to pay for education or consumption is consistent with the life-cycle model. However, beware that when you borrow to consume now, you repay by forgoing consumption later. Moreover, when you borrow you will incur borrowing costs, which means that the amount of repayment will exceed what you borrow – perhaps by a lot. We will return to discuss borrowing and repayment in future modules.

How much saving is required? The amount is highly individualized, determined by an optimization process called the economic life-cycle model. *The Life-Cycle Model* is the subject matter of Module 2.

Assets, Liabilities and Net Worth

When people say, "I wish I had a lot of money," they are probably describing a desire for wealth. Wealth is an accumulation of value, the same way that the volume of water in a swimming pool is an accumulation.

Assets are what you own that have value because they can generate income or be sold. Many assets increase in value, but money does not – it only stores value and facilitates exchange. Wealth held in investments is put to productive use, so it is expected to increase in value. However, money is required to make exchanges, and not all investments (or all assets) can be easily converted to money. *Liquid assets* are easily converted to money. For example, bonds or stocks that can immediately be sold for cash are considered liquid assets. *Illiquid assets* are more difficult to convert to money. For example, while houses or businesses can be sold, it may take several months or longer to find a buyer, negotiate a price, and receive payment. Houses or businesses are illiquid assets.

*Liabilities*or *debts* are what you owe; the terms liabilities and debts are interchangeable. Whereas assets generate future income or can be sold, liabilities represent claims that you will need to pay at some future time, and might also cost you future income (in the form of interest expenses). Examples of liabilities include credit cards, student loans, car loans, and home mortgages. Whereas assets can be sold or can generate income, you can think of liabilities as *negative assets*, since they represent a claim against future income.

Another name for a household's wealth is the *net worth*. Net worth is the accounting measure that is the difference between assets and liabilities. Net worth is calculated as:

$$Net\ worth = Assets - Liabilities$$

Or simply:

$$NW = A - L$$

Assets, liabilities, and net worth figure into a financial statement called the balance sheet. Balance sheets are discussed in more detail in section 1.4 below.

1.3 Introduction to Spreadsheets

A *spreadsheet* is a type of document that stores numerical values and performs calculations on them using customizable formulas that describe the calculations to perform. A spreadsheet automatically recalculates values based on changes to the inputs. Spreadsheets will play an important role in our quantitative analysis throughout this book.

A *spreadsheet program* is a computer application for creating interactive models using numbers and equations. The most popular spreadsheet program on the market is Microsoft Excel (for both Windows and Mac). Apple has a spreadsheet program called Numbers, which comes installed with new Mac computers. Alternatives such as OpenOffice[1] and Google Spreadsheets[2] are available for free.

All of these spreadsheet programs have the same general set of features, including inputting text and numbers, building formulas (i.e., calculations that use the data that you've entered), built-in functions to perform mathematical, statistical, and financial calculations, and the ability to create graphs and charts. There are some minor differences in the user interface between these programs, but you can probably pick up on those differences on your own. When in doubt, you can always search the World Wide Web for a specific feature.

Throughout this book, new spreadsheet features will be introduced in boxes like this one.

Using Spreadsheets: Cells, Rows, and Columns

A spreadsheet is organized into horizontal *rows*, which are labeled by numbers (e.g., row 3), and vertical *columns*, which are labeled by letters (e.g., column B). A *cell* is the intersection of a row and a column, and is labeled by the combined row-column index (e.g., cell C3). Cells are the building blocks of all spreadsheets. All data is entered into a cell, and any formula or calculation is computed in a cell.

Using Spreadsheets: Data Entry and Formatting

To enter data into a cell, you must first select it. You can select a cell by single clicking on it with the mouse, or by using the arrow keys to navigate to that cell. Once a cell is selected (as is cell B3 in Figure 1.1), you may type text, numbers, or a formula into that cell.

Figure 1.1 Spreadsheet Example

	A	B	C
1			
2		**ASSETS**	
3		Cash	$ 180
4		Checking Account	$ 2,680
5		Certificates of Deposit	$ 18,790
6		Investment Accounts	$ 218,762
7		Car	$ 19,200
8		House	$ 412,800
9			
10		**TOTAL ASSETS**	**$ 672,412**

[1] http://www.openoffice.org/
[2] https://developers.google.com/chart/interactive/docs/spreadsheets

Using Spreadsheets How-to: Arithmetic Operations

The magic of a spreadsheet is that cells can contain formulas that automatically recalculate whenever the data in the sheet changes. To create a formula, we must always begin typing in a cell with the equals sign ("="). We can then type an equation using literal numbers, arithmetic operators, and references to other cells.

The *arithmetic operators* in excel are: + (addition), - (subtraction), * (multiplication), and / (division).

A *reference* is a way to refer to the value contained in another cell. We create a reference by typing the cell's label, for example "B5" (without quotation marks).

Suppose we want to compute the sum of cells B3 and B4, and show the result in cell B5. To do so, we would first select cell B5, and then enter the formula "=B3+B4" (without quotation marks). When we hit enter, the formula is entered into the cell and the result is calculated immediately. Further, if we change the data in either of the input cells (i.e., B3 or B4), the formula will recalculate the value in cell B5 immediately.

Using Spreadsheets: The SUM Function

Oftentimes, we need to add values from many cells together. In Excel, the built-in *SUM* function calculates the sum of a series of many cells. For example, the formula "=SUM(B3:B8)" (without quotation marks) will calculate the sum of all values in column B, from row 3 to row 8.

Online Tutorials

There are many great resources on the World Wide Web for learning spreadsheets. This website: http://www.gcflearnfree.org/excel2010 has some excellent Excel Tutorials.

YouTube Example Videos

As an ongoing effort, I am developing some tutorial videos to demonstrate some common spreadsheet features. This is not an all-inclusive set, but you can see a playlist containing my example videos here:http://tinyurl.com/nd4rxap.

1.4 Modeling Financial Health using Spreadsheets

Now that we've introduced the terminology of financial literacy, let's step back and look at two specific examples of fictitious households. We will assume values for assets and liabilities, and construct a balance sheet to summarize the financial health of each household.

Case Study: Dr. West and Mr. East

To illustrate the concepts about financial health, we will use a common case study throughout this module. Consider two fictional neighbors who live in very similar houses across the street from each other in a Boston suburb. Both are age 40, married, have 2 kids (ages 5 and 3), and each is the sole income earner in his or her household (i.e., their spouses do not work).

Dr. West graduated from college at age 22, medical school at age 26, and completed her residency at age 29. After her residency, she worked as a junior doctor in an established private practice, and later in her own practice as a family physician. She is now the medical director of her practice, with a salary of $160,000. Dr. West and her family just bought a house and moved into the neighborhood this year. Dr. West drives a brand-new Mercedes sedan.
Mr. East is a small business owner who has a small-engine repair business. When he graduated from high school at age 18, he took a summer job mowing lawns for an established landscape maintenance company. As part of his job, he learned how to repair and maintain all kinds of power equipment including mowers, blowers, chainsaws, etc. Some landscape customers started to ask him to repair their equipment. He recognized the demand and began a side business. He now owns a small engine repair shop with 2 employees, and takes a salary of $85,000 from his business. Mr. East and his family have lived on the street for 15 years. Mr. East drives a 3-year old Chevrolet pickup truck.

By knowing only their occupations, household composition, salary, and what kind of car they drive, many people will assume that Dr. West must be in better financial fitness than Mr. East. However, do we really have enough information to make a judgment about the financial fitness of Dr. West and Mr. East? One cannot judge financial fitness from outside appearances.

The Balance Sheet

A *balance sheet* is a statement of a household's wealth or net worth at one point in time. Assets and liabilities are listed separately so each can be totaled. *Liquidity* describes how easily assets can be converted to cash. Assets are listed with the most liquid assets (e.g., cash, bank deposits) on top, and the least liquid assets (e.g., investments in marketable securities, houses) at the bottom. Liabilities are listed with short-term liabilities, those due more imminently, listed at the top (e.g., credit cards and car loans), and long-term liabilities (e.g., student loans, mortgage loans) at the bottom. Figure 1.2 and Figure 1.3 show examples of balance sheets for Dr. West and Mr. East, respectively.[3]

[3] In Figure 1.2 and Figure 1.3, assets are listed as positive numbers because they "add" to net worth. Liabilities are listed as negative numbers because they "subtract" from net worth. The charts in Figure 1.2 and Figure 1.3 show absolute values for easy comparison.

Figure 1.2 Balance Sheet for Dr. West

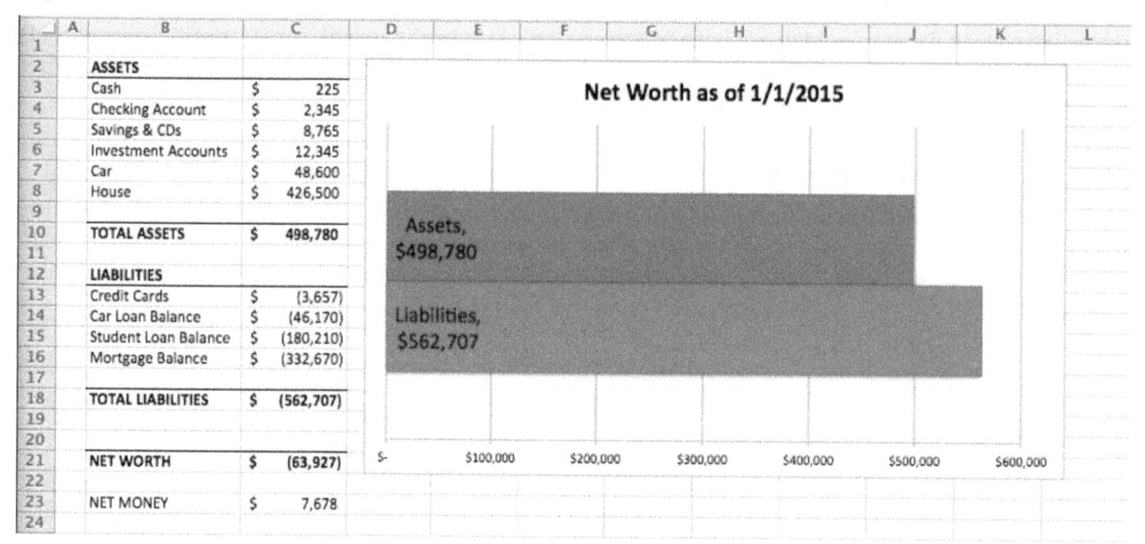

ASSETS		
Cash	$	225
Checking Account	$	2,345
Savings & CDs	$	8,765
Investment Accounts	$	12,345
Car	$	48,600
House	$	426,500
TOTAL ASSETS	**$**	**498,780**
LIABILITIES		
Credit Cards	$	(3,657)
Car Loan Balance	$	(46,170)
Student Loan Balance	$	(180,210)
Mortgage Balance	$	(332,670)
TOTAL LIABILITIES	**$**	**(562,707)**
NET WORTH	**$**	**(63,927)**
NET MONEY	$	7,678

Figure 1.3 Balance Sheet for Mr. East

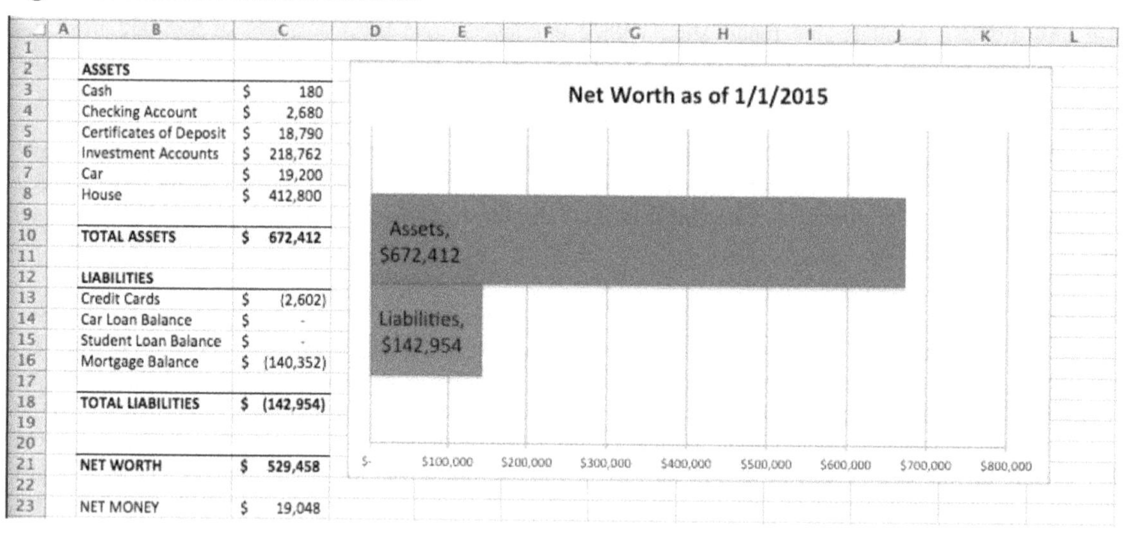

ASSETS		
Cash	$	180
Checking Account	$	2,680
Certificates of Deposit	$	18,790
Investment Accounts	$	218,762
Car	$	19,200
House	$	412,800
TOTAL ASSETS	**$**	**672,412**
LIABILITIES		
Credit Cards	$	(2,602)
Car Loan Balance	$	-
Student Loan Balance	$	-
Mortgage Balance	$	(140,352)
TOTAL LIABILITIES	**$**	**(142,954)**
NET WORTH	**$**	**529,458**
NET MONEY	$	19,048

We can discuss the money holdings of Dr. West and Mr. East. Dr. West has$225 of cash in her purse. She has $2,345 in a checking account, and $8,765 between her savings account and a certificate of deposit at her local bank that serves as her emergency funds. The Wests owe $3,657 on a credit card, which they pay in full at the end of each month so as not to accrue any interest charges. The Wests' net money holdings are:

$$\$225 + \$2,345 + \$8,765 - \$3,657 = \$7,678.$$

Mr. East has $180 of cash in his wallet, a checking account balance of $2,680, and $18,790 in emergency funds held in a certificate of deposit at his bank. The Easts owe $2,602 on a charge card, which they pay off in full at the end of each month. The Easts' net money holdings are:

$$\$180 + \$2,680 + \$18,790 - 2,602 = \$19,048.$$

Consequently, Mr. East has more money than Dr. West.

Net Worth

Dr. West's net worth is presented in Figure 1.2 in cell C21. Recall that $NW = A - L$. In this example, Dr. West's net worth is:

$$\$498{,}780 - \$562{,}707 = -\ \$63{,}927.$$

Therefore, Dr. West has a negative net worth.

Mr. East's net worth is presented in Figure 1.3 in cell C21. In this example, Mr. East's net worth is:

$$\$672{,}412 - \$142{,}954 = \$529{,}458.$$

Mr. East has a substantial positive net worth, almost $600,000 higher than Dr. West's net worth.

Valuing Physical Assets

The typical balance sheet follows generally accepted accounting principles for determining the value of assets and liabilities. In the case of financial assets it is easy to determine their values: simply look at your bank or brokerage statement for the amount labeled "total balance" or "total account value." Marketable securities' (e.g., stocks and bonds) prices are quoted throughout the business day, and mutual fund prices are available at the end of each trading day. Similarly, for financial liabilities, one can examine his or her credit card, car loan, or student loan statement and look for the amount labeled "outstanding balance."
Physical assets present an additional challenge. Is your 2002 Honda Civic worth the $16,452 (including taxes and registration fees) that you paid for it 10 years ago? If you tried to sell it today, it wouldn't fetch nearly that price. Businesses usuallyadjust the value of an asset on their balance sheetto reflect the "using up" of the asset over time. *Depreciation* is the expense which describes the "using up" of an asset over time. The *book value* is an accounting term for the price paid, less any accumulated depreciation.

Buyers and sellers of used vehicles are familiar with the Kelley Blue Book, which is a report of recent sales prices of used cars and trucks.[4] Don't be deceived by the unassuming name "Kelley Blue Book." The recent sales prices reported inthe KBB describe the market of buyers and sellers of used cards. *Market value* reflects the amount that a buyer could expect to pay (or a seller could expect to receive) for buying (selling)an asset in the market.

Economic Net Worth

Typically, the balance sheet shows the values of tangible assets and liabilities, such as property, money, investments, and loans. We saw that with physical assets there is some degree of interpretation involved in determining the appropriate measure of value. Based on their conventional net worth, it appears that Dr. West is relatively poor, with a negative net worth, and that Mr. East is relatively wealthy, with a net worth of over half a million dollars. To many, this conclusion feels counter-intuitive, because of the substantial difference between their lifetime incomes. Conventional accounting practices tend to ignore the value of Dr. West's future income. The conventional net worth does not completely reflect the neighbors' true financial situations.

To keep the accountants happy, we will use the term *conventional net worth* to describe the net value of *conventional assets* (financial and physical) minus *conventional liabilities*. However, this does not account for intangible assets and liabilities. In this section, we will examine how the *economic net worth* is the total net value of one's resources, including both tangible and intangible assets and liabilities.

Intangible Assets

Recall that an asset is something you own that produces an income or can be sold in exchange for money.The *economic value* of an asset is the value of its expected income or sale price.*Intangible assets*are assets because they produce an income. However, unlike conventional assets, intangible assets cannot be readily traded in the marketplace. The values of intangible assets are not included on the conventional balance sheet, but since they have economic value we must include them in evaluating a household's complete financial condition.

[4] www.kbb.com.

Human Capital

We know that Dr. West's future earnings will outpace Mr. East's earnings by a substantial amount each year due to her higher salary ($160,000 per year versus $85,000 per year). To account for the lifetime earnings, we will estimate each neighbor's human capital. *Human capital* is the estimated value of one's future labor income. Human capital is an intangible asset that you can convert to income by working.

For example, consider an estimate of Dr. West's human capital. At age 40, Dr. West expects that she will continue working for 25 more years, and expects to earn $160,000 in "today's dollars" per year for the rest of her career.[5] As a first approximation, we can calculate Dr. West's human capital as:

$$\text{HC} = \text{Expected Salary per year} \times \text{Number of Working Years}$$
$$\text{HC} = \$160{,}000/\text{year} \times (65 - 40)\text{ years} = \$4{,}000{,}000$$

Similarly, Mr. East's human capital is approximately $\$85{,}000/year \times 25\, years = \$2{,}125{,}000.$

For most young people, human capital represents their single largest asset, and from an economic decision-making perspective, it would be unwise to exclude it from any financial plan.
In general, human capital is a declining asset because as you age, you "use up" your human capital and have fewer years of work remaining. When young, you have many years of work ahead of you. Recall that the calculation for human capital is salary multiplied by years of work – the more years of work ahead, the higher the value.

In the above example, Dr. West, age 40, has 25 years of work salary income remaining, so her human capital is worth $\$160{,}000/year \times 25\, years = \$4{,}000{,}000$. At age 50, she will have 15 years of work remaining, so her human capital would be worth $\$160{,}000/year \times 15\, years = \$2{,}400{,}000$. Finally, when she retires at age 65, Dr. West does not expect any future income from her human capital, i.e., human capital expires at retirement. Figure 1.4 depicts the evolution of Dr. West's human capital as she ages.

Figure 1.4 Human Capital Declines Over Time

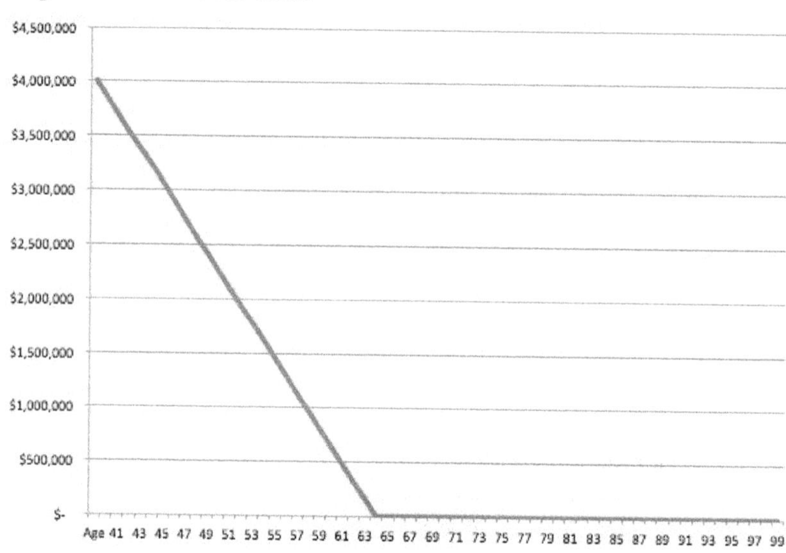

It's important to note that the inevitable using up of one's human capital is the incentive to accumulate financial and physical assets during one's working years. By systematically converting human capital to financial capital, we can provide for our continued consumption once we no longer receive a salary income. This will be discussed further in Module 3, where the life-cycle model is explained in more detail.

[5] While her salary will likely increase over time, most of that increase will be due to inflation, or the general rise of prices. We say "today's dollars" to refer to the purchasing power of her current salary, which we expect will be approximately stable over time.

Other Intangible Assets

In addition to human capital, other intangible assets include pensions and annuities. *Pensions* are ongoing incomes paid to retirees by their former employers. Most pensions provide an income for life, but do not have a fixed asset value and are not readily bought and sold they physical or financial assets could be sold.

Lifetime annuities(sometimes simply called *annuities*) are financial products sold by insurance companies, which provide an income for life. Annuities provide protection against outliving one's assets. Annuities will be discussed further in 1. *Social Security benefits*, which most workers receive in retirement, are a form of annuity paid by the United States government.

While pensions, annuities, and Social Security benefits are generally not included in calculating a household's net worth, these intangible assets are extremely important in the life-cycle model and determining a household's sustainable standard of living.

Figure 1.5 and Figure 1.6 show revised balance sheets for both neighbors, economic net worth including human capital.

Figure 1.5Dr. West's Economic Net Worth (Including Human Capital)

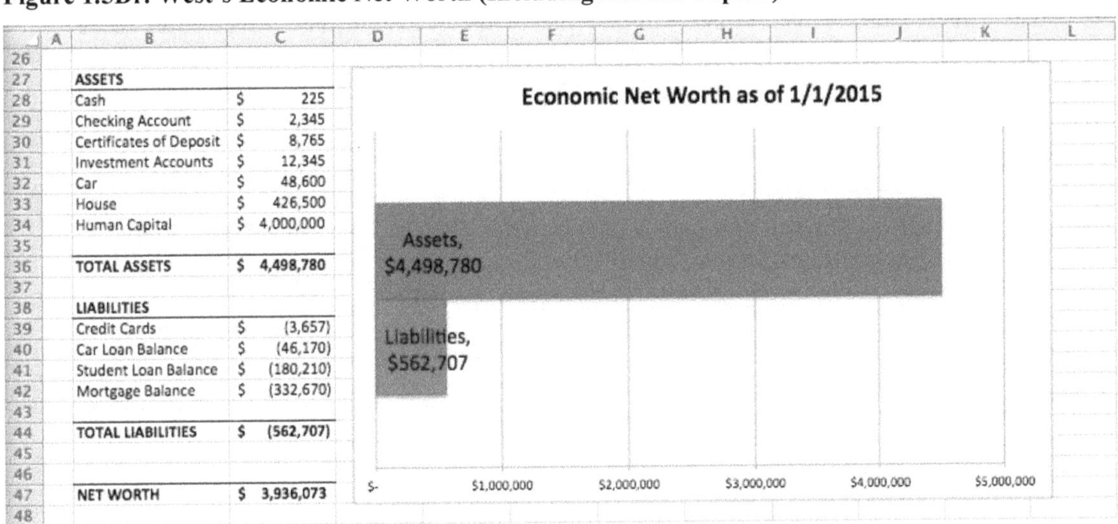

ASSETS	
Cash	$ 225
Checking Account	$ 2,345
Certificates of Deposit	$ 8,765
Investment Accounts	$ 12,345
Car	$ 48,600
House	$ 426,500
Human Capital	$ 4,000,000
TOTAL ASSETS	**$ 4,498,780**
LIABILITIES	
Credit Cards	$ (3,657)
Car Loan Balance	$ (46,170)
Student Loan Balance	$ (180,210)
Mortgage Balance	$ (332,670)
TOTAL LIABILITIES	**$ (562,707)**
NET WORTH	**$ 3,936,073**

Figure 1.6 Mr. East's Economic Net Worth (Including Human Capital)

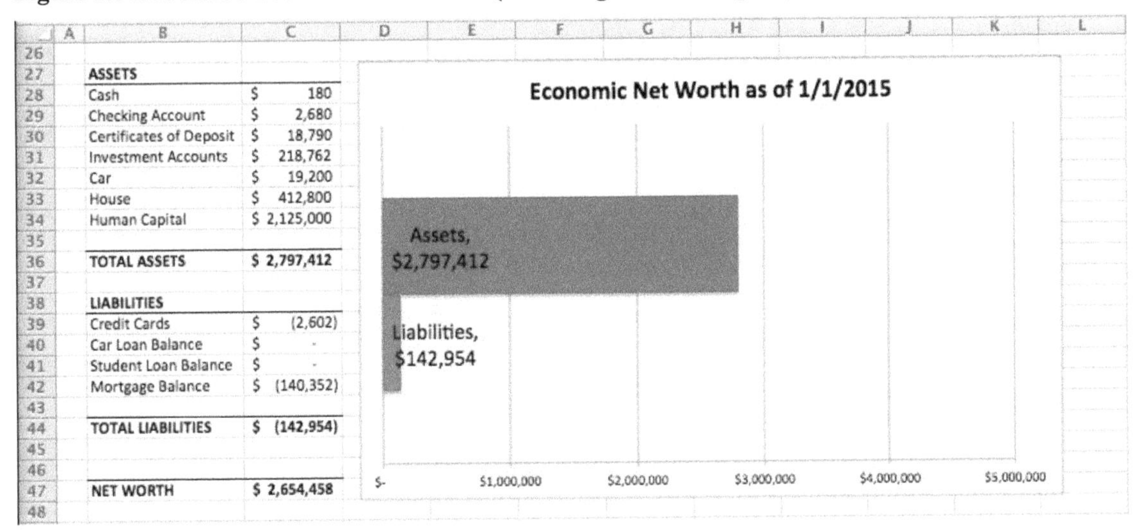

ASSETS	
Cash	$ 180
Checking Account	$ 2,680
Certificates of Deposit	$ 18,790
Investment Accounts	$ 218,762
Car	$ 19,200
House	$ 412,800
Human Capital	$ 2,125,000
TOTAL ASSETS	**$ 2,797,412**
LIABILITIES	
Credit Cards	$ (2,602)
Car Loan Balance	$ -
Student Loan Balance	$ -
Mortgage Balance	$ (140,352)
TOTAL LIABILITIES	**$ (142,954)**
NET WORTH	**$ 2,654,458**

After reviewing Figure 1.5 and Figure 1.6, we see that the conventional net worth presented in Figure 1.2 and Figure 1.3 really did not represent the actual financial fitness of each household. One way to think about this is that the conventional net worth describes the running total of what has occurred, financially, so far in the history of the household. Dr. West has accumulated $498,780 of conventional assets and $562,707 of liabilities to-date, which explains her conventional net worth of –$63,927. By including human capital, the economic net worth describes both the past and the future financial condition of the households. Dr. West's economic net worth is approximately 3.9 million dollars. This is the amount of total resources available for the rest of her life, from which she can plan her future consumption.

We will return to using the economic net worth in module 2, when we introduce the economic life-cycle (LC) model.

1.5 Summary

The applied mathematics of personal finance includes a diverse set of mathematical tools with a shared purpose: to help model a household's financial situation. Unlike other math books, this book is a mix of personal finance topics, mathematical tools, and applied modeling using spreadsheets.

Financial literacy is about understanding the vocabulary of financial matters, and about making optimal financial decisions. In this module we have introduced most of the terminology used to measure the financial health of an individual or household. In the case study of Dr. West and Mr. East, we have seen that outer appearances may be deceiving, and that one cannot judge a household's economic condition by income or assets alone.

1.6 Review Questions

1. Define personal finance and financial literacy.
2. Define the principal indicators of financial health (assets, debt, net worth, income, human capital).
3. Briefly explain each of these concepts: wealth, money, income, consumption, and savings.
4. Define and explain each of these concepts: physical assets, financial assets, and intangible assets.
5. Explain the difference between checks/debit cars, and credit cards, and why none of these are classified as money.
6. Explain the concept of human capital:
    a. How might we estimate the value of one's human capital?
    b. How does human capital figure into economic net worth?
    c. How does human capital evolve over time?
7. What is a spreadsheet, and what does it do?
8. Explain these elements of a spreadsheet:
    a. Rows, columns, and cells
    b. Formulas
    c. References

2 Modeling Systems of Equations and the Economic Life-Cycle Model

Learning Objectives

- Discuss equations and systems of equations in the context of building a numerical model.
- Introduce the life-cycle model as a general framework with which to understand household economic decision-making.
- Introduce the budget constraint equation to explain the relationship between income, consumption, and savings.
- Develop a simplified two-period life-cycle model with which to introduce the notation to account for saving and borrowing in the economic life-cycle.
- Build a spreadsheet model consisting of a series of interrelated equations.
- Define borrowing constraints as a reason you might not be able to maximize your economic happiness.

2.1 Systems of Equations and the Life-Cycle Model

Suppose you are 22 years old, have just graduated from college, and started your first job. Your salary of $44,000 per year is more income than you have ever had before. How should you allocate your income between spending and saving to achieve the maximum amount of happiness throughout your life?

How should you evaluate the financial decisions you will face, about what kind of work to do, where to live, and how to invest? Economic science has a framework with which to quantify your decisions.

A *financial plan* describes how much an individual (or household) should spend and save each year. The *economic life-cycle (LC) model* provides a conceptual framework for determining how much of your income and assets you should spend and save each year, given your total lifetime resources and your needs for consumption. The LC model is a *system of equations*, wherein the *output* of one equation becomes the *input* to the next equation. In a system of equations, all of the expressions must be *recalculated together* – this is where the mathematics and spreadsheets come into play. This module provides a deeper introduction to the concepts of the LC model, to which we will return many times throughout this book, as well as how to implement a model involving a system of equations as a spreadsheet.

2.2 History and Overview of the LC Model

Over decades of work, financial economists have developed an integrated framework to evaluate all personal economic decisions that incorporates spending, saving, borrowing, and insuring. These decisions are based on a holistic understanding of household composition, lifespan, total resources, and fixed obligations. This model is based on the *life-cycle hypothesis*[6] proposed by Nobel laureate Franco Modigliani and his student Richard Brumberg in 1954. Modigliani and Brumberg explained that the reason people saved was as a precaution: to be able to continue consuming at the same level during times of limited income (i.e., in the event of job loss or during old age).

The central concern of life-cycle economics is to understand how people should best allocate their resources over their lifespan. The motivating question is *why do people save*? We can explain saving as a function of three main motivations:

- *Saving for retirement.* Retirement savings are used to build an accumulation of financial assets from which to sustain your standard of living when you are old and can no longer earn income from work.

- *Saving for emergencies.* An emergency fund provides ready access to money in the event of an unforeseen disruption in income (i.e., unemployment) or financial disaster (e.g., major accident or theft).

---

[6] Modigliani, Franco, and Richard H. Brumberg, 1954, "Utility analysis and the consumption function: an interpretation of cross-section data."

- *Saving for major purchases.* A major purchase like a house or car will require you to prepare by accumulating at least part of the purchase price as a down payment.

The life-cycle hypothesis explains the reason that households accumulate assets (by saving) during the income-producing years and spend down assets (by dissaving) during the retirement years (see Figure 2.1).

Figure 2.1 Income, Consumption, and Savings over the Economic Life-Cycle

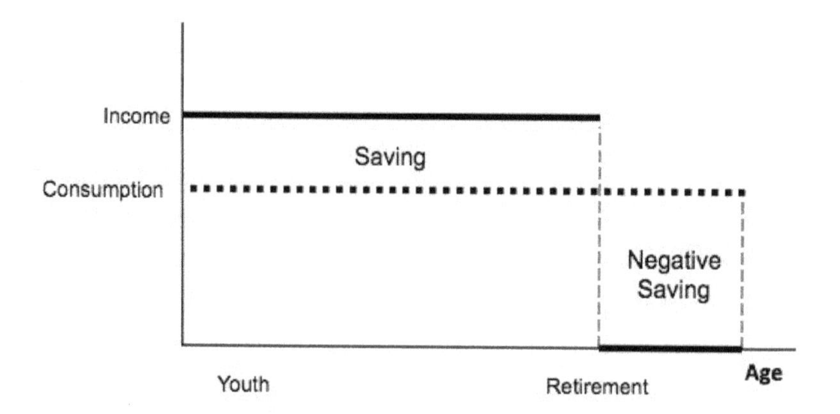

The life-cycle hypothesis describes the income, saving and consumption behavior of households. From young adulthood to retirement, a typical household receives a fairly constant annual income from working, and during retirement no income is received from working. While the labor income is received only during the working years, the household must consume through its entire lifetime (i.e., until death). Thus, households accumulate assets during the working years to provide resources for consumption during the non-working years.

The implication of this pattern of income, saving, and consumption is that household wealth is accumulated over the income producing years, reaches its maximum level at retirement, and is consumed during the retirement years (Figure 2.2).

Figure 2.2 The Level of Household Wealth over the Economic Life-Cycle

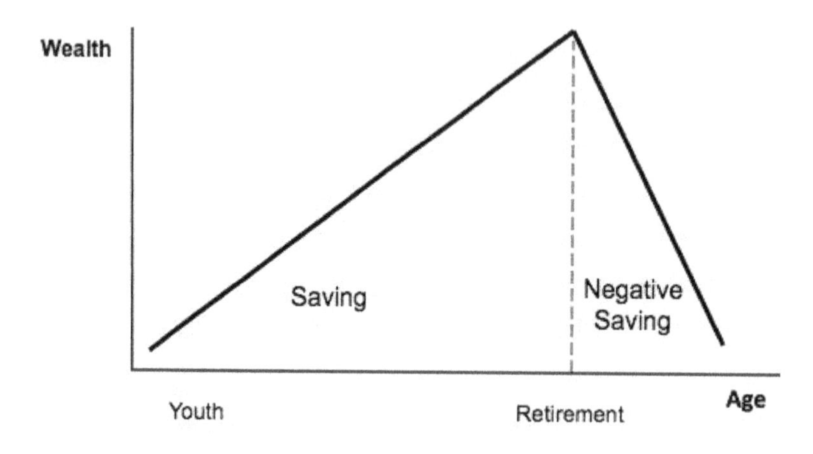

The life-cycle hypothesis provides an explanation of household saving and consumption behavior. Over decades, economists have developed the life-cycle model, which is an integrated framework for personal financial decision-making based on the life-cycle hypothesis.

Great Minds in Applied Mathematics/Personal Finance: Franco Modigliani

Franco Modigliani (1918-2003) was a Professor of Economics at MIT from 1962 until his death.

In the early 1950s, the conventional explanation of why households save was based on their marginal propensity to consume. The *marginal propensity to consume* (MPC) describes how a household would consume given an additional dollar of income. Under the MPC, households with lower income were said to consume a higher fraction of their income and households with higher income were said to consume a lower fraction of their income. In turn, this would mean that households with low incomes would save almost nothing and households with high incomes would save a great deal. According to this conventional wisdom, the goal of saving is to leave a legacy to future generations.

Modigliani rejected this conventional wisdom, arguing that the motivation for saving did not depend on the level of household income. Instead, he explained saving as a precautionary mechanism which enabled households to continue smooth consumption during times of lower income, similar to how firms build-up inventory to prevent shocks to sales during times of limited production. Modigliani introduced the *life-cycle hypothesis* to explain the level of wealth accumulated during a household's working years in preparation for consuming in excess of income during its retirement years. Modigliani described an economic life-cycle in which young people with low incomes would borrow to increase their consumption; middle-aged people with high income would save (repay borrowing and build up assets); old people with low income would dissave. In essence, the life-cycle model provides a theoretical basis for consumption smoothing – spreading your economic resources over your entire lifetime.[7]

Modigliani also made substantial contributions to corporate finance theory (along with fellow Nobel Laureate Merton Miller) about the capital structure of firms, arguing that the value of a firm is independent of its debt-to-equity ratio. The Modigliani-Milller model has become a cornerstone of modern corporate finance theory.

According to fellow Nobel Laureate Robert Merton, at his Nobel acceptance speech, Modigliani explained the life-cycle theory by saying 'when people are young, they save money, and when they are old, they consume their savings.' Franco Modigliani was awarded the Nobel Memorial Prize in Economics in 1985.

Why Should We Use the Life-Cycle Model?

The LC model explains households' propensity to save when young and consume their savings when old. Fundamental questions we can answer with the life-cycle model include:

- What is the level of consumption (i.e., spending) that a household can sustain until its maximum age of life?
- How many assets must be accumulated during the working years?
- How much should workers save each year, to accumulate enough assets prior to retirement?
- At what rate should households spend down their assets (i.e., dissaving) during retirement?

The LC model uses the facts about an individual's or household's assets, debts, income and fixed expenses as input data and then calculates the household's sustainable standard of living.The *sustainablestandard of living* is the amount of spending on all goods and services that a household can afford until its maximum age of life[8], without running out of resources. Once we know the sustainable standard of living we can use this to create a viable and sustainable financial plan: how much to spend and save each year.

2.3 A Two-Period Life-Cycle Model

Oftentimes, we can explain the conceptual intuition behind a complicated problem by describing a similar but simpler problem. One of the main ideas of the LC model is *consumption smoothing*, which means spending one's economic resources smoothly over time. To begin to explain consumption smoothing, and the role of saving and

[7] Modigliani, Franco, Adventures of an Economist, 2001.

[8] In this module, we assume that the maximum age of life is 100 years. This is a reasonable assumption, as only about 2% of the population will live beyond 100 years of age. In fact, an individual's age at death is uncertain, and in Module 7 (For Whom the Bell Tolls) we will address this uncertainty by estimating the probability of survival to a specific age. We will revisit the LC model to account for uncertain longevity at that time.

dissaving in the economic life-cycle, we now turn to a two-period LC model[9] for saving and consumption. After we establish the notation and arithmetic with a two-period model, we will develop an 80-period model (i.e., one period for each year of your life).

In the two-period LC model, you live and consume for two periods: youth and old age. During the period of youth, you work and receive labor income. During the period of old age, you do not work and must consume only from your accumulated assets. Using a two-period LC model provides a reasonable approximation of consumption for real life: imagine that each period in this model represents approximately 40 years of adulthood. Beginning around age 20, you work for about 40 years and beginning around age 60, you stop working and live for approximately another 40 years in retirement (to your maximum age of life).

Simplifying Assumptions

We will begin by making several assumptions to keep the arithmetic as simple as possible. For now we will assume:

- No taxes or transfer payments.
- No real return on invested wealth (i.e., if you save money when young, you get back the same amount of wealth when old).
- No uncertainty about earnings or longevity.

We will address each of these assumptions in the modules ahead, but to begin we want to convey the conceptual framework of the LC model without forward references.

Budget Constraints

A *budget constraint* is an equation that describes how your income must be allocated over time. During the first period, you must allocate your work income (W) between consumption when young (Cy) and saving (S) for old age. Thus, we can write the budget constraint for the first period as:

$$W = Cy + S$$

Alternatively, we can rewrite this budget constraint in terms of consumption. The amount of consumption when young is equal to income minus savings:

$$Cy = W - S$$

The budget constraint for old age is much simpler; the only resources available for consumption when old (Co) are the savings (S) from the period of youth. We can write the budget constraint for the period of old age as:

$$Co = S$$

The Lifetime Budget Constraint

Now we can combine the budget constraints for both periods. Since consumption when old (Co) is equal to savings (S), by substitution we arrive at a lifetime budget constraint:

$$W = Cy + Co$$

In other words, your income from working must support consumption when young *and* consumption when old. By examining this equation, it is clear that there are many possible values of Cy and Co that would satisfy the lifetime budget constraint.

[9] We develop a two-periodlife-cycle model to explain the conceptual framework while keeping the arithmetic as simple as possible. Later in this module, we will develop an 80-period life-cycle model, although the model can be implemented for any number of periods and its insight and interpretation will remain the same.

Figure 2.3 Plotting the Lifetime Budget Constraint for a two-period LC

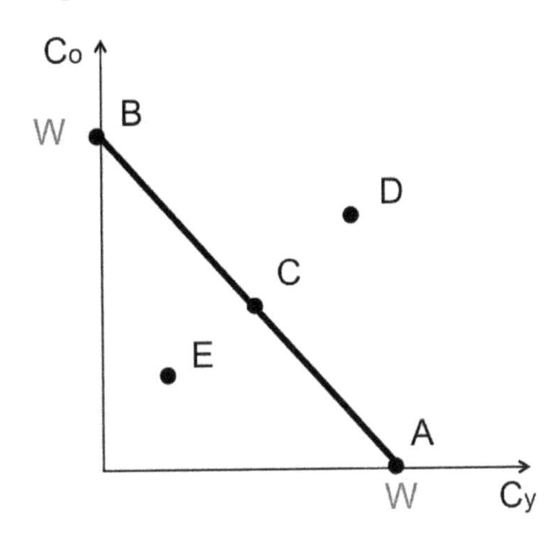

Figure 2.3 illustrates the tradeoff between consumption when young (the horizontal axis) and consumption when old (the vertical axis). We can describe the labeled points as follows:

- Point A represents consuming everything when young and starving when old.
- Point B represents starving when young and consuming everything when old.
- Point C represents a balanced approach of consuming the same amount in both periods.
- Point D is infeasible. It is beyond the bound of the lifetime budget constraint (i.e., attempting to spend more than one's total lifetime resources). An individual who attempts to achieve this level of consumption will end up under-saving during his or her working life and will therefore suffer a drastic reduction in standard of living at retirement. Put another way, if you consume most of your lifetime income when young, you cannot expect to have enough wealth to sustain the same level of consumption during old age. The person who plans to consume at the level of point D will likely find himself or herself actually consuming close to point A: starving when old.
- Point E is inefficient. It is within the bound of the lifetime budget constraint, but represents an individual who does not consume all of his or her available resources. This might be explained by altruism (wanting to leave an inheritance for one's descendants), over-saving (i.e., under-consuming), or working too much given one's low desire for consumption.

Case Study: Wealthy Uncle Elwyn

Let's suppose that you just turned 21 years old, and studied ancient philology[10] in college. You are shocked to discover upon graduation that there are no jobs in this field and you have no expected labor income. Fortunately your wealthy old Uncle Elwyn, who had no children of his own, has passed away and left you a small but significant fortune – for example, $10 million. You receive this inheritance at age 21 and you expect to live to age 100 (a total of 80 years). We will continue to assume that there are no taxes or benefits and that you can borrow or lend as much as you like with 0% real interest rates[11] for borrowing and investment.

How would you consume this wealth over your lifetime? One idea is to plan for your lifetime consumption in phases, with one level of consumption when you are young (age 21-60) and another level of consumption when you are old (age 61-100). For example, if you wait to age 61 before you begin to use Uncle Elwyn's inheritance you could spread the $10 million over 40 years by spending $\$10{,}000{,}000 / 40\,years = \$250{,}000\,per\,year$.

---

[10] Philology is the study of language in its written form and ancient philology includes the study of non-text writing such as hieroglyphs. There are many degree programs in philology.

[11] In this module, we make the simplifying assumption that there is no return on investment, and no change in the purchasing power of money. We will address return on investment and inflation in Module 4 (*Thinking in Present Value*).

Alternatively, you could spend $100,000 per year from age 21 to 60, and $150,000 per year from age 61 to 100. Figure 2.4 shows some alternatives.

Figure 2.4 How to Spend Uncle Elwyn's Wealth: Annual Consumption Choices

Choice	When Young	When Old
A	$ -	$ 250,000
B	$ 50,000.00	$ 200,000.00
C	$ 100,000.00	$ 150,000.00
D	$ 150,000.00	$ 100,000.00
E	$ 200,000.00	$ 50,000.00
F	$ 250,000.00	$ -

Given a finite amount of lifetime wealth, there are in fact many choices of how to consume this wealth. Figure 2.5 shows graphical representations of just two of these consumption plans.

Figure 2.5 How to spend Uncle Elwyn's wealth? Some Feasible Consumption Plans

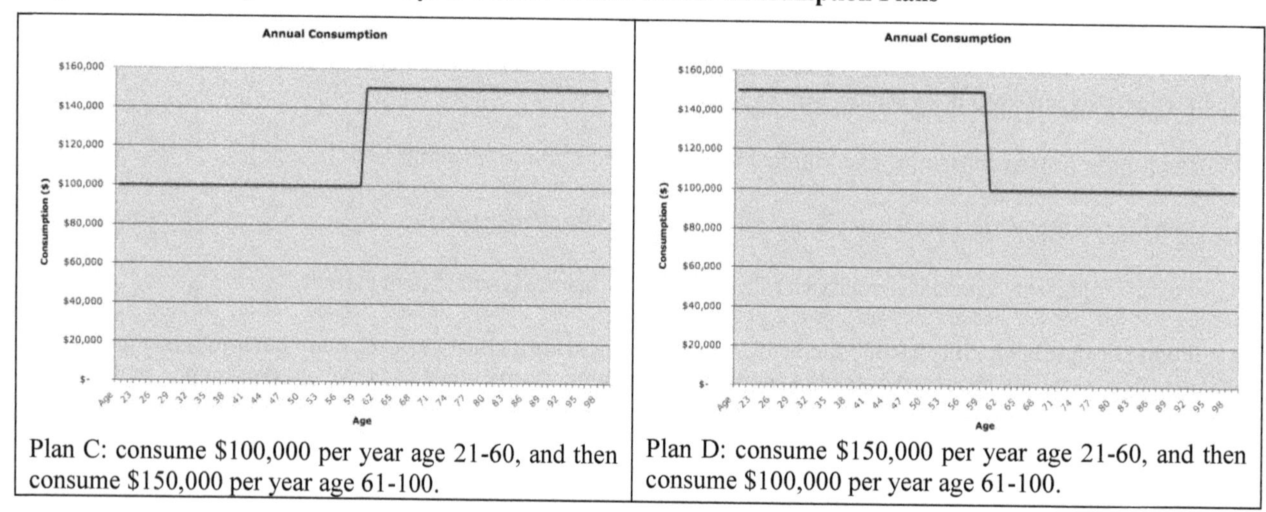

Plan C: consume $100,000 per year age 21-60, and then consume $150,000 per year age 61-100.

Plan D: consume $150,000 per year age 21-60, and then consume $100,000 per year age 61-100.

Which of these consumption plans is the best? They are both viable options, so the choice depends on person preference (i.e., which one will give you the most happiness).

How Economists Measure Happiness

Economists measure happiness using a theoretical model of utility maximization.[12] In economics, *utility* is a measure of happiness. Briefly, economics states that more utility is better and rational people should make choices that maximize their utility. In general, you increase your utility by consuming more, and decrease utility by having an unexpected decrease in your consumption. The LC model depends on maximizing utility across multiple time periods: now and in the future. We make the simplifying assumption that people will be happiest (i.e., have the most utility) by having a sustainable standard of living, i.e., smoothing consumption.

[12] A discussion of the theory of utility maximization is beyond the scope of this book. If you're curious, you can learn a lot from Wikipedia. Read about *marginal utility*, *economicsatiation*, and *diminishing marginal utility*. Watch out: economists are closeted mathematicians!

2.4 Consumption Smoothing

How do households choose how much to consume each year?Economist Franco Modigliani argued that rational consumers want to maximize their utility of consumption across time. The shorthand term for this approach is *life-cycle consumption smoothing*, which means to spread your consumption smoothly over your lifetime. Modigliani identified this desire to consume the same amount each year as an explanation for why households save some of their income when young, and consume their wealth when old.

Consumption Smoothing in aTwo-PeriodLife-Cycle Model

In our two-period LC model consumption smoothing is achieved by consuming the same amount when young as when old. Recall that the two-period lifetime budget constraint for a household that works only when young is $W = Cy + Co$.

Let $\overline{C}$ be the amount of smooth consumption, such that $\overline{C} = Cy = Co$.Total lifetime wealth is consumed over the two periods, so:

$$W = Cy + Co = C + C = 2C.$$

We can solve for $\overline{C} = W / 2$.

Consider a numerical example to further illustrate consumption smoothing with the two-period LC model's lifetime budget constraint. Suppose $W = 100$:

$$\overline{C} = \frac{W}{2} = \frac{100}{2} = 50$$

Now that we know $\overline{C} = 50$ we plug this value into the budget constraint for youth and arrive at the amount of savings required when young:

$$S = W - Cy = 100 - 50 = 50.$$

By saving 50 when young the budget constraint when old is $Co = S = 50$. In this example we have smoothed consumption for 2 periods and the lifetime budget constraint has given us the outcome variables of $\overline{C}$ (how much to consume each period) and S (how much to save in each period to achieve this level of consumption).

A Two-PeriodLife-Cycle Model with Income in Both Periods

As we continue to make our lifetime-budget constraint more realistic, the next iteration is to assume that individuals will receive income not only when young but *also* when old. Perhaps you will receive income when old from part time work, a pension plan or Social Security old age benefits. We can revise the previous two-period lifetime budget constraint to account for income in both periods. When young, income can be consumed or saved:

$$Wy = Cy + S$$

We can re-arrange to:

$$Cy = Wy - S$$

When old, consumption can be funded by income when old (Wo) and by any savings that were accumulated when young and brought into old age:

$$Co = Wo + S$$

Now, we combine the two period budget constraints, again using S as the common term, and arrive at this lifetime budget constraint:

$$Cy + Co = Wy + Wo$$

As with our earlier lifetime budget constraint, the interpretation remains the same: lifetime consumption is bounded by lifetime wealth.

Saving to Smooth Consumption

In the two-period LC model with income in both periods, it is possible to shift consumption through time in two ways: from youth to old age (save when young and dissave when old), or from old age to youth (borrow when young and repay when old). Figure 2.6 illustrates this saving and dissaving graphically.

Figure 2.6Shifting Consumption from Youth to Old Age

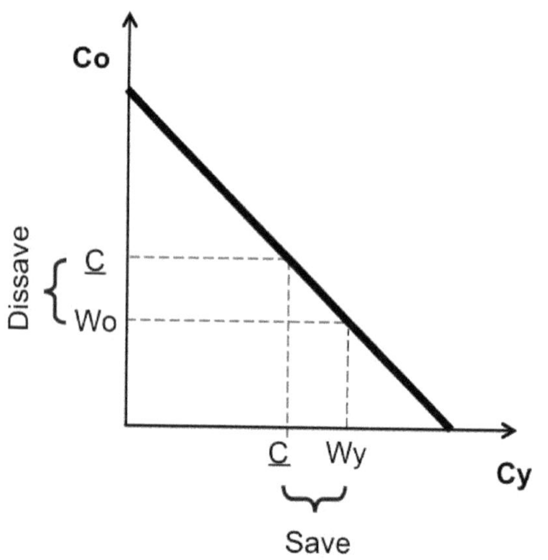

Consider a numerical example to further illustrate smoothing consumption by saving when young. Suppose $Wy = 100$, and $Wo = 50$. It would be rational to consume the same amount in both periods. Let $\overline{C}$ be the amount of smooth consumption such that $\overline{C} = Cy = Co$.

$$2\overline{C} = Wy + Wo = 100 + 50 = 150$$

$$\overline{C} = \frac{Wy + Wo}{2} = \frac{100 + 50}{2} = 75$$

To achieve this level of consumption, savings when young would be:

$$S = Wy - Cy = 100 - 75 = 25.$$

Recall from Module 1 that assets are an accumulation of savings. The savings from the first period are accumulated (added to) assets. Assuming that we begin youth with no assets, saving 25 in the period of youth results in having assets of 25 at the in the period of old age.

Consumption when old would be:

$$Co = Wo - S = 50 - (-25) = 50 + 25 = 75.$$

Notice that the savings when old are -25. Negative saving (or dissaving) requires withdrawing or using up assets. Since we had 25 of assets at the beginning of the period of old age, we can withdraw 25 to fund consumption when old. This results in no assets remaining at the end of life.

Borrowing to Smooth Consumption

When there is income in both periods it is also possible to borrow to smooth consumption. In this case, you would borrow to be able to consume more than income when young and repay the borrowing out of income when old. Figure 2.7 illustrates this borrowing and repayment graphically.

Figure 2.7 Shifting Consumption From Old Age to Youth

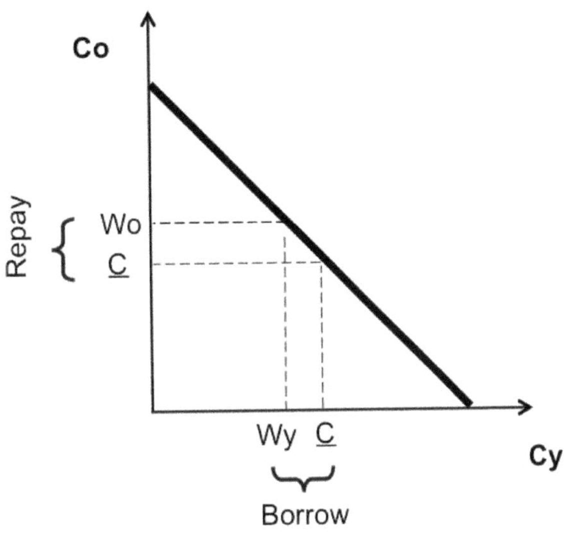

Consider a numerical example with $Wy = 100$, and $Wo = 150$. It would still be rational to consume the same amount in both periods. Let $\overline{C}$ be the amount of smooth consumption such that $\overline{C} = Cy = Co$. Using the known amount of income, we can solve for smooth consumption as:

$$2\overline{C} = Wy + Wo = 100 + 150 = 250$$

$$\overline{C} = \frac{Wy + Wo}{2} = \frac{100 + 150}{2} = 125$$

To achieve this level of consumption, one would have to dissave when young. Since there are no initial assets from which to dissave one would need to borrow to smooth consumption.

$$S = Wy - Cy = 100 - 125 = -25.$$

In old age, one would have to repay the amount borrowed when young, so consumption when old would be:

$$Co = Wo + S = 150 + (-25) = 125.$$

2.5 *Constructing an 80-Period Life-Cycle Model*

The two-period LC model is a useful abstraction to explain the concepts of consuming now versus later and to introduce the budget constraint. In the real world we earn income in each of many years and need to consume in all years, up to our maximum age of life.

A more realistic LC model would be flexible enough to take into consideration annual earnings and find annual spending. Assuming the maximum age of life to be age 100, the appropriate number of periods in the model would be 100 minus the current age. For a young person, an 80-period LC model might be appropriate (i.e., age 21 to age 100), wherein wages are earned in the working years until retirement, and then no wages are earned in the years of retirement. It would be extremely tedious to write out all 80 period's worth of annual budget constraints and to calculate each year's savings and assets. Thankfully, we can use a spreadsheet to create an 80-period model with ease.

Case Study of the 80-period Life-Cycle Model: Roxanne

As an example, let's consider Roxanne, a 21-year-old who will begin working at a salary of $44,000 per year. She expects to work until age 65, so she will earn 45 years of salary. This gives her an initial human capital of $45\,years \times \$44{,}000 = \$1{,}980{,}000$ but no other assets. We will continue to assume no taxes or transfer payments, no real return on savings, no uncertainty, and that she will live to age 100. How much should she consume each year?

We can determine the amount of smooth consumption by taking total human capital and dividing by the number of years of consumption. Roxanne's smooth annual consumption would be:

$$\bar{C} = \frac{\$1{,}980{,}000}{80\,years} = \$24{,}750\,per\,year$$

s

To achieve this smooth standard of living Roxanne will need to save during her 45 working years when here income is greater than $24,750 and spend down her savings during her 35 non-working years when her income is less than $24,750. Roxanne's annual budget constraint during each of her working years is:

$$C_i = W_i - S_i$$

Notice the subscript **i**, such as in $\mathbf{S_i}$. The subscript $i$refers to a numeric index. Since we will have the same equation for each year, we will use the index to differentiate between one year's equation and another year's equation. For example, $\mathbf{S_{21}}$refers to thesavings at age 21.

In each year,Roxanne will earn wages *W*. She will consume the amount *C* and save the amount *S*. The annual budget constraint shows that her savings in each of the working years will be:

$$S_i = W_i - \bar{C} = \$44{,}000 - \$24{,}750 = \$19{,}250.$$

Figure 2.8 shows an implementation of the 80-period LC model as a spreadsheet in Microsoft Excel, with some lines hidden for brevity.

Figure 2.8 An Excel implementation of an 80-period Life-Cycle Model

	A	B	C	D	E	F	G	H
1	Human Capital	$ 1,980,000		Annual Budget Constraint: Si = Wi - Ci				
2	Num Years	80		Annual Assets Calculation: Ai = Ai-1 + Si				
3	Compsumption	$ 24,750	per year					
4								
5	Age	Income	Consumption	Savings	Financial Assets	Human Capital	Economic Wealth	Saving or
6		W	C	S	A	HC	ENW	Dissaving?
7	21	$ 44,000.00	$ 24,750	$ 19,250	$ 19,250	$ 1,936,000	$ 1,955,250	saving
8	22	$ 44,000.00	$ 24,750	$ 19,250	$ 38,500	$ 1,892,000	$ 1,930,500	saving
9	23	$ 44,000.00	$ 24,750	$ 19,250	$ 57,750	$ 1,848,000	$ 1,905,750	saving
10	24	$ 44,000.00	$ 24,750	$ 19,250	$ 77,000	$ 1,804,000	$ 1,881,000	saving
11	25	$ 44,000.00	$ 24,750	$ 19,250	$ 96,250	$ 1,760,000	$ 1,856,250	saving
12								
51	64	$ 44,000.00	$ 24,750	$ 19,250	$ 847,000	$ 44,000	$ 891,000	saving
52	65	$ 44,000.00	$ 24,750	$ 19,250	$ 866,250	$ -	$ 866,250	saving
53	66	$ -	$ 24,750	$ (24,750)	$ 841,500	$ -	$ 841,500	dissaving
54	67	$ -	$ 24,750	$ (24,750)	$ 816,750	$ -	$ 816,750	dissaving
55								
85	97	$ -	$ 24,750	$ (24,750)	$ 74,250	$ -	$ 74,250	dissaving
86	98	$ -	$ 24,750	$ (24,750)	$ 49,500	$ -	$ 49,500	dissaving
87	99	$ -	$ 24,750	$ (24,750)	$ 24,750	$ -	$ 24,750	dissaving
88	100	$ -	$ 24,750	$ (24,750)	$ -	$ -	$ -	dissaving

In Figure 2.8, there are columns for the period (i.e., age), annual income (*W*), annual consumption (*C*) and annual savings (*S*). The financial assets (net worth) column is an accumulation of savings from all previous years. The additional columns showing human capital (*HC*) and economic net worth (*ENW*) are included to show the rate at which human economic capital will decline but these are not necessary for the implementation of the annual budget constraint.

Using Spreadsheets: Absolute References

The *absolute reference* is the most important feature you need to master to be able to make financial spreadsheets work correctly. An absolute reference allows you to put a formula in one cell, and then apply that formula to other cells by dragging or copy/paste.

For example, if you have a value that is calculated in cell B3, which is needed in each row of the spreadsheet (e.g., the level of annual consumption). The default behavior when you drag or copy the formula is to reference the *next row*, i.e., D3. However, if you want to always reference cell B3, you need to use an absolute reference.

To create an absolute reference, you must edit a formula to add a dollar sign ($) before both the row and column indices. In this example, you would want to change the formula to reference B3 andhit <enter> to save the formula. Once you've done this you can safely copy/paste the formula to any row, any column and it will always refer to the value in cell B3.

In the spreadsheet example in Figure 2.8, we use the same expression (calculation) for each year's consumption. *We will need to use an absolute reference to refer to the cell containing the amount of annual consumption.*

Assets are Accumulated Savings

In each of her working years, Roxanne will save $19,250 of earnings, which will be added to her accumulated financial assets. To see how this works, Let A_i be the assets in year i. The value of assets in year i will be:

$$A_i = A_{i-1} + Si$$

In words, this means that at the end of each year, the amount of assets A_i is calculated as the previous year's assets A_{i-1} plus this year's savings Si.

For example, at age 21, her financial assets begin at $0 and Roxanne saves $19,250. At the end of the year, Roxanne's assets are $19,250. At age 22, Roxanne's financial assets are the amount of assets she had at the end the previous year ($19,250), plus her new savings of $19,250, for a total of $38,500. Thus, the column showing Roxanne's financial assets (i.e., net worth) is the running-total accumulation of her savings from all previous periods.

In each of Roxanne's retirement years, she will consume by spending down her accumulated financial capital. Recall the annual budget constraint was $C_i = W_i - S_i$. When old, income is $0 so the annual budget constraint simplifies to $C_i = - S_i$. To smooth consumption, Roxanne will have negative savings of:

$$C_i = -S_i = \$24{,}750$$

Negative saving is called dissaving, which reduces the amount of accumulated assets. To smooth consumption, Roxanne will dissave $24,750 in each year of retirement.

Annual Income and Consumption

Figure 2.9 shows Roxanne's annual income and consumption. Even though her income happens only during the working years, her consumption is smooth for all years. Roxanne must save $19,250 during each working year and dissave $24,750 during each non-working years.

Figure 2.9 Roxanne's Annual Income and Consumption

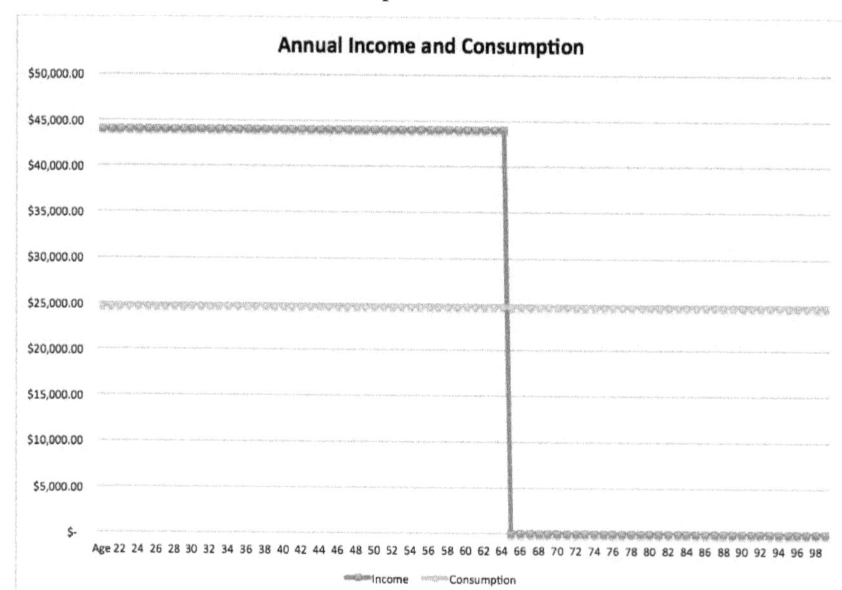

Accumulated Financial Assets

Figure 2.10 shows Roxanne's accumulation of financial assets (net worth). She begins at age 22 with no financial assets and gradually accumulates financial assets by saving each year. During her non-working years Roxanne consumes by spending down her financial assets. Roxanne's financial capital rises each year, peaks at the start of her retirement and declines until reaching $0 at her maximum age of life.

Figure 2.10 Roxanne's Accumulated Financial Assets

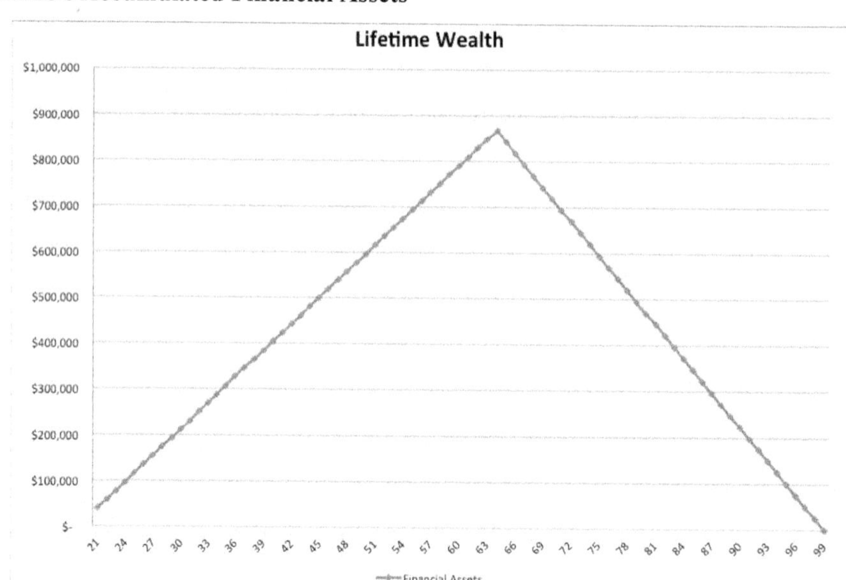

Recall that we said that Roxanne's human capital at age 21 was $1,980,000. As she earns each year's income, that amount is deducted from her human capital. As a result, her human capital declines at a rate of $44,000 per year, for each year from age 21 to age 65, when it expires worthless. Figure 2.11 shows Roxanne's human capital, financial assets, and her total economic net worth (human capital plus financial assets).

Figure 2.11 Roxanne's Human Capital, Financial Assets, and Economic Net Worth

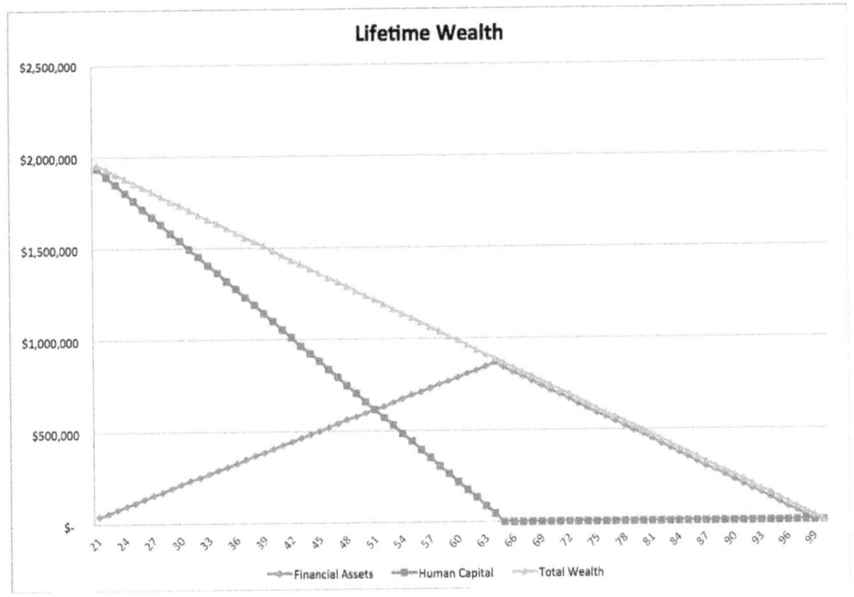

In Figure 2.11, we see that the consumption smoothing plan enables Roxanne to consume her lifetime wealth smoothly over her maximum lifespan. The annual savings required by her lifetime budget constraint when young enables Roxanne to accumulate enough financial capital to sustain her consumption until her maximum age of life.

Recall that savings is income that is not consumed, and that for most young people income is earned by working. In essence, saving is the process of converting human capital to financial capital, and financial capital is used to sustain consumption once the human capital is used up.

2.6 *Extending the LC Model*

The initial presentation of the LC model in Excel made several simplifications, and we will begin to unwind these in this section and the modules that follow.

An 80-Period Life-CycleModel with Uneven Wages

Lifetime consumption smoothing does not require flat annual wages. For example, suppose that Roxanne begins working at age 21 at an entry-level income of $44,000 per year, and then gets a promotion to a management position at age 31, wherein she will earn $55,000 per year until retirement at age 65. To keep things simple, we will continue to assume no taxes and transfer payments, no real investment return on wealth and no uncertainty.

In this case, Roxanne's human capital would be:

$$\$44{,}000\ \times\ 10\,years\ +\$55{,}000\ \times\ 35\,years =\ \$2{,}365{,}000$$

And Roxanne's smooth annual consumption would be:

$$\bar{C} = \frac{\$2{,}365{,}000}{80\,years} = \$29{,}563/year$$

To implement her plan for smooth consumption, we re-arrange the annual budget constraint to solve for the level of savings required during her first 10 years of work:

$$S\ =\ W - C\ =\ \$44{,}000 - \$29{,}563\ =\ \$14{,}437$$

When her income increases at age 31, her saving will adjust to:

$$S = W - C = \$55{,}000 - \$29{,}563 = \$25{,}437.$$

Even though Roxanne would be earning more she would be consuming the same amount per year. As income changes, the amount of savings will change as is necessary to achieve the goal of smoothing consumption. Figure 2.12 shows the spreadsheet implementation of the consumption smoothing model for Roxanne's revised earnings profile.

Figure 2.12 An 80-period Consumption-Smoothing Model with Uneven Earnings

	A	B	C	D	E	F	G	H
1	Human Capital	$ 2,365,000		Annual Budget Constraint: Si = Wi - Ci				
2	Num Years	80		Annual Assets Calculation: Ai = Ai-1 + Si				
3	Compsumption	$ 29,563	per year					
4								
5	Age	Income	Consumption	Savings	Financial Assets	Human Capital	Economic Wealth	Saving or
6		W	C	S	A	HC	ENW	Dissaving?
7	21	$ 44,000.00	$ 29,563	$ 14,438	$ 14,438	$ 2,321,000	$ 2,335,438	saving
8	22	$ 44,000.00	$ 29,563	$ 14,438	$ 28,875	$ 2,277,000	$ 2,305,875	saving
9	23	$ 44,000.00	$ 29,563	$ 14,438	$ 43,313	$ 2,233,000	$ 2,276,313	saving
10								
16	29	$ 44,000.00	$ 29,563	$ 14,438	$ 129,938	$ 1,969,000	$ 2,098,938	saving
17	30	$ 44,000.00	$ 29,563	$ 14,438	$ 144,375	$ 1,925,000	$ 2,069,375	saving
18	31	$ 55,000.00	$ 29,563	$ 25,438	$ 169,813	$ 1,870,000	$ 2,039,813	saving
19	32	$ 55,000.00	$ 29,563	$ 25,438	$ 195,250	$ 1,815,000	$ 2,010,250	saving
20	33	$ 55,000.00	$ 29,563	$ 25,438	$ 220,688	$ 1,760,000	$ 1,980,688	saving
21								
52	64	$ 55,000.00	$ 29,563	$ 25,438	$ 1,009,250	$ 55,000	$ 1,064,250	saving
53	65	$ 55,000.00	$ 29,563	$ 25,438	$ 1,034,688	$ -	$ 1,034,688	saving
54	66	$ -	$ 29,563	$ (29,563)	$ 1,005,125	$ -	$ 1,005,125	dissaving
55	67	$ -	$ 29,563	$ (29,563)	$ 975,563	$ -	$ 975,563	dissaving
56								
85	96	$ -	$ 29,563	$ (29,563)	$ 118,250	$ -	$ 118,250	dissaving
86	97	$ -	$ 29,563	$ (29,563)	$ 88,688	$ -	$ 88,688	dissaving
87	98	$ -	$ 29,563	$ (29,563)	$ 59,125	$ -	$ 59,125	dissaving
88	99	$ -	$ 29,563	$ (29,563)	$ 29,563	$ -	$ 29,563	dissaving
89	100	$ -	$ 29,563	$ (29,563)	$ -	$ -	$ -	dissaving

In Figure 2.12, we see that the level of consumption is smooth even though income is not smooth. To implement this series of equations, we must begin by (1) finding the level of consumption and then (2) solve for the amount of savings required to satisfy each year's annual budget constraint.

Figure 2.13 Roxanne's Revised Annual Income and Consumption

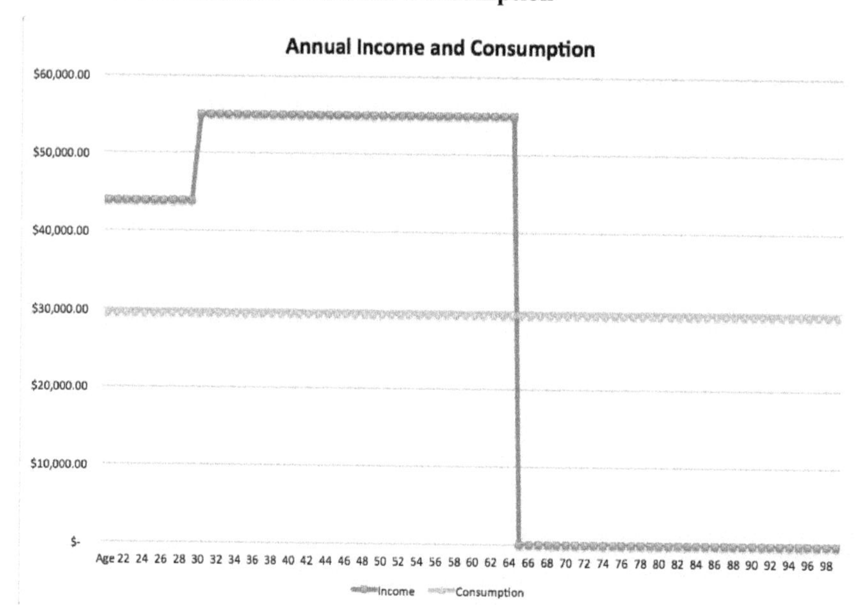

Figure 2.13 makes it clear that it is possible to achieve smooth lifetime consumption even with uneven annual wages. Saving makes it possible to achieve smooth consumption despite uneven income. We can experiment with the inputs in the spreadsheet to achieve a consumption-smoothing plan for any income profile.

Incorporating Initial Assets and Bequests

The previous numerical examples assumed that human capital was the only asset. Incorporating initial assets or debts and final bequests (leaving assets at the end of life) is a rather simple extension. The level of initial assets and final bequests are incorporated into finding the level of annual consumption. More generally, the level of smooth annual consumption is:

$$\bar{C} = \frac{human\ capital + initial\ assets - bequests}{years\ of\ consumption}$$

For example, if Roxanne had initial assets of $100,000, and she did not want to leave any bequests, her level of smooth consumption would be:

$$\bar{C} = \frac{\$1{,}980{,}000 + \$100{,}000 - \$0}{80\ years} = \$26{,}000/year$$

Similarly, if Roxanne had initial assets of $100,000, and planned to leave a bequest of $250,000 at the end of her life, her annual consumption would be:

$$\bar{C} = \frac{\$1{,}980{,}000 + \$100{,}000 - \$250{,}000}{80\ years} = \$22{,}875/year$$

Finally, the annual budget constraint is calculated using this level of smooth consumption to find the amount of savings required in each year.

Case Study: Borrowing Against Uncle Elwyn's Inheritance

Recall theexample in which we discussed receiving an inheritance from your wealthy Uncle Elwyn. After considering several choices of how to consume the wealth over your lifetime, we learned that a rational person would maximize his utility by consuming the wealth smoothly over his or her lifetime.

Now let's change the scenario as follows:suppose your wealthy Uncle Elwyn bequeaths to you $10,000,000, but stipulates that you should receive the wealth at age 61 (40 years from now). How will you allocate the wealth over your lifetime?

In a world without borrowing, you would have no choice: you would have to wait until age 61 before you could begin to consume Uncle Elwyn's $10 million. You could apply consumption smoothing from age 61 to age 100, consuming $250,000 per year. Figure 2.14 shows a graph of this consumption plan.

Figure 2.14 Waiting Until Age 61 To Smooth Consumption

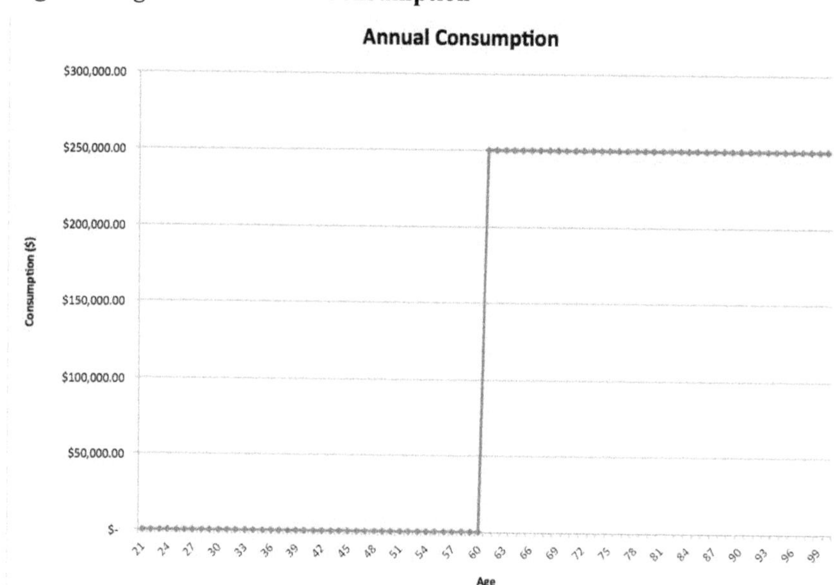

The consumption plan depicted in Figure 2.14could be described as "starving when young" and "splurging when old." Instead, you want to maximize your utility by smoothing consumption, and this consumption plan fails to achieve that goal.

Now suppose you live in a world where borrowing is permitted and does not incur any borrowing costs (i.e., no interest). In this scenario, you would begin borrowing immediately in anticipation of receiving your inheritance. To smooth your consumption over your lifetime, you would want to consume $125,000 per year ($10,000,000 divided by 80 years).

Consuming Uncle Elwyn's inheritance smoothly over time would entail borrowing each year from age 21 to 60. When you receive the inheritance at age 61, you will repay all the borrowing, and then spend down your remaining wealth until age 100. Figure 2.15depicts the net worth you would have with this consumption plan.

Figure 2.15Net Worth When Consuming Uncle Elwyn's Inheritance Smoothly Over Time

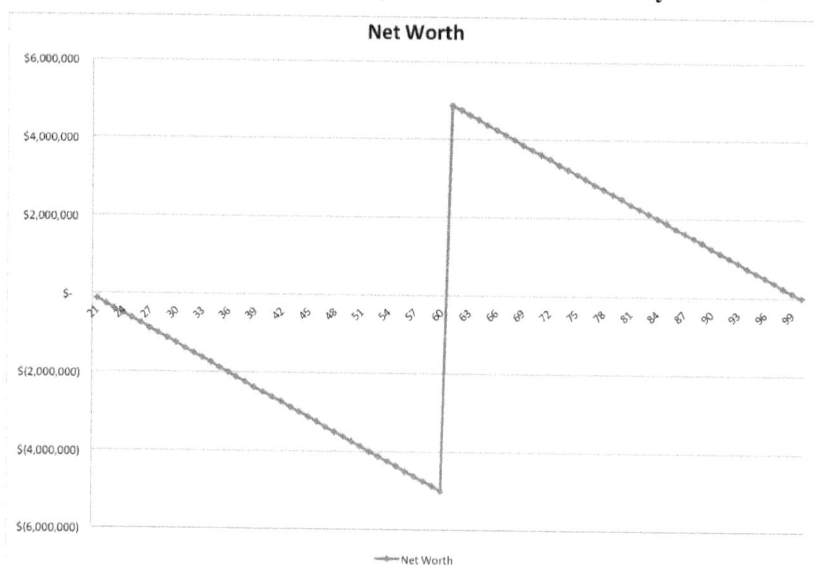

The net worth diagram looks quite daunting: going deeply into debt only to repay it later. However, if the inheritance were guaranteed to arrive on schedule this would be perfectly rational. In this case, borrowing would be a necessary part of maximizing your lifetime utility.

The example of borrowing in anticipation of a substantial inheritance is a bit extreme. Nonetheless, borrowing to smooth consumption is a common and desirable way to increase one's standard of living. Many households borrow to smooth consumption by using installment purchase agreements (e.g., a car loan, mortgage, etc.). These cases call for consuming in the present out of income that is yet to be earned.

2.7 Borrowing Constraints

The theory behind consumption smoothing is that individuals (or households) will maximize their utility by smoothing their consumption. In some circumstances, smoothing consumption would require borrowing against future income.

In general, lenders do not trust borrowers to repay. *Collateral* is an asset pledged to secure the promise of repayment on a loan. In the case that the borrower cannot repay the loan, the lender can repossess the collateral asset. For example, when a lender makes a mortgage loan against a house, the house is pledged as collateral to secure the loan. If the borrower defaults, the lender will repossess the house and sell it to recoup the loan.

Lenders are especially wary of unsecured loans (i.e., loans without collateral), and fear they will not be repaid. Generally, it is not possible to borrow against an intangible asset (for example human capital) to smooth consumption.

A household is *borrowing constrained* when consumption smoothing is not possible because it would require borrowing. Empirical evidence shows that as many as $2/3$ of households are borrowing constrained prior to retirement and thus they cannot smooth consumption.

We can illustrate the effects of borrowing constraints using the two-period model. When a household is borrowing constrained, consumption when young is limited to income when young. As a result, the budget constraint plots as a vertical line at the level of income when young (Figure 2.16).

Figure 2.16A Two-PeriodLifetime Life-Cycle Model UnderBorrowing Constraints

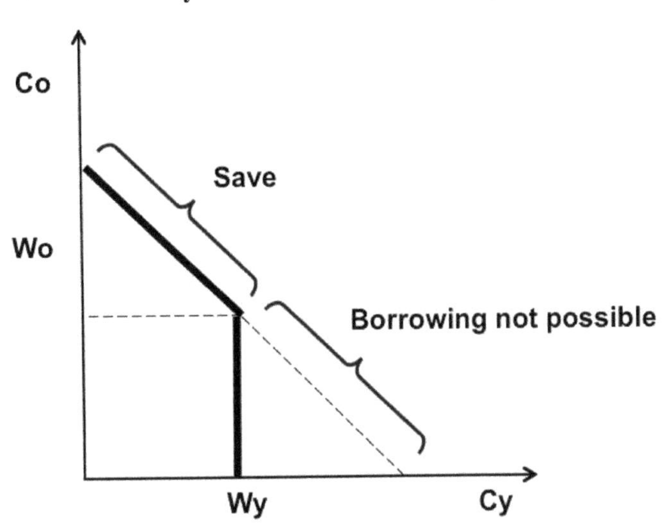

Now, consider a numerical example: let $Wy = 100$, and $Wo = 150$. Let $\bar{C}$ be the amount of smooth consumption such that $\bar{C} = Cy = Co$.

$$2\bar{C} = Wy + W0 = 100 + 150 = 250$$

$$\bar{C} = \frac{Wy + Wo}{2} = \frac{100 + 150}{2} = 125$$

Given the budget constraint when young, we can solve for the amount of savings required to achieve smooth consumption:

$$S = Wy - Cy = 100 - 125 = -25.$$

Smoothing consumption would require negative savings (borrowing) of 25 when young. However, since borrowing is not possible, consumption when young is bounded by income when young: $Cy = Wy = 100$. Consuming all of Wy when young implies that savings when young is: $S = Wy = -Cy = 0$. Finally, given savings when young of 0, consumption when old is:

$$Co = Wo + S = 150 + 0 = 150.$$

Implications of Borrowing Constraints

One implication of borrowing constraints is that some households will not be able to maximize their economic happiness. As we move on to evaluating personal financial decisions we will give preference to choices that do not lead to borrowing constraints. We will also be interested in finding ways to alleviate borrowing constraints or reduce their effect on our level of consumption.

The second implication of borrowing constraints is that it introduces additional computational difficulties. First of all, we cannot assume that consumption smoothing is always possible and we will need to check. If borrowing constraints are present we will need to take the "smoothest path without borrowing."[13]

[13] When it is not possible to smooth consumption, it might still be possible to at least prevent declines in consumption (which would negatively affect utility). Module 1 (Borrowing Constraints and Dynamic Programming) provides a further discussion of borrowing constraints, and smoothing consumption subject to borrowing constraints.

2.8 Summary

In this module, we introduced the economic life-cycle model to explain an individual's (or a household's) accumulation of wealth over its life-cycle. The LC model is a system of equations that models the relationship among income, consumption, savings and assets, and enables us to find the sustainable level of consumption for an individual or household over its lifetime.

The life-cycle model provides a framework for understanding the relationship between income, savings, and consumption. When income and consumption occur in different periods, or have different time horizons, savings is used to transfer wealth through time. A budget constraint describes how to allocate your income between consumption and savings. Alternatively, a budget constraint shows how you will fund your consumption, from income or from dissaving your assets.

We introduced a simplified two-period lifetime to illustrate the budget constraint for each period and a lifetime budget constraint. The lifetime budget constraint relates consumption when young to consumption when old and provides a frontier of possible choices about when to consume. After explaining the main ideas using the two-period model, we developed an 80-period model to illustrate how the LC model and how the annual budget constraint applies to real-life.

Economists use utility as a measure for happiness, and assume that rational people will choose to maximize their utility through time: rational people will want to smooth consumption over their lifetimes at the highest sustainable level. Alternatively, if you are satisfied with lower sustainable consumption you could reduce your work (and thus your lifetime income) accordingly. With consumption smoothing as a goal, saving or dissaving are tools used to support the sustainable standard of living.

When income is earned in multiple periods, it is possible that income when old will be higher than income when young. In this case, you would want to borrow against your future income and use those funds to consume when young. However, it is generally not possible to borrow against future income without collateral, a condition called borrowing constraints. When a household is borrowing constrained it is not possible to smooth consumption and the best you can do is to consume at the highest level without borrowing.

2.9 Review Questions

1. The big idea: explain why we have the economic life-cycle model? What does it help us understand?
2. Explain the terms income, consumption, and savings in the context of the economic life-cycle model.
3. What is utility, and why is utility important in the economic LC model?
4. Write and explain the two-period lifetime budget constraint for a household that lives for 2 periods and works only when young.
5. Draw a graph of the two-period lifetime budget constraint, and label the feasible and infeasible choices for consumption.
6. Explain by giving a numerical example: what is lifetime consumption smoothing?
7. Write the two-period lifetime budget constraint for a household with income in both periods (when young and when old). Explain how savings and borrowing would figure into utility maximization.
8. Explain the concept of borrowing constraints. What is the implication of borrowing constraints for maximizing utility?
9. Within a spreadsheet, what is an absolutely reference? Why would you need to use an absolute reference?

2.10 Application/Practice

Jane St. Clair is 31 years old and expects to work until age 65 and live to age 100. She currently earns $30,000 per year, but expects to have a rising income with $1,000 more each year until age 65, when she will earn $65,000.

1. Using a spreadsheet program (i.e., Excel or Google Docs), construct a 70-period LC model, similar to the one included in this module. Assume no taxes or benefits, no return on investment, and no risk.
    a. What is her sustainable level of annual consumption?
    b. How much does she need to save at age 30?
    c. Is she borrowing constrained?
    d. How much assets must she accumulate by age 66 when she retires?

2. Modify your spreadsheet to account for Jane having $50,000 in initial assets at age 30.
    a. What is her sustainable level of annual consumption?
    b. How much does she need to save at age 30?
    c. Is she borrowing constrained?
    d. How much assets must she accumulate by age 66 when she retires?

3. Modify your spreadsheet to account for Jane having $50,000 in initial assets at age 30, and planning to leave a bequest of $200,000 at age 100.
    a. What is her sustainable level of annual consumption?
    b. How much does she need to save at age 30?
    c. Is she borrowing constrained?
    d. How much assets must she accumulate by age 66 when she retires?

3 Borrowing Constraints And Dynamic Programming

Learning Objectives

- Revisit the concept of borrowing constraints; identify several practical reasons that households might be borrowing constrained; and demonstrate the effect of borrowing constraints on smoothing consumption.
- Review the notation and equations used to describe the relationship between income, consumption, and savings.
- Demonstrate that given a household's assets and income, there exist two optimal financial plans: *smooth consumption* and the *smoothest consumption without borrowing.*
- Frame the problem of consumption smoothing as a case of overlapping sub-problems, in which a solution from one problem is an input to the next problem.
- Describe the computational time required to find the optimal financial plan with borrowing constraints.
- Introduce dynamic programming, and provide numerical examples to illustrate using a dynamic program to find the optimal financial plan with borrowing constraints.

3.1 *Introduction: Borrowing Constraints Revisited*

In this module, we revisit the problem of borrowing constraints, which were introduced at the end of Module 2 (*The Life-Cycle Model*). Rational households like to maximize their sustainable smooth standard of living, but in some instances smoothing consumption would require borrowing against future income to fund current consumption. Solving the lifetime budget constraint when borrowing constraints exists proves to be a difficult problem. This module will introduce the computational technique called dynamic programming, which will help us solve for the smoothest lifetime consumption without borrowing.

Causes of Borrowing Constraints

There are many reasons that households could be borrowing constrained, including:

- *Expectations of a rapidly rising income*. If income when old will be substantially higher than income when young, smoothing consumption would require borrowing against income when old to fund consumption when young.

- *Expecting a large pension income later in life*. Income to be received when old cannot be used to fund consumption when young without borrowing.

- *Expecting a large inheritance later in life*. Again, income to be received when old cannot be used to fund consumption when young.

- *Paying off debts too quickly*. Paying off debts is a form of saving. However, debt repayment counts as a non-discretionary expense that must be paid off the top, before smoothing consumption. If too much current income is required to repay debts, less current income is available for smoothing consumption. While borrowing to smooth consumption is usually not possible, repaying debts more slowly is sometimes possible.

- *Saving too quickly*. Saving to build an emergency fund or for a down payment on a house requires that you accumulate assets that will not be used to smooth consumption. Accumulating these funds too quickly will lead to borrowing constraints, but might be rational because of the benefits (i.e., having an emergency fund to protect against loss of income).

- *Over-contributing to tax-deferred retirement accounts* (e.g., 401k, IRA). Tax-deferred retirement accounts have special terms that prohibit early withdrawal. Once funds have been contributed to the retirement account, they cannot be used to smooth consumption. Retirement accounts will be discussed in more detail in Module 12 (*Saving for Retirement*).

Notation Used in This Module

One note about notation: in Module 2, we described a two-period budget constraint with income when young (Wy), income when old (Wo), consumption when young (Cy), and consumption when old (Co). In this section, we will discuss 2-, 3-, 4-, and n-period budget constraints, so we will adjust the terminology to describe income in each period as W_1, W_2, etc., and consumption in each period as C_1, C_2, etc. In this module, we will continue to refer to smooth consumption as $\bar{C}$, such that $\bar{C} = C_1 = C_2$.

Vector Notation

We can refer to the set of possible incomes for the 2 periods as the vector W, which contains values for each of W_1 and W_2. For example, if we write $W = [150, 50]$, this is equivalent to writing $W_1 = 150$ and $W_2 = 50$. When we expand to a 3 or 4 period model, we will use a vector of incomes such as $W = [50, 40, 60, 50]$, i.e., $W_1 = 50$, $W_2 = 40$, $W_3 = 60$, and$W_4 = 50$.

Similarly, we will use vector notation to describe the level of consumption in each of multiple periods by writing $C = [45, 45, 55, 55]$ to mean $C_1 = 45$, $C_2 = 45$, $C_3 = 55$ and $C_4 = 55$.

Assumptions In This Module

To facilitate calculations and focus attention on the important issues of computational complexity and dynamic programming, we will assume for simplicity that the real interest rate $r = 0$[14]. Further, we will assume for the time being that there are no taxes and benefits.

3.2 *Borrowing Constraints Revisited*

Life-cycle planning starts out easy. Consider a one-period model for lifetime income and consumption, in which a household lives and works for one period. Recall that we use the variable W for income and C for consumption. In this model (depicted graphically in Figure 3.1), all income is consumed. By definition, consumption is smooth.

Figure 3.1 A One-Period Life-Cycle, with W_1 = 100.

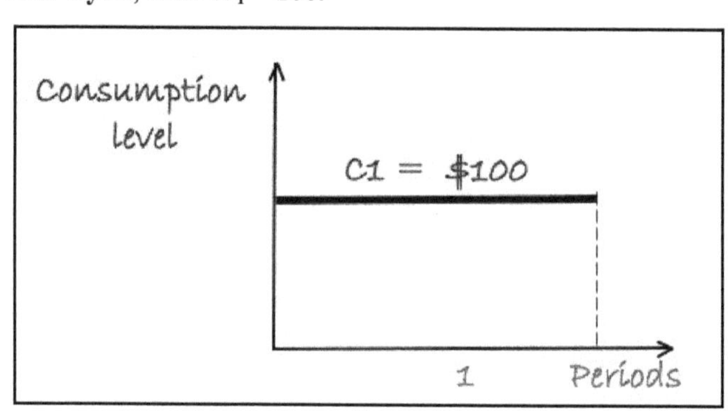

Now consider a two-period model with lifetime income of 200. Let W_1 and W_2 be the income in periods 1 and 2, respectively, and let C_1 and C_2 be the level of consumption in periods 1 and 2, respectively.

The two-period lifetime budget constraint is $W_1 + W_2 = C_1 + C_2$. In other words, lifetime income equals lifetime consumption. Suppose that $W_1 = W_2 = 100$, in which case it's easy to smooth consumption: $C_1 = C_2 = W_1 = W_2 = 100$. Figure 3.2 shows this graphically.

[14] We will address real interest rates in Module 4. When $r > 0$, the budget constraint would be expressed as a present value, i.e., $W_1 + W_2/(1+r) = C_1 + C_2/(1+r)$. By assuming a real rate of return of $r = 0$, we can solve the equation with addition, which keeps the examples tidy. For actual cases with $r > 0$, we would need to be careful to only add together values which occur in the same time period, or to discount/compound as appropriate to move amounts between periods.

Figure 3.2 A Two-Period Life-Cycle Model, with $W_1 = W_2 = 100$.

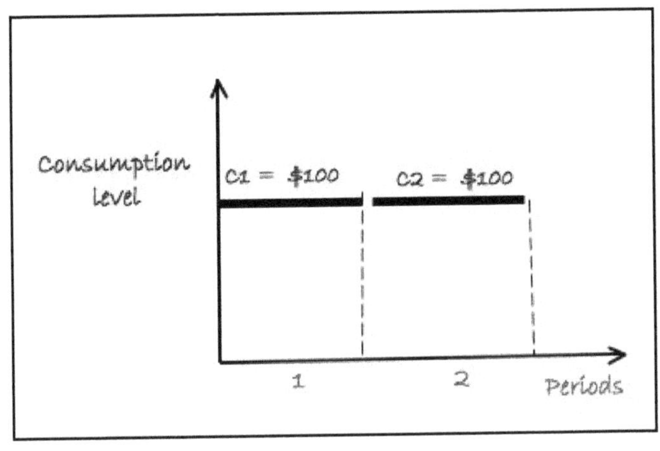

What about cases when $W_1 \neq W_2$? Clearly, it is possible to smooth consumption in cases in which $W_1 > W_2$: we solve the two-period budget constraint to find the value of savings (S) required to smooth consumption:

$$\begin{aligned} C_1 &= W_1 - S_1 \\ C_2 &= W_2 + S_2 \end{aligned}$$

Recall that the savings is the amount of income that is not consumed. In any given period *i*, savings is defined by the annual budget constraint $S_i = W_i - C_i$, i.e., income that is not consumed. In the 2-period example above, savings from period 1 are accumulated as assets, and are available for consumption in period 2. When we know income and consumption, we can determine the amount of savings required to satisfy the annual budget constraint.

Let $\bar{C}$ be the level of smooth consumption, such that:

$$\bar{C} = \frac{W_1 + W_2}{2}$$

and solve for the level of savings, such that:

$$S_1 = W_1 - \bar{C}.$$

We will find that the amount of savings from period 1 will be the correct amount to satisfy the budget constraint for period 2, i.e., $\bar{C} = W_2 + S_2$.

Smoothing Consumption: Numeric Example

Suppose $W_1 = 150$ and $W_2 = 50$. We can solve for smooth consumption as:

$$\bar{C} = \frac{W_1 + W_2}{2} = \frac{150 + 50}{2} = 100$$

Then, we can find the amount of savings to satisfy the annual budget constraint in each period.

$$S_1 = W_1 - \bar{C} = 150 - 100 = 50.$$

In this case, we would save 50 in the first period (i.e., accumulate 50 of assets), and dissave from these assets to sustain smooth consumption in the second period.

Borrowing Constraints

The challenge arrives when $W_1 < W_2$, for example if $W_1 = 50$ and $W_2 = 150$. It would be rational to smooth consumption, but smoothing consumption would require borrowing against future income, which is usually not possible. Therefore, when $W_1 < W_2$, (income when young is less than income when old), the household is borrowing constrained.

3.3 Consumption Smoothing Under Borrowing Constraints

A Two Period Life-Cycle Model

Consider a two-period model with $W = [50, 150]$. Total lifetime income is still 200, so at first glance we might think that $\bar{C} = C_1 = C_2 = 100$ (Figure 3.3, graph A).

Let savings in period 1 be: $S_1 = W_1 - \bar{C} = 50 - 100 = -50$. Saving a negative amount is called dissaving, and if there are no initial assets it's called borrowing. In this case we would want to borrow 50 in period 1, and repay 50 in period 2. However, *borrowing is not allowed.* The next-best alternative is to consume at the level that provides the *smoothest path without borrowing*, shown in Figure 3.3, graph B. In this case, all income is consumed in the period in which it was earned.

Figure 3.3 A Two-Period Life-Cycle Model, with $W = [50, 150]$.

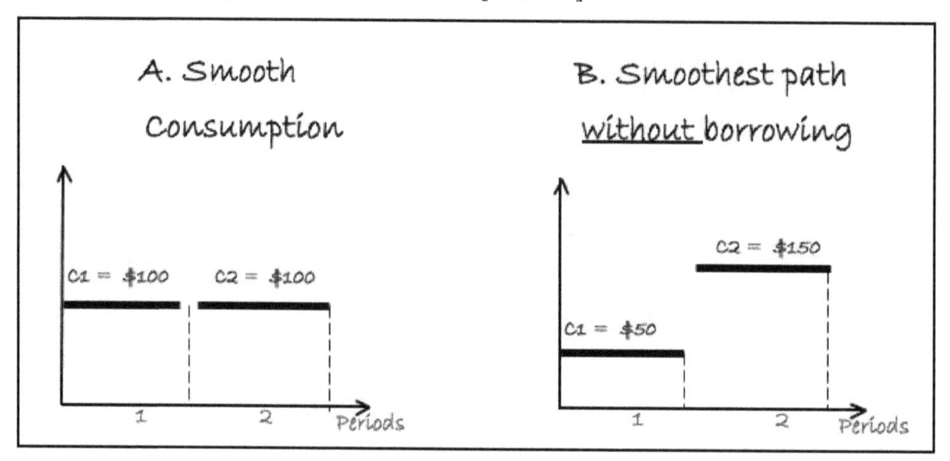

For a two-period model, there are two consumption plans to consider: consumption smoothing, and the smoothest path without borrowing. As an exercise, try some examples, picking values for W_1 and W_2 to confirm this.

A Three Period Life-Cycle Model

What about a three-period model? How many consumption plans are possible? Depending on the levels of income W_1, W_2, and W_3, there are 4 possible consumption paths (see Figure 3.4).

Figure 3.4 A 3-Period Life-Cycle Model Presents 4 Possible Consumption Plans.

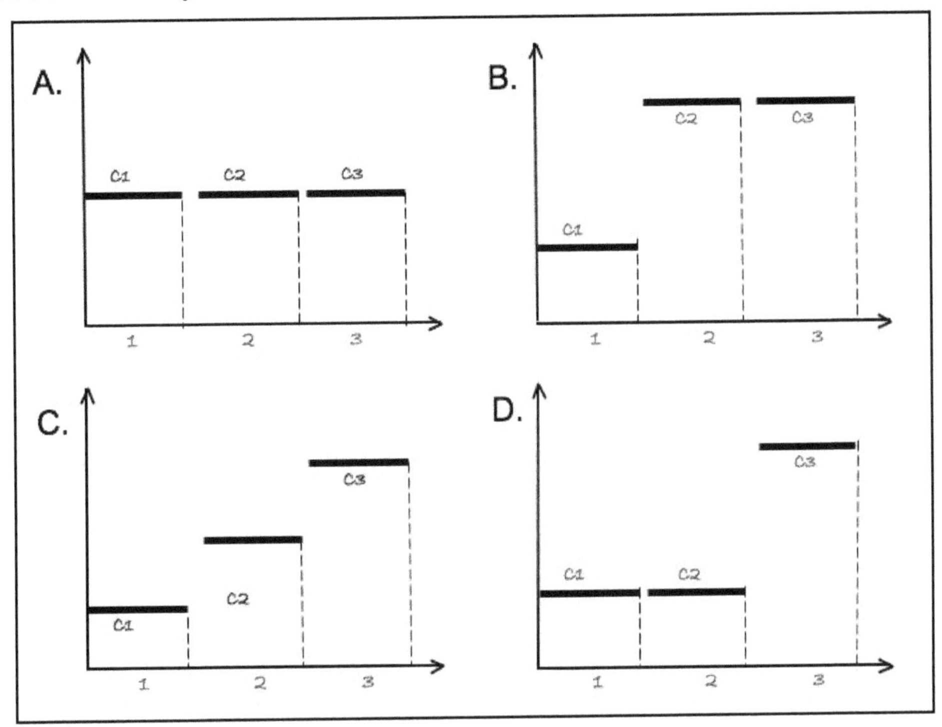

Figure 3.4 graphically explains how a three-period model results in having four alternative consumption plans:

- In Graph A, $W_1 \geq W_2$, $and W_2 \geq W_3$, and consumption smoothing is possible for all periods. To smooth consumption, save the portion of income in period(s) 1 (and 2), in order to consume at a smooth level in all periods. Example: $W_1 = 100$, $W_2 = 50$, and $W_3 = 50$.

- In Graph B, $W_1 < W_2$, $but\ W_2 \geq W_3$, and it is not possible to smooth consumption in the first period. The smoothest path without borrowing is to consume all income in period 1, but to smooth consumption in periods 2 and 3 (by saving in period 2 and consuming that savings in period 3). Example: $W_1 = 50$, $W_2 = 100$, and $W_3 = 100$.
-

- In Graph C, $W_1 < W_2$, $and W_2 < W_3$, and it is not possible to smooth consumption. The smoothest path without borrowing is to consume all income in each period. Example: $W_1 = 20$, $W_2 = 50$, and $W_3 = 80$.

- Finally, in Graph D, $W_1 \geq W_2$, $but\ W_2 < W_3$. In this case, it is possible to smooth consumption between the first two periods, but it is not possible to smooth consumption across all three periods. The smoothest path without borrowing requires saving in period 1, consume all assets in period 2 (thus smoothing consumption across 2 periods), and finally consuming all income in period 3. Example: $W_1 = 100$, $W_2 = 50$, and $W_3 = 100$.

To further illustrate the choices in Figure 3.4, let's consider some numerical examples.

Example 1

Consider a three-period example with incomes $W = [50, 50, 50]$. We can set the level of consumption to:

$$\bar{C} = \frac{W_1 + W_2 + W_3}{3} = \frac{50 + 50 + 50}{3} = 50$$

In this case, we can easily set $\bar{C} = 50$, and consume each period's income without borrowing or saving (i.e., Figure 3.4, panel A).

Example 2

Consider a three-period example with incomes $W = [100, 70, 70]$. We could find that smooth consumption would be:

$$\bar{C} = \frac{W_1 + W_2 + W_3}{3} = \frac{100 + 70 + 70}{3} = 80$$

To smooth consumption in period 1, we consume $C_1 = 80$, which leads to savings of $S_1 = W_1 - C_1 = 100 - 80 = 20$. That is, at the end of period 1 we would have 20 in assets. In period 2, we could would consume $C_2 = 80$,, which leads to savings of $S_2 = W_2 - C_2 = 70 - 80 = -10$. Since we have assets of 20, we *can* dissave 10 to help smooth consumption (i.e., without borrowing). The same consumption plan applies for period 3 (i.e., dissave 10). At the end of period 3, there are no assets remaining. The level of consumption is smoothed for all years (i.e., Figure 3.4 panel A again).

Example 3

Consider a three-period example with incomes of $W = [60, 100, 50]$. We would like to set consumption to be:

$$\bar{C} = \frac{W_1 + W_2 + W_3}{3} = \frac{60 + 100 + 50}{3} = 70$$

When we solve the budget constraint for period 1, we find it requires savings of $S_1 = W_1 - C_1 = 60 - 70 = -10$. This amount of consumption is infeasible, because *we cannot borrow to smooth consumption*. The next best alternative is to consume $C_1 = 60$, and to try to smooth consumption in periods 2 and 3.

Taken together, periods 2 and 3 look like a 2-period model in which we have high income followed by low income. We can in fact consume $C_2 = C_3 = (W_2 + W_3)/2 = (100 + 50)/2 = 75$ in each period, since this does not require borrowing. In period 2, we set of $S_2 = W_2 - C_2 = 100 - 75 = 25$. We bring 25 of assets into period 3. In period 3, we consume 75, which requires saving of $S_3 = W_3 - C_3 = 50 - 75 = -25$. Since we began period 3 with 25 in assets, we can dissave 25 to smooth consumption. In this case, the smoothest plan without borrowing is $C = [60, 75, 75]$ (Figure 3.4, Panel B).

Example 4

Consider a three-period example with incomes of $W = [50, 75, 100]$. The smooth level of consumption would be:

$$\bar{C} = \frac{W_1 + W_2 + W_3}{3} = \frac{50 + 75 + 100}{3} = 75$$

When we solve the budget constraint for period 1, we find it requires savings of $S_1 = W_1 - C_1 = 50 - 75 = -25$. Consuming 75 in period 1 would require borrowing 25, which violates the borrowing constraint. The next-best alternative is to consume 50 in period 1, and save 0.

Next, we try to smooth consumption between periods 2 and 3:

$$\bar{C} = \frac{W_2 + W_3}{2} = \frac{75 + 100}{2} = 87.5$$

We verify by using the annual budget constraint to solve for $S_2 = W_2 - C_2 = 75 - 87.5 = -12.5$. Again, we've encountered the borrowing constraint and cannot smooth consumption. Thus, we consume 75 in period 2. Finally, we consume 100 in period 3. The smoothest plan without borrowing is $C = [50, 75, 100]$ (Figure 3.4, Panel C).

Example 5

Consider a three-period example with incomes of $W = [80, 60, 100]$. The level of smooth consumption would be:

$$\bar{C} = \frac{W_1 + W_2 + W_3}{3} = \frac{80 + 60 + 100}{3} = 80$$

In the first period, we would want to consume 80 and have savings of $S_1 = W_1 - C_1 = 80 - 80 = 0$. The annual budget constraint is satisfied because no borrowing is required to be able to smooth consumption.

In the second period, we would want to consume 80, but this would require borrowing. The most we could consume in period 2 without borrowing would be 60. A rational person would not want to consume 80 in period 1 and then have consumption decreases to 60 in period 2. It would make more sense to consume 70 each in periods 1 and 2. This would require saving of $S_1 = W_1 - C_1 = 80 - 70 = 10$, which would provide assets of 10 from which to smooth consumption in period 2, whereof $S_2 = W_2 - C_2 = 60 - 70 = -10$. Finally, we could consume all 100 in period 3. The smoothest plan without borrowing is $C = [70, 70, 100]$ (Figure 3.4, Panel D).

Try some additional examples of incomes for W_1, W_2, and W_3. You will find that no matter the incomes, with a 3-period model the only choices are the four shown in Figure 3.4.

A Four Period Life-Cycle Model

Consider a four-period life-cycle. Figure 3.5 shows all of the possible consumption plans in a four-period model.

Figure 3.5 A Four-Period LC Model Presents 8 Possible Consumption Plans

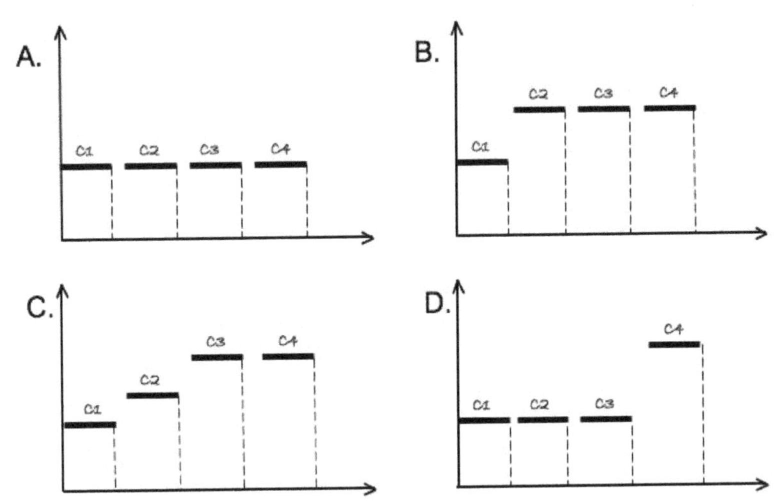

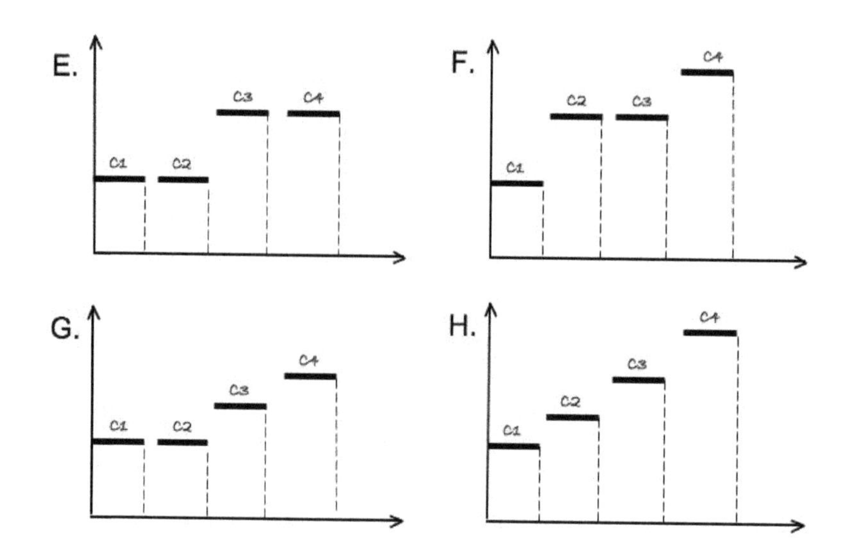

You should experiment with some numerical examples for W_1, W_2, W_3, and W_4 to verify your understanding of these consumption plans presented in Figure 3.5. No matter the sequence of incomes, with 4 periods there are exactly 8 possible consumption plans that can satisfy the borrowing constraints.

3.4 Computational Time

We have just demonstrated that as we increase the number of periods in a LC model, the number of possible consumption plans grows exponentially. For a one-period LC model, there is only one consumption plan. In a two-period LC model, there are two possible plans. For a three-period LC model, there are four possible plans, and for four periods there are eight possible plans. In general we can say that for a LC model with n periods, there will be 2^{n-1}possible consumption plans. Figure 3.6 shows this pattern graphically for up to 12 periods.

Figure 3.6 Under the "Try-All" Approach, the Number of Possible Consumption Plans Grows Exponentially with the Number of Periods

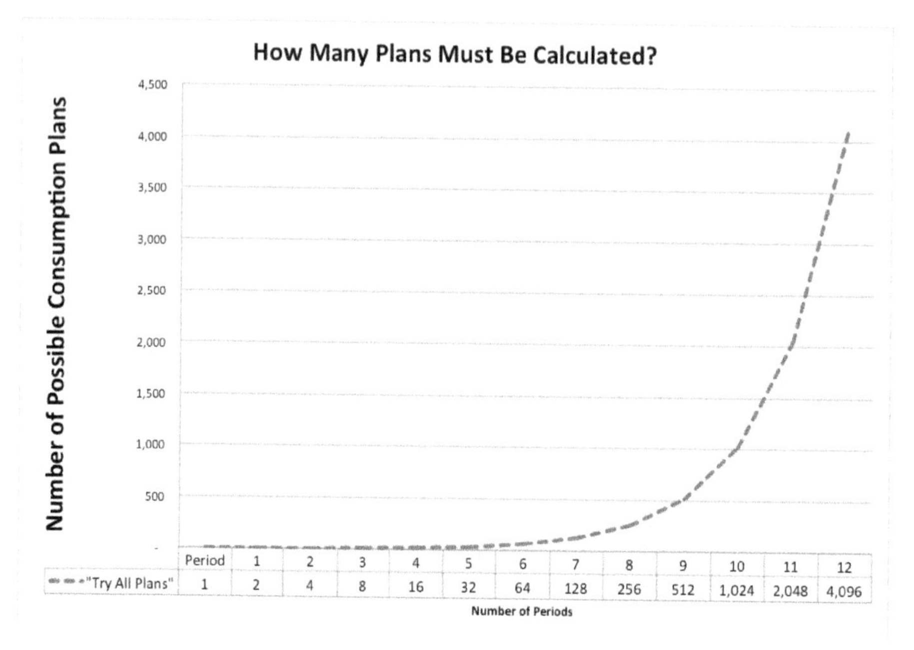

As we see in Figure 3.6, a 10-period LC model would require calculating $2^9 = 512$ possible consumption plans to find the smoothest path without borrowing. A 20-period LC would require calculating $2^{19} = 524{,}288$ possible consumption plans.

Computer scientists measure the efficiency of an algorithm by describing the *computation time* (the number of operations) required to carry out the algorithm as a function of the size of the data. In the case of the try-all approach to finding the smoothest path without borrowing, the worst-case scenario would require solving 2^{n-1}different consumption plans to pick the "best" one for *n* periods in the LC.

As *n* gets very large, the number of possible consumption plans becomes almost unimaginably large. For example, in an 80-period model, there would be 2^{79} possible plans. $2^{79} = 604{,}462{,}909{,}807{,}315{,}000{,}000{,}000$possible consumption plans. To get a feel for how large this number is, the age of the entire universe is estimated to be about 13.75 billion years, which is about 433,600,000,000,000,000 seconds. If we had a computer that was able to calculate 1,500,000 plans per second and had started calculating at the time of the Big Bang, it would be finishing up just about now. This kind of problem is described as being *computationally complex*: the computational requirements of solving all of these consumption plans by the try-all approach are simply infeasible.

3.5 Dynamic Programming

Another approach to finding the smoothest path without borrowing is called dynamic programming. *Dynamic programming*[15] is a divide-and-conquer strategy, which involves solving a complex problem by breaking it down into a series of simpler but overlapping sub-problems. The characteristic of *overlapping sub-problems* is present in consumption smoothing, because each period's budget constraint of how much to consume and save in each period, given the level of income and assets depends on the solutions to all of the later periods' budget constraints. The smallest problem is solved first, and each increasingly more difficult problem is solved in sequence. To this end, the dynamic program for consumption smoothing begins at the *last period* and works backwards to the *initial period.*

Great Minds in Applied Mathematics/Personal Finance: Paul Samuelson

Paul Samuelson (1915-2009) was Professor of Economics at the Massachusetts Institute of Technology.

Samuelson was famous as an economics generalist, with influential publications in international trade, investment, and economics theory. In his book *Foundations of Economic Analysis*, based on his Harvard Ph.D. Dissertation, Samuelson showed that virtually all economic behavior is based on maximization subject to constraints. In doing so, Samuelson brought a new level of mathematical rigor to economics, which had previously been previously been subject to verbal descriptions of events and behaviors. Samuelson's famous *Economics* textbook, first published in 1948, was for decades the best-selling economics textbook, with 19 different editions (most recent in 2010). It has been translated into many 41 different languages.[16]

In a 1969 paper about investment portfolio selection[17], Samuelson demonstrated the use of dynamic programming to solve the series of related decisions required to select the optimal level of consumption throughout one's lifetime. In the 1969 paper, he considered a stochastic model (involving an uncertain but effectively random rate of return on investment), but dynamic programming is also a useful technique in the static (non-random) context to help solve for the highest, smoothest level of consumption under borrowing constraints.

Paul Samuelson was awarded the Nobel Memorial Prize in Economics in 1970 for "for the scientific work through which he has developed static and dynamic economic theory and actively contributed to raising the level of analysis in economic science."[18]

---

[15] The term *dynamic programming* was originally used in the 1940s by Richard Bellman to describe the process of solving problems of this nature, i.e., a series of interrelated equations.
[16] http://en.wikipedia.org/wiki/Economics_(textbook)
[17] Samuelson, Paul A. "Lifetime Portfolio Selection by Dynamic Stochastic Programming." Review of Economics and Statistics, August 1969
[18] "The Sveriges Riksbank Prize in Economic Sciences in Memory of Alfred Nobel 1970." Nobelprize.org. 18 Nov 2012. http://www.nobelprize.org/nobel_prizes/economics/laureates/1970/

Consumption Functions

The basic strategy of using dynamic programming for consumption smoothing involves creating a function to determine the level of consumption for each period. In mathematics, a *function* takes one or several inputs called parameters, applies an equation or process using those parameters, and produces an output or result. A *consumption function* determines the level of consumption, given the available resources (income and assets). The consumption function is written in the form:

$$C_i(W_i, A_i)$$

Let the subscript *i* represent a period in the LC model, such as $1 \leq i \leq n$, where *n* is the number of periods in the LC model. In words, the amount of consumption in period *i* (C_i) depends on the amount of income (W_i) and assets (A_i) in that period.

Absent a bequest motive, a rational person would want to consume all resources during his or her finite lifetime, so there is no reason to have resources left over.[19] For a one period LC model (or the last period of a multi-stage LC), the goal would be to consume all income and all remaining assets:

$$C(W, A) = W + A$$

Decision Rules

For each period other than the last period, the consumption function implements the following two-pronged decision rule:

1. Try to smooth consumption for this period and all subsequent periods, so long as it does not require borrowing. To do so, we try to set the level of consumption such that:

$$C_i(W_i, A_i) = C_{i+1}(W_{i+1}, A_{i+1})$$

   a. Find the amount of smooth consumption, $\bar{C}$.
   b. Calculate the amount of saving required, i.e., $S_i = W_i - C_i$
   c. Calculate the following period's assets, i.e., $A_{i+1} = A_i + S_i$.
   d. Check that $A_{i+1} \geq 0$. If $A_{i+1} \geq 0$, we are able to smooth consumption for the subsequent (remaining) periods.
   e. Recalculate subsequent years' consumption functions to make use of the new amount of assets.

2. If $C_i(W_i, A_i) = C_{i+1}(W_{i+1}, A_{i+1})$ resulted in $A_{i+1} < 0$ (i.e., it required borrowing), it is not possible to smooth consumption.

   a. Set $C_i(W_i, A_i) = W_i + A_i$.
   b. Calculate the amount of saving required, i.e., $S_i = W_i - C_i$.
   c. Calculate the following period's assets, i.e., $A_{i+1} = A_i + S_i$.
   d. If $A_{i+1} > 0$, recalculate subsequent years' consumption functions.
      If $A_{i+1} = 0$, there is no need to recalculate subsequent years' consumption functions.

In words: try to set this year's consumption equal to next year's consumption. Recalculate subsequent year's consumption if needed. However, if smoothing consumption would require borrowing, we will not smooth consumption and instead consume at the highest possible level without borrowing.

---

[19] In many cases, individuals may have a bequest motive to leave assets to their family or charitable organizations when they die. We can adapt this methodology to include bequests, so that we consume at the highest smoothest possible level while leaving enough assets at the end of life to fund the bequest..

It is important to reiterate that the reason for accumulating assets is to be able to maintain a smooth level of consumption from one year to the next. In the event that we cannot smooth consumption, we can at least use the assets to have the smoothest level of consumption without borrowing.

When building a complete solution for n periods, we would have n consumption functions, one for each period. For example, in an 80-period model (i.e., for a 20-year old), there would be 80 consumption functions. Several numerical examples will follow in the next section.

Running Time Analysis

Previously we described the try-all approach as having a computational running time of 2^{n-1}, i.e., it would require trying all 2^{n-1}discrete consumption plans to find the smoothest path without borrowing.

By using the dynamic programming approach, the maximum number of consumption plans to calculate is substantially smaller. In each consumption function, there are 2 choices: select either the smoothest path, or the smoothest path without borrowing. For n periods in the model, each consumption function is calculated at least once. In any given period, if consumption cannot be smoothed for a remaining periods, it might be necessary to recalculate all subsequent consumption values.

For example, in a 3-period model, the solution to C_2 might require recalculating C_3, and the solution to C_1 might require recalculating C_2 and C_3. This worst-case scenario would be that each year's consumption function required recalculating every subsequent year's consumption values. For an n-period LC model, the maximum number of times the consumption functions must be calculated is:

$$1 + 2 + 3 + \cdots + (n-1) = \frac{n(n-1)}{2} = \frac{n^2 - n}{2}$$

We can estimate that the dynamic program will require approximately n^2 calculation steps. This might seem like a large number of calculation steps, but recall that under the "try-all" approach, the worst-case scenario of how many consumption plans to try requires 2^{n-1} computation steps.

Mathematically, $n^2 < 2^n$ for all n. Regardless the number of periods in the life-cycle, the dynamic programming algorithm will be faster than the try-all algorithm. Figure 3.7 provides a graphical illustration of this comparison.

Figure 3.7 With Dynamic Programming, the Number of Possible Consumption Plans Grows with the Square of the Number of Periods

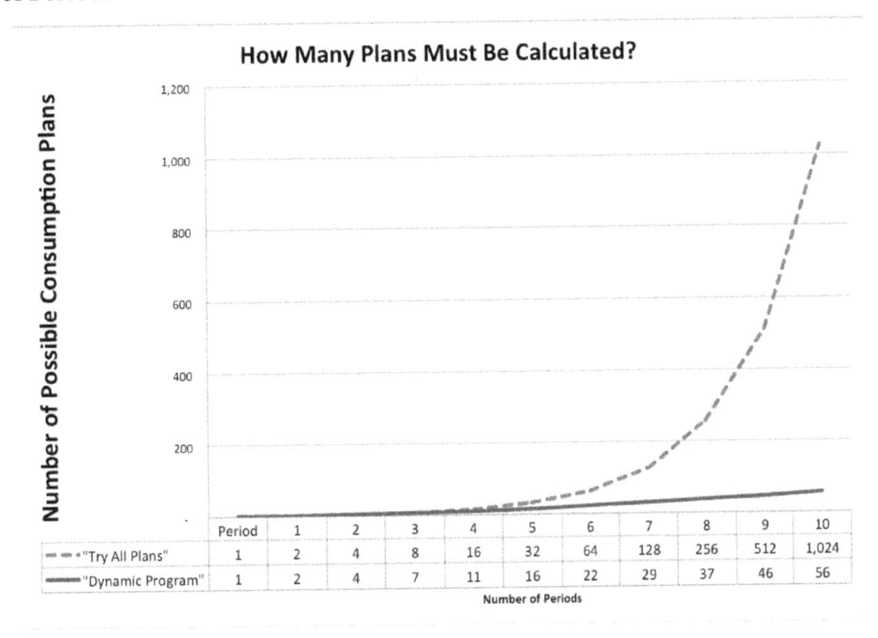

Period	1	2	3	4	5	6	7	8	9	10
"Try All Plans"	1	2	4	8	16	32	64	128	256	512	1,024
"Dynamic Program"	1	2	4	7	11	16	22	29	37	46	56

For an 80-period LC model, the number of computation steps required would be:

$$\frac{n(n-1)}{2} = \frac{80 \times 79}{2} = 3{,}160$$

With modern computers, this relatively small number of computations is easily solved in a fraction of a second.

3.6 *Dynamic Programming: Some Numerical Examples*

In this section, we develop some simple numerical examples to illustrate how dynamic programming works. As in our consumption smoothing spreadsheet, we need to take inputs for the level of income in each period. Our goal is to understand how to smooth consumption under the condition of borrowing constraints.

Finding the Smoothest Path Without Borrowing

To arrive at the smoothest path without borrowing, the consumption functions for each period must be solved together (because they are overlapping sub-problems). *The simplest strategy is to work backwards starting from the last period.*

For each period, we begin by taking all future income, and any current assets, and attempting to allocate this amount over the remaining number of periods. Since we assume no interest or taxes, we can simply use addition (adding all future income) and division (divide by number of periods remaining).

As a general form, the level of smooth consumption $\bar{C}$beginning in period *x* of an *n*-period model would be:

$$\bar{C} = \frac{A_x + \sum_{i=x}^{n} W_i}{n - x + 1}$$

In words, the amount of smooth consumption over periods *x* through *n* is the sum of the assets available in period *x*, plus all income from periods *x* through *n*, divided by the number of periods remaining $(n - x + 1)$.

In each of the examples that follow, we begin with the set of incomes as a vector W. We will create three outcome vectors C, S, and A, for consumption, savings, and assets. Once we know the amount of income (W_i) and the amount of consumption (C_i) for a given period *i*, we can solve for savings $(S_i = W_i - C_i)$ in that period. Also, recall that assets are the cumulative amount of savings from all previous periods. In each consumption function, we know the assets at the start of the period $(A_i)$and we calculate the amount of savings (S_i). Thus, we can find the assets to bring into the next period as $A_{i+1} = A_i + S_i$.

A Three-Period Life Cycle

Consider a three-period LC model with incomes of $W = [30, 50, 40]$ and no initial assets. To begin, we set $A = [0, 0, 0]$. We will discover how much assets we need to accumulate after we determine the smoothest feasible level of consumption. Let's write the vectors $C = [?, ?, ?]$ and $S = [?, ?, ?]$ to create place-holders for the consumption and saving values we find with the dynamic plan.

Finding Consumption for Period 3

In the dynamic program, we begin by first solving for consumption in last period. Recall that in a one-period model, the plan is simple: consume all income and assets. The consumption function for period 3 is:

$$C_3(W_3, A_3) = W_3 + A_3$$

Given $W_3 = 40$ and $A_3 = 0$ we set $C_3 = 40$. We can write the partial solution as $C = [?, ?, 40]$, $S = [?, ?, 0]$. The assets vector remains $A = [0, 0, 0]$.

Finding Consumption for Period 2

Now we begin to work backwards and find the plan for period 2. We can effectively treat this as a two-period LC model. The goal is to smooth consumption between periods 2 and 3, so long as it doesn't require borrowing. We use the income for periods 2 and 3, and assets at the start of period 2, and try to smooth consumption. As of period 2, the incomes are $W_2 = 50$ and $W_3 = 40$. We also have $A_2 = 0$. With two periods remaining, we can find the level of smooth consumption to be:

$$\bar{C} = \frac{W_2 + A_2 + W_3}{2} = \frac{50 + 0 + 40}{2} = 45$$

Is it feasible to consume 45 in each period, without violating the borrowing constraint? To find out, we plug values into the periodic budget constraint and check whether borrowing would be required. Given $W_2 = 50$ and $C_2 = 45$, we find:

$$S_2 = W_2 - \bar{C} = 50 - 45 = 5$$

Recall that $A_2 = 0$, so we can calculate the next period's assets as:

$$A_3 = A_2 + S_2 = 0 + 5 = 5$$

Since $A_3 > 0$, we have satisfied the borrowing constraint: we are able to smooth consumption without borrowing. In period 3, the budget constraint is:

$$S_3 = W_3 - \bar{C} = 40 - 45 = -5$$

In words: to smooth consumption, we must dissave 5. Since we began period 3 with $A_3 = 5$, we are able to dissave 5 and arrive at ending assets of 0. We update the partial solution is updated to $C = [?, 45, 45]$, $S = [?, 5, -5]$ and $A = [0, 0, 5]$.

Finding Consumption for Period 1

We continue to work backwards and find the plan for period 1. The goal is to smooth consumption for all 3 periods. The decision rule is to set $C_1(W_1, A_1) = C_2 = C_3$, so long as it will result in $A_2 \geq 0$ (i.e., no borrowing). However, if smoothing consumption would require borrowing, then set $C_1 = W_1 + A_1$ (i.e., consume at the highest level without borrowing).

We use the income for periods 1, 2 and 3, and assets at the start of period 1, and try to smooth consumption. As of period 1, we know the values of all future incomes is $W = [30, 50, 40]$, and know that$A_1 = 0$. With 3 periods remaining, we find the level of smooth consumption to be:

$$\bar{C} = \frac{W_1 + A_1 + W_2 + W_3}{3} = \frac{30 + 0 + 50 + 40}{3} = 40$$

Is it feasible to consume 40 in each period, without violating the borrowing constraint? To find out, we plug values into the periodic budget constraint and check whether borrowing would be required. Given $W_1 = 30$ and$\bar{C} = 40$, we find:

$$S_1 = W_1 - \bar{C} = 30 - 40 = -10$$

Recall that $A_1 = 0$, so we calculate the next period's assets as:

$$A_2 = A_1 + S_1 = 0 + (-10) = -10$$

Since $A_2 < 0$, we have encountered the borrowing constraint. We cannot smooth consumption without borrowing, and borrowing is not allowed. The next best alternative is to consume at the highest level without borrowing:

$$C_1 = W_1 + A_1 = 30 + 0 = 30$$

We can thus find the level of savings as:

$$S_1 = W_1 - C_1 = 0$$

Since $S_1 = 0$, we do not change the level of A_2. Therefore, we *do not* need to recalculate the consumption functions for periods 2 or 3. The complete solution can be shown as $C = [30, 45, 45]$, $S = [0, 5, -5]$, and $A = [0, 0, 5]$. Figure 3.8 illustrates the consumption plan.

Figure 3.8 The Consumption Plan for the 3-Period Life-Cycle

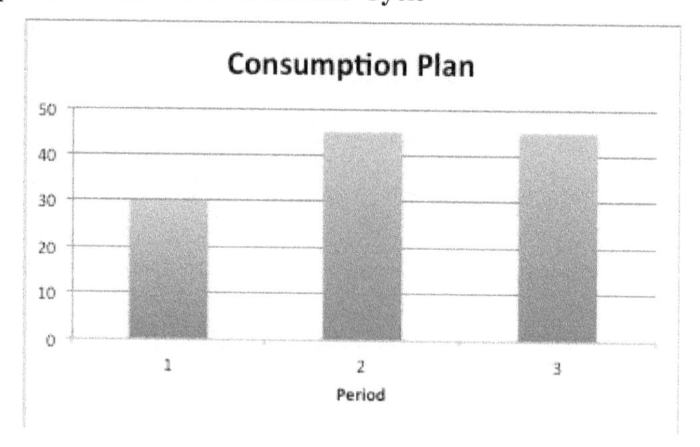

Compare the graph in Figure 3.8 to the graph in Figure 3.4 Panel B. This consumption plan is the smoothest path without borrowing.

A Four-Period Life Cycle

Consider a four-period LC model with incomes of $W = [37, 45, 50, 40]$ and no initial assets. To begin, we set $A = [0, 0, 0, 0]$. Let's write the vectors $C = [?,?,?,?]$ and $S = [?,?,?,?]$ to create place-holders for the consumption and saving values we find with the dynamic plan.

Finding Consumption for Period 4

In the dynamic program, we begin by first solving for consumption in last period. The consumption function for period 4 is:

$$C_4(W_4, A_4) = W_4 + A_4$$

Given $W_4 = 40$ and $A_4 = 0$ we find in $C_4 = 0$.
The partial solution can be shown as $C = [?,?,?,40], S = [?,?,?,0]$, and $A = [0, 0, 0, 0]$.

Finding Consumption for Period 3

Now we begin to work backwards and find the plan for period 3. As of period 3, the incomes are $W_3 = 50$ and $W_4 = 40$. We also have $A_3 = 0$. With two periods remaining, we can find the level of smooth consumption to be:

$$\bar{C} = \frac{W_3 + A_3 + W_4}{2} = \frac{50 + 0 + 40}{2} = 45$$

Is it feasible to consume 45 in each period, without violating the borrowing constraint? Given the income of $W_3 = 50$ and $C_3 = 45$, we find:

$$S_3 = W_3 - \bar{C} = 50 - 45 = 5.$$

Recall that $A_3 = 0$, so we can calculate the next period's assets as:

$$A_4 = A_3 + S_3 = 0 + 5 = 5.$$

Since $A_4 > 0$, we have satisfied the borrowing constraint: we are able to smooth consumption without borrowing. In period 4, the budget constraint is:

$$S_4 = W_4 - \bar{C} = 40 - 45 = -5.$$

In words: to smooth consumption, we must dissave 5. Since we entered period 4 with $A_4 = 5$, we can dissave 5 and arrive at ending assets of 0.
The partial solution can be shown as $C = [?, ?, 45, 45]$, $S = [?, ?, 5, -5]$, and $A = [0, 0, 0, 5]$.

Finding Consumption for Period 2

We continue to work backwards and find the plan for period 2. The goal is to smooth consumption for all 3 periods. The decision rule is to set $C_2(W_2, A_2) = C_3 = C_4$, so long as it will result in $A_3 \geq 0$ (i.e., no borrowing). However, if smoothing consumption would require borrowing, then set $C_2 = W_2 + A_2$(i.e., consume at the highest level without borrowing).

As of period 2, we know the values of all future incomes are $W_2 = 45$, $W_3 = 50$ and $W_4 = 40$, and know that $A_2 = 0$. With 3 periods remaining, we find the level of smooth consumption to be:

$$\bar{C} = \frac{W_2 + A_2 + W_3 + W_4}{3} = \frac{45 + 0 + 50 + 40}{3} = \frac{135}{3} = 45$$

Is it feasible to consume 45 in each period, without violating the borrowing constraint? Given the income of $W_2 = 45$ and $C_2 = 45$, we find:

$$S_2 = W_2 - \bar{C} = 45 - 45 = 0$$

Since $S_2 = 0$, no dissaving is required and we have not encountered a borrowing constraint. Further, since $S_2 = 0$, we do not change the level of A_3 nor do we need to recalculate the consumption functions for periods 3 or 4. The partial solution can be shown as $C = [?, 45, 45, 45]$, $S = [?, 0, 5, -5]$, and $A = [0, 0, 0, 5]$.

Finding Consumption for Period 1

We continue to work backwards and find the plan for period 1. The goal is to smooth consumption for all 4 periods. The decision rule is to set $C_1(W_1, A_1) = C_2 = C_3 = C_4$,, so long as it will result in $A_2 \geq 0$ (i.e., no borrowing). However, if smoothing consumption would require borrowing, then set $C = W_1 + A_1$(i.e., consume at the highest level without borrowing).

As of period 1, we know the values of all future incomes is $W_1 = 37$, $W_2 = 45$, $W_3 = 50$ and $W_4 = 40$, and know that $A_1 = 0$. With 3 periods remaining, we find the level of smooth consumption to be:

$$\bar{C} = \frac{W_1 + A_1 + W_2 + W_3 + W_4}{4} = \frac{37 + 0 + 45 + 50 + 40}{4} = \frac{172}{4} = 43$$

Is it feasible to consume 43 in each period, without violating the borrowing constraint? Given income of $W_1 = 37$ and $C_1 = 43$, we find:

$$S_1 = W_1 - \bar{C} = 37 - 43 = -6.$$

Recall that $A_1 = 0$, so we calculate the next period's assets as:

$$A_2 = A_1 + S_1 = 0 + (-6) = -6.$$

Since $A_2 < 0$, we have encountered the borrowing constraint. We cannot smooth consumption without borrowing, and borrowing is not allowed. The next best alternative is to consume at the highest level without borrowing:

$$C_1 = W_1 + A_1 = 37 + 0 = 37.$$

We can thus find the level of savings as:

$$S_1 = W_1 - C_1 = 0.$$

Since $S_1 = 0$, we do not change the level of A_2. Therefore, we do not need to recalculate the consumption functions for periods 2, 3, and 4. The complete solution can be shown as $C = [37, 45, 45, 45]$, $S = [0, 0, 5, -5]$, and $A = [0, 0, 0, 5]$. Figure 3.9 illustrates the consumption plan.

Figure 3.9 The Consumption Plan for the 4-Period Life-Cycle

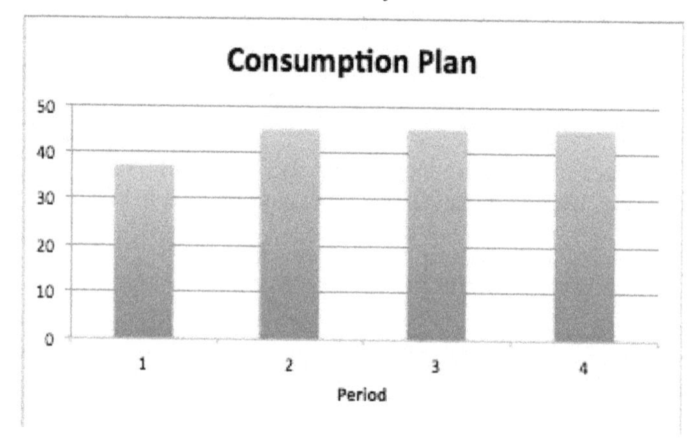

Compare the graph in Figure 3.9 to the graph in Figure 3.5 Panel B. This consumption plan is the smoothest path without borrowing.

Incorporating Initial Assets

The two previous examples (both the 3-period and 4-period dynamic programs) assumed income in each period, but no initial assets. Now we consider a four-period example with initial assets of 100. Let $W = [20, 40, 60, 40]$, and let the vector of assets be $A = [100, 0, 0, 0]$ before we begin the dynamic program. We create vectors $C = [?, ?, ?, ?]$ and $S = [?, ?, ?, ?]$ as place-holders for the consumption and saving values to be calculated with the dynamic program.

Finding Consumption for Period 4

Beginning in period 4, we have $W_4 = 40$ and $A_4 = 0$. We set $C_4 = W_4 + A_4 = 40 + 0 = 40$. The partial solution can be shown as $C = [?, ?, ?, 40]$, $S = [?, ?, ?, 0]$, and $A = [100, 0, 0, 0]$.

Finding Consumption for Period 3

In period 3, we try to smooth consumption across 2 periods:

$$\bar{C} = \frac{W_3 + A_3 + W_4}{2} = \frac{60 + 0 + 40}{2} = 50$$

We calculate savings in period 3 as:

$$S_3 = W_3 - \bar{C} = 60 - 50 = 10.$$

This savings in period 3 is added to the assets in period 4, so $A_4 = 10$. We recalculate the consumption function for period 4 as:

$$C_4 = W_4 + A_4 = 40 + 10 = 50.$$

The annual budget constraint is satisfied with $A_4 = 10$, which allows dissaving 10 in period 4. The partial solution can be shown as $C = [?,?,50,50]$, $S = [?,?,10,-10]$, and $A = [100,0,0,10]$.

Finding Consumption for Period 2

In period 2, we try to smooth consumption across 3 periods (rounded to nearest whole dollar):

$$\bar{C} = \frac{W_2 + A_2 + W_3 + W_4}{3} = \frac{40 + 0 + 60 + 40}{3} = \frac{140}{3} = 47$$

We calculate the periodic savings required to smooth consumption as:

$$S_2 = W_2 - \bar{C} = 40 - 47 = -7$$

To smooth consumption across periods 2, 3, and 4 would require dissaving 7 in the second period. However, assets at the start of period 2 are 0, so dissaving 7 would require borrowing, which is not possible. We have encountered the borrowing constraint in period 2 and we must instead consume the maximum amount possible without borrowing in period 2, i.e., $C_2 = 40$, and set $S_2 = 0$. Since $S_2 = 0$, A_3 will remain 0, and we do not need to recalculate the consumption functions for periods 3 and 4. The partial solution can be shown as $C = [?,40,50,50]$, $S = [?,0,10,-10]$, and $A = [100,0,0,10]$.

Finding Consumption for Period 1

In period 1, we try to smooth consumption across 4 periods. Notice that we are now finally taking into account the initial assets ($A_1 = 100$), which we will use to smooth consumption. With four periods in which to consume, we find the level of smooth consumption to be:

$$\bar{C} = \frac{W_1 + A_1 + W_2 + W_3 + W_4}{4} = \frac{20 + 100 + 40 + 60 + 40}{4} = \frac{260}{4} = 65$$

We calculate the periodic savings required to smooth consumption as:

$$S_1 = W_1 - \bar{C} = 20 - 65 = -45.$$

With $A_1 = 100$, and $S_1 = -45$ we calculate $A_2 = A_1 + S_1 = 100 - 45 = 55$. Since $A_2 > 0$, there is no borrowing constraint in period 1.

Next, we re-calculate the consumption function for the second period with $W_2 = 40$, $A_2 = 55$, and $\underline{C} = 65$. The level of savings required in period 2 is:

$$S_2 = W_2 - \bar{C} = 40 - 65 = -25$$

Since $A_2 = 55$, it is possible to dissave 25. This leaves $A_3 = A_2 + S_2 = 55 - 25 = 30$. Since $A_3 > 0$, there is no borrowing constraint in period 2.

Next, we recalculate the consumption functions for periods 3 and 4. In period 3, we begin with $W_3 = 60$, $A_3 = 30$, and $\bar{C} = 65$. We can now find the amount of savings required to satisfy the periodic budget constraint:

$$S_3 = W_3 - \bar{C} = 60 - 65 = -5.$$

Thus, we must dissave 5 in period 3, which we can do since $A_3 = 30$. This leaves $A_4 = A_3 + S3 = 30 - 5 = 25$.

Finally, in period 4,we begin with $W_4 = 40$, $A_4 = 25$, and $\bar{C} = 65$. To satisfy the periodic budget constraint, we need saving of:

$$S_4 = W_4 - \bar{C} = 40 - 65 = -25.$$

Since $A_4 = 25$, we can dissave 25 and end up with 0 of regular assets at the end of the plan. The complete solution can be shown as $C = [65, 65, 65, 65]$, $S = [-45, -25, -5, -25]$, and $A = [100, 55, 30, 25]$.Figure 3.10 illustrates the consumption plan.

Figure 3.10 The Consumption Plan for the 4-Period Life-Cycle with Initial Assets

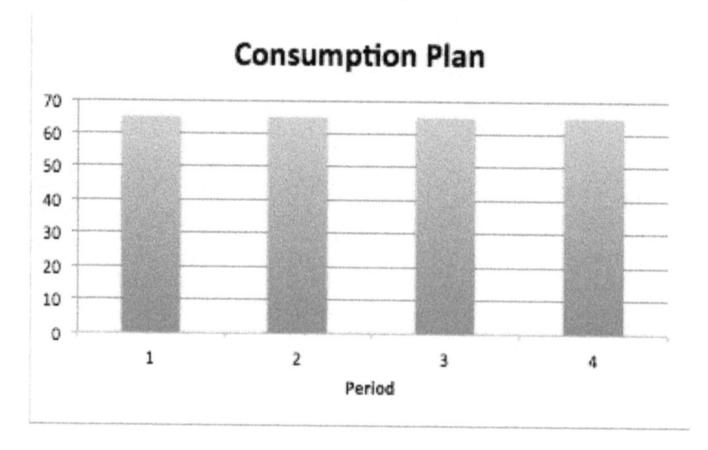

A Four Period Life-Cycle with Multiple Periods of Borrowing Constraints

In the previous examples, we have seen cases in which the income path led to some years of borrowing constraints, followed by some years of being able to smooth consumption. Another subtle case of borrowing constraints could arise, wherein a household could face borrowing constraints at multiple periods of its life cycle. Further, it might be possible to smooth consumption for some years at one level and then be able to smooth consumption in later years at a higher level.

As a numerical example, consider the following vector of incomes $W = [60, 40, 70, 50]$, with no initial assets. As in the previous examples, we begin by finding the level of consumption in the final period and working backwards to the first period.

Finding Consumption for Period 4

In period 4, $W_4 = 50$ and $A_4 = 0$. Thus $C_4 = W_4 + A_4 = 50 + 0 = 50$. The partial solution can be shown as $C = [?, ?, ?, 50]$, $S = [0, 0, 0, 0]$, and $A = [0, 0, 0, 0]$.

Finding Consumption for Period 3

In period 3, we find that total resources are: $W_3 + A_3 + W_4 = 70 + 0 + 50$. With two periods in which to consume, we find the level of smooth consumption to be:

$$\bar{C} = \frac{W_3 + A_3 + W_4}{2} = \frac{70 + 0 + 50}{2} = 60$$

We calculate the savings required in period 3 as:

$$S_3 = W_3 - \bar{C} = 70 - 60 = 10.$$

This savings in period 3 adjusts the assets in period 4, so $A_4 = 10$. Finally, we recalculate the consumption function for period 4 as:

$$C_4 = W_4 + A_4 = 50 + 10 = 60.$$

The annual budget constraint is satisfied with $A_4 = 10$, which allows dissaving 10 in period 4. The partial solution can be shown as $C = [?, ?, 60, 60]$ and $A = [0, 0, 0, 10]$, with $S = [?, ?, 10, -10]$.

Finding Consumption for Period 2

In period 2, we find that the total resources are: $W_2 + A_2 + W_3 + W_4 == 40 + 0 + 70 + 50 = 160$. With three periods in which to consume, we find the level of smooth consumption to be (cents omitted):

$$\bar{C} = \frac{W_2 + A_2 + W_3 + W_4}{3} = \frac{40 + 0 + 70 + 50}{3} = \frac{160}{3} = 53$$

We calculate the periodic savings required to smooth consumption as:

$$S_2 = W_2 - \bar{C} = 40 - 53 = -13$$

To smooth consumption across periods 2, 3, and 4 would require dissaving 13 in period 2. However, assets at the start of period 2 are 0, so dissaving 13 would require borrowing and borrowing is not possible. Since we have encountered the borrowing constraint, we must instead consume the maximum amount possible in period 2 without borrowing: $C_2 = 40$ and $S_2 = 0$. Since $S_2 = 0$, A_3 will remain 0, and we do not need to recalculate the consumption functions for periods 3 and 4. The partial solution can be shown as $C = [?, 40, 60, 60]$, $S = [?, 0, 10, -10]$, and $A = [0, 0, 0, 10]$.

Finding Consumption for Period 1

Having calculated the level of consumption for periods 2, 3, and 4 and having encountered the borrowing constraint in period 2, we now calculate consumption for period 1. To smooth consumption, we would set the level of consumption at:

$$\bar{C} = \frac{W_1 + A_1 + W_2 + W_3 + W_4}{4} = \frac{60 + 0 + 40 + 70 + 50}{4} = \frac{220}{4} = 55$$

We check this consumption plan by plugging C into each period's budget constraint to verify that it can be done without borrowing. In the first period, the periodic budget constraint would require saving:

$$S_1 = W_1 - \bar{C} = 60 - 55 = 5$$

We would enter the second period withassets $A_2 = 5$. In the second period, the periodic budget constraint would require saving:

$$S_2 = W_2 - \bar{C} = 40 - 55 = -15$$

However, since $A_2 = 5$, it is not possible to dissave 15. Our program encountered another borrowing constraint: we cannot smooth consumption from periods 1 through 4, since it would require borrowing in period 2.

In this case, we can still smooth consumption across periods 1 and 2. The level of smooth consumption for periods 1 and 2 is:

$$\bar{C} = \frac{W_1 + A_1 + W_2}{2} = \frac{60 + 0 + 40}{2} = \frac{100}{2} = 50$$

To verify that this is a feasible plan, we check the periodic budget constraint for each period. In period 1, we find that savings is:

$$S_1 = W_1 - \bar{C} = 60 - 50 = 10$$

Recall that saving in the first period adds to the value of assets that are brought into the second period. We recalculate $A_2 = A_1 + S_1 = 0 + 10 = 10$. In the second period, the budget constraint would require savings of:

$$S_2 = W_2 - \bar{C} = 40 - 50 = -10$$

Since $A_2 = 10$, it *is* possible to dissave 10 without borrowing.

The complete solution can be shown as $C = [50, 50, 60, 60]$, $S = [10, -10, 10, -10]$, and $A = [0, 10, 0, 10]$. Figure 3.11 illustrates the consumption plan.

Figure 3.11 The Consumption Plan for the 4-Period Life-Cycle with Two Periods of Borrowing Constraints

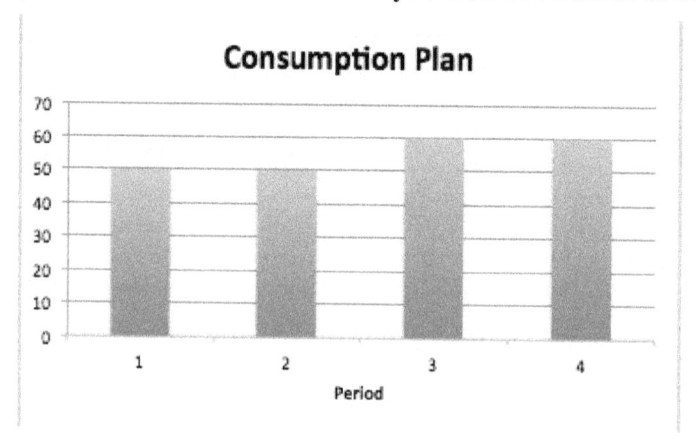

Compare the graph in Figure 3.11 to the graph in Figure 3.5 Panel E. This consumption plan is the smoothest path without borrowing.

3.7 Violations of Borrowing Constraints

The discussion in this module has focused on explaining how to choose the highest level of consumption without borrowing against future income. However, some real-life situations violate borrowing constraints. For example, credit cards, student loans, and payday loans all violate borrowing constraints in one way or another. The common characteristic of these loans is that they are unsecured loans (i.e., not backed by collateral). The borrowers wish to borrow against future income to consume in the present, and in each case the lenders make loans based on future income *despite the borrowing constraints*.

Unsecured loans that are not backed by collateral are generally risky to lenders. The lenders that make these kinds of loans know that some borrowers will not be able to repay in full or on time. Since they are in business to make a profit, lenders making unsecured loans do so by charging unreasonably high interest rates, having onerous contract terms, or organizing special arrangements in which their loans are guaranteed by a third party.

For example, credit cards charge extremely high interest rates, currently around 15% per year (compared to around 4% for a home mortgage loan or car loan). Payday loans charge even higher interest rates (or fees that behave like interest rates), some approaching 100% per year or more. Student loans generally require a third-party guarantor (e.g., the government secures the repayment of the loan, or a parent must co-sign to accept responsibility for repayment). In addition, student loans lack basic consumer protections, as they cannot be refinanced or discharged in bankruptcy. In general, loans that violate borrowing constraints are a bad deal for borrowers.

As we revisit borrowing constraints in future modules, we will examine many personal financial decisions that lead to being borrowing constrained as well as decisions that might alleviate borrowing constraints.

3.8 Summary

Life-cycle planning would be easy if we could know with certainty our future income, and could easily borrow against future income or assets to smooth consumption in the present. Borrowing constraints arise when future income is higher than current income or when additional assets will be available in the future but are not accessible currently. Under conditions of borrowing constraints, it is often not possible to smooth consumption. Rational individuals who would want to maximize their utility by consuming smoothing cannot do so.

Finding the smoothest consumption plan without borrowing is not arithmetically challenging, but when the number of periods becomes large the problem becomes computationally complex. For a realistic 80-period LC model, it would be computationally infeasible to find the smoothest path without borrowing by a brute-force try-all strategy, as the number of consumption plans to try would approach 2^n.

Dynamic programming provides a strategy to solve the consumption-smoothing problem in a reasonable amount of time, even with borrowing constraints. A dynamic program works by creating a series of overlapping equations that must be solved together. A consumption function is a plan for how much to consume and save in a certain period, given your income and assets in that period and the income in future periods. The basic decision rule is simple: try to smooth consumption if that does not require borrowing. Otherwise, consume at the highest level without borrowing. By working backwards, it is possible to find the smoothest path without borrowing in a reasonable amount of computation time. The dynamic program would require as an upper bound of calculating about $(n^2 - n)/2$ consumption plans.

Some lenders do offer unsecured loans in violation of borrowing constraints. While these loans might help alleviate borrowing constraints, they are generally a bad deal for consumers.

3.9 Review Questions

Conceptual Questions

1. Briefly define borrowing constraints. How can you determine if a household is borrowing constrained?
2. Identify and explain 3 reasons why a household might be borrowing constrained.
3. What is the advantage of using dynamic programming to find the level consumption as compared to the try-all approach? Give a mathematical reasoning.
4. Why are unsecured loans generally unfavorable to consumers (i.e., high interest rates)?

Numerical Questions

In each question that follows, use dynamic programming to find the smoothest level of consumption without borrowing for each period. Find consumption and savings for each period.

5. Let W = [100, 90, 110] and use initial assets of 0.
6. Let W = [100, 115, 95] and use initial assets of 0.
7. Let W = [75, 85, 100, 80] and use initial assets of 0.
8. Let W = [100, 70, 100, 80] and use initial assets of 0.
9. Let W = [50, 100, 200, 100] and use initial assets of 100.
10. Let W = [80, 100, 80, 50] and use initial assets of 30. Plan to leave final assets (i.e., a bequest) of 40.

4 Thinking in Present Value

Learning Objectives

- Illustrate the power of exponential growth and decay through the time value of money: the relationship between present values, interest rates, and future values.
- Explain the algebra of calculating future values and present values.
- Explain the limits of exponential growth and compounding.
- Consider changes in the price level, inflation, and purchasing power.

4.1 Introduction: Savings, Investment, and Interest

In the previous modules, we discussed the role of savings in the economic life-cycle. *Savings* is income that is not consumed. Households use savings to move wealth through time, for example, from the period of work (when young) to the period of non-work (when old). An *investment* is made to grow the amount of savings over time. Households make investments by lending their savings to another agent in the economy (an individual, household, or firm). The person lending the money is making the investment, and the borrower is taking on a debt (a promise to repay). The borrower's liability is the investor's asset.

Recall that money is a means of exchange, but it is not the only store of value. Money can be used for immediate consumption or to buy assets that can store value for future consumption. *Interest* is the "rental fee" for a loan of money, and is paid from the borrower to the lender (investor). Interest helps quantify the time-preference of having money for consumption. To the lender, interest is the incentive to forgo consumption, i.e., payment for deferred gratification. To the borrower, interest is the cost of borrowing, i.e., how much future consumption must be forgone to have money for consumption now.

Financial Intermediation

Households with excess money (potential lenders) and households with a shortage of money (potential borrowers) are free to make borrowing/lending arrangements directly with each other. However, borrowers might not be trustworthy to repay in full and on time, and lenders might not want to deal with the hassle of collection.

A bank is a firm that provides *financial intermediation* services between households or firms within the economy. Households that have excess money can make deposits at banks, for example by purchasing a Certificate of Deposit. A *Certificate of Deposit* (CD) is a deposit account in which the depositor agrees to leave the funds in place for a fixed period of time. Common time periods for CDs range from 3 months to 5 years.

Banks usually advertise the rates offered on CDs in the window of the bank building, and on the bank's website. Interest rates are not static, but rather change often. When you buy a CD, you will get a fixed rate for the period of time of the CD (i.e., for 3 months or 5 years), but the rates the bank will offer change periodically, perhaps even daily. A discussion of the different interest rates and what affects the level of interest rates will be discussed at the end of this module.

The Cash Flow Timeline

A simple way to explain the concept of a time deposit is with a cash flow timeline. The cash flow timeline shows time periods on the top, and dollars on the bottom. The cash flow timeline always starts at "today" (or time 0), and extends to the right for future periods (months or years from now). By accounting convention, cash outflows are shown as negative numbers, and cash inflows are shown as positive numbers. Mathematically, only the absolute value of the amounts will matter.

For example, suppose you invest $1,000 in a 1-year Certificate of Deposit at your local savings bank. The bank offers to pay you 10% interest on your deposit for one year.[20] You give the bank $1,000 (called the principal) today, and the bank promises to return your $1,000 in one year along with interest. At an interest rate of 10%, your

---

[20] At the time of this writing in August 2013, the interest rates offered by banks are at a historical low point, and rates of return on Certificates of Deposit are in the 1% range. Historically higher rates have been more common, and in the first examples we use a 10% rate for illustrative purposes.

investment would earn interest of $\$1{,}000 \times 10\% = \$100$ for the year. Figure 4.1 shows a cash flow timeline for this investment in a 1-year CD.

Figure 4.1 Cash Flow Timeline for a 1-Year Investment.

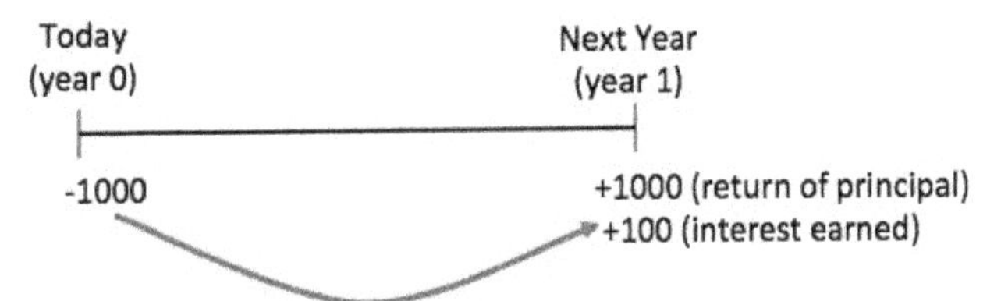

In Figure 4.1, the deposit is made today, so you would have a cash outflow of $1,000. The bank will return your $1,000 of principal one year from now, along with $100 of interest. The $1,000 return of principal and the $100 of interest are shown as positive numbers, since you will have cash coming to you at that time.

More generally, we refer to amounts of money today as ***present values***, and the amounts to be received in the future as ***future values***. Figure 4.2 shows the cash flow timeline with the dollars amounts as above, as well as the present value (labeled PV) and future value (labeled FV).

Figure 4.2 Cash Flow Timeline: Present Value and Future Value

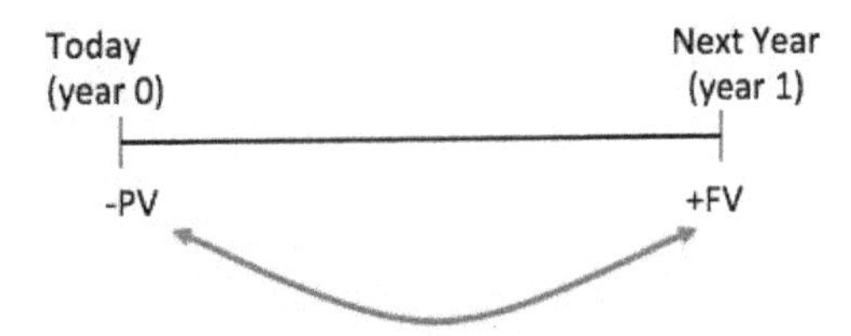

Let's extend the previous example: suppose you reinvest your funds with the bank for a second year, also at 10% interest. After 1 year, you would have had $1,000 of principal plus $100 of interest, for a total of $1100 to reinvest. Figure 4.3 shows the cash flow timeline for this investment of $1100 beginning 1 year from now.

Figure 4.3 Cash Flow Timeline: Reinvesting for a Second Year

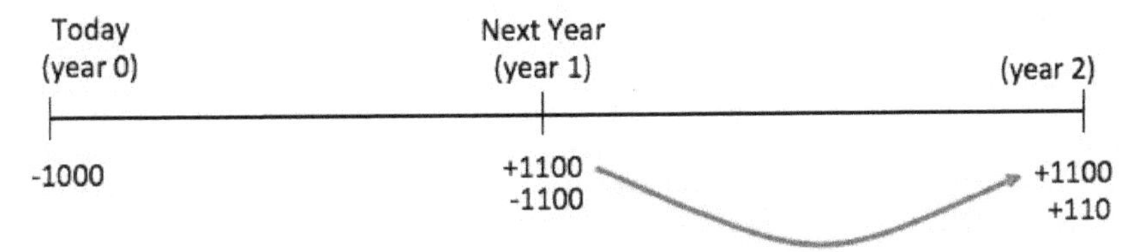

In Figure 4.3, the $1,100 cash inflow from the first year's investment is exactly offset by a cash outflow of $1,100 (the amount invested for the second year). At the end of the second year, you would receive a cash inflow of $1,100 for the return of the principal invested for the year, and an interest payment of $\$1100 \times 10\% = \$110$. At the end of the 2-year investment, you would receive from the bank a total of $1,210.

With a 2-year investment to a Certificate of Deposit, we don't need to think about the offsetting cash inflow and cash outflow happening 1 year from today. What really matters is how much you invest today, and how much you receive back at the end of the 2-year investment. Figure 4.4 shows a cash flow timeline for the complete 2-year investment.

Figure 4.4 Cash Flow Timeline for a 2-year Investment

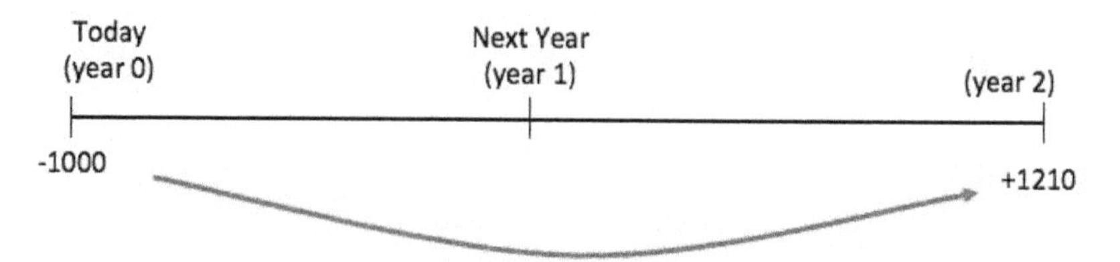

Compound Interest

We can refine our definition of interest as follows: *Simple interest* is the interest earned on the principal. *Compound interest* is the interest earned on previously earned interest.

The previous example illustrated the effect of investing for multiple years. Beginning with principal of $1,000, 10% interest for the first year was $100. In the second year, $1,100 was reinvested at 10% interest. The interest earned in the second year is $\$1,100 \times 10\% = \$110$. Of this amount, $100 was simple interest and $10 was compound interest. The table in Figure 4.5 carries out this example for thirty years.

Figure 4.5 Calculation of Compound Interest at 10% Interest for 30 years.

Year	Beginning Amount	Interest Earned	Ending Amount	Compound Interest
1	$ 1,000.00	$ 100.00	$ 1,100.00	$ -
2	$ 1,100.00	$ 110.00	$ 1,210.00	$ 10.00
3	$ 1,210.00	$ 121.00	$ 1,331.00	$ 21.00
4	$ 1,331.00	$ 133.10	$ 1,464.10	$ 33.10
5	$ 1,464.10	$ 146.41	$ 1,610.51	$ 46.41
6	$ 1,610.51	$ 161.05	$ 1,771.56	$ 61.05
7	$ 1,771.56	$ 177.16	$ 1,948.72	$ 77.16
8	$ 1,948.72	$ 194.87	$ 2,143.59	$ 94.87
9	$ 2,143.59	$ 214.36	$ 2,357.95	$ 114.36
10	$ 2,357.95	$ 235.79	$ 2,593.74	$ 135.79
11	$ 2,593.74	$ 259.37	$ 2,853.12	$ 159.37
12	$ 2,853.12	$ 285.31	$ 3,138.43	$ 185.31
13	$ 3,138.43	$ 313.84	$ 3,452.27	$ 213.84
14	$ 3,452.27	$ 345.23	$ 3,797.50	$ 245.23
15	$ 3,797.50	$ 379.75	$ 4,177.25	$ 279.75
16	$ 4,177.25	$ 417.72	$ 4,594.97	$ 317.72
17	$ 4,594.97	$ 459.50	$ 5,054.47	$ 359.50
18	$ 5,054.47	$ 505.45	$ 5,559.92	$ 405.45
19	$ 5,559.92	$ 555.99	$ 6,115.91	$ 455.99
20	$ 6,115.91	$ 611.59	$ 6,727.50	$ 511.59
21	$ 6,727.50	$ 672.75	$ 7,400.25	$ 572.75
22	$ 7,400.25	$ 740.02	$ 8,140.27	$ 640.02
23	$ 8,140.27	$ 814.03	$ 8,954.30	$ 714.03
24	$ 8,954.30	$ 895.43	$ 9,849.73	$ 795.43
25	$ 9,849.73	$ 984.97	$ 10,834.71	$ 884.97
26	$ 10,834.71	$ 1,083.47	$ 11,918.18	$ 983.47
27	$ 11,918.18	$ 1,191.82	$ 13,109.99	$ 1,091.82
28	$ 13,109.99	$ 1,311.00	$ 14,420.99	$ 1,211.00
29	$ 14,420.99	$ 1,442.10	$ 15,863.09	$ 1,342.10
30	$ 15,863.09	$ 1,586.31	$ 17,449.40	$ 1,486.31
	Total Interest	$ 16,449.40	Compound Interest	$ 13,449.40

As we see in Figure 4.5, the column labeled Interest Earned is calculated as the Beginning Amount times the interest rate. The column Ending Amount is the sum of the Beginning Amount plus the Interest Earned in the year. The last column labeled Compound Interest is the amount of compound interest earned, i.e., the amount of interest in excess of the simple interest. The exciting thing to notice about the compound interest is not only that the compound interest is increasing, but also that the compound interest is increasing at an increasing rate. The graph in Figure 4.6 illustrates the compounding effect.

Figure 4.6 Principal, Simple Interest, and Compound Interest, Calculated with 10% Annual Interest

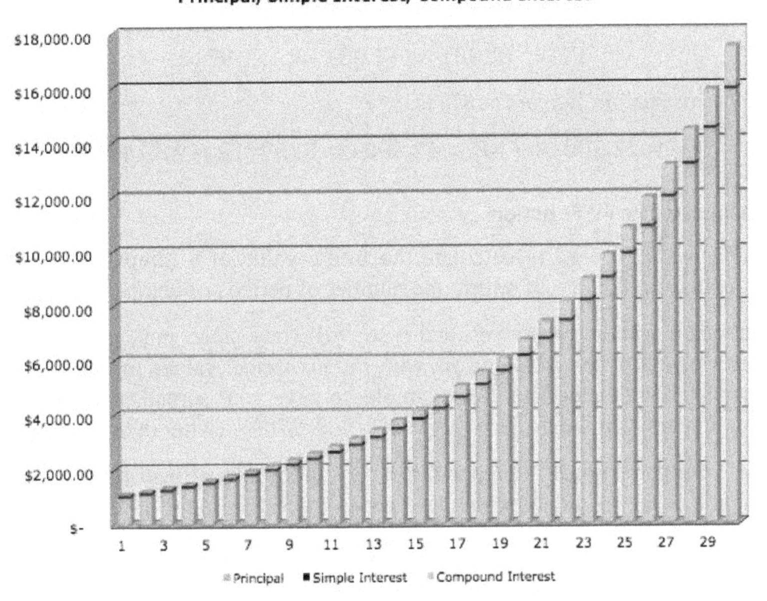

It has often been repeated that the famous physicist and Nobel Laureate Albert Einstein answered "Compound Interest," when asked to name the most powerful force in the universe.[21]

4.2 The Future Value of a Lump Sum

We refer to a single amount of present value or future value as a *lump sum*. When an investment is made for one year, we can find the future value at the end of the year by finding out the amount of interest earned in the year, and adding that to the present value (the return of principal).

For an investment of $1,000 at 10% interest, the future value at the end of the year is:

$$FV = \$1{,}000 + \$1{,}000 \times 10\% = \$1{,}000 + \$100 = \$1{,}100$$

More generally, the future value of a 1-year investment is:

$$FV = PV + PV \times i = PV(1 + i)$$

When we invested the $1100 for a second year, we found its future value was:

$$FV = \$1{,}100 \times (1 + i) = \$1{,}100 \times (1.10) = \$1{,}210$$

In general, we can calculate the future value of a lump sump PV after compounding for n years as:

$$FV = PV \times (1 + i)^n$$

Notice the exponent n accounts for the number of periods of compounding. The *future value factor* describes the value in the future of $1 worth of present value invested at the interest rate i for n years. The future value factor is given by:

$$FVF = (1 + i)^n$$

To find a future value of any lump-sum amount today, we can multiply the present value by the future value factor:

[21] The famed scientist probably never said this. See http://www.snopes.com/quotes/einstein/interest.asp.

$$FV = PV \times FVF$$

For example, the future value factor for a 3-year investment at 8% is:

$$FVF = (1.08)^3 = (1.08)^3 = 1.259712$$

The future value of $2,500 invested for 3-years at 8% is:

$$FV = \$2{,}500 \times FVF = \$2{,}500 \times 1.259712 = \$3{,}149.28$$

1.1.1.1 Using Spreadsheets: The FV Function

The FV function implements the algebra to calculate the future value of a lump sum today. The function takes parameters (changeable inputs) for the rate of return, the number of periods of compounding, and the present value.

To use the FV function, begin with an empty cell and type "=FV(rate, nper, pmt, pv)" (without quotation marks), and replace the parameters named *rate*, *nper*, and *pv* with the numerical values to use in the calculation. We will discuss the *pmt* parameter later in this module. For example, to have your spreadsheet calculate the future value of $2,500 invested at 8% for 3 years, you would type "=FV(8%, 3, 0, 2500)" (without quotation marks).

It is also possible to have your spreadsheet reference other cells in the spreadsheet for the inputs to the FV function. Figure 4.7 shows an example of the FV function that references other cells in the spreadsheet.

Figure 4.7 Using the FV Function to Calculate the Future Value of a Lump Sum

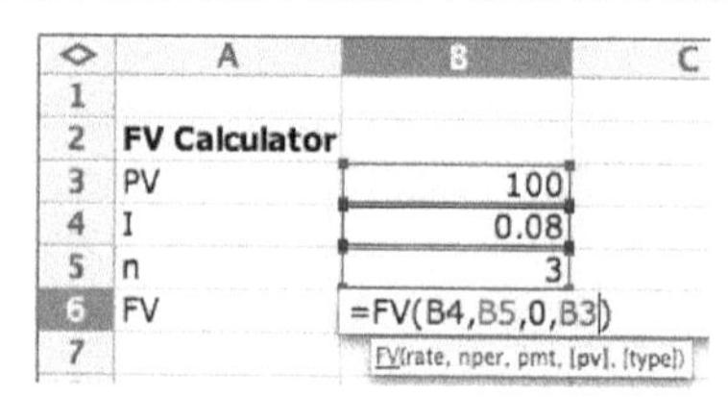

	A	B	C
1			
2	FV Calculator		
3	PV	100	
4	I	0.08	
5	n	3	
6	FV	=FV(B4,B5,0,B3)	
7		FV(rate, nper, pmt, [pv], [type])	

In the example in Figure 4.7, we could implement the FV function by typing the "=FV(", and then clicking on the cell B4 for the rate, typing the comma (","), clicking on the cell B5 for the number of periods, typing the next comma, typing 0 (for *pmt*), typing the next comma, and finally clicking on the cell B3 for the present value.

Note: the FV function includes an optional parameter for "type." This is a distinction that will be explained in the section on annuities below, and we can ignore it for now (or give the type of 0).

4.3 *The Present Value of a Lump Sum*

The previous section addressed the question, how much will we have in the future, if we invested *x* dollars today at the interest rate *i* for *n* periods. In this section, we will address the question, how much would we have to invest today, to achieve the amount *x* in the future?

Amounts to be Received in the Future Must be Discounted to the Present

Consider this deal: Suppose you were offered $100 for a cup of coffee with the stipulation that the coffee must be provided now, and the $100 would be paid to you in 50 years. Is this a good deal for a potential coffee purchaser to make?

Given the choice between receiving money now or later, we would prefer to receive it now. When considering amounts to be received in the future (even one day in the future), these amounts must be discounted to present value to make any kind of meaningful comparison. The present value of an amount to be received in the future must be worth less than the nominal future value, because of the time value of money.

Consider a potential coffee purchaser, who has $2 in cash right now. When trying to decide whether she should take the deal (buy the coffee now, and receive $100 in 50 years), she must also consider several alternatives: She could enjoy the coffee herself (e.g., consume her resources now). Perhaps she would invest it and earn a return on investment, and have more than $100 in the future. She might also be concerned that if she takes this deal, there is the risk that she might not get repaid (investment risk) or that the cost of a cup of coffee in 50 years might be more than $100 (inflation risk reduces purchasing power; inflation will be discussed later in this module).

Calculating Present Values

In a previous section, we established the simple arithmetic to calculate the future value of a sum of money invested today at a known interest rate. What about finding the present value of an amount to be received in the future? Figure 4.8 shows a cash flow timeline for this type of problem.

Figure 4.8 Cash Flow Timeline: Finding the Present Value of Funds to be Received in the Future

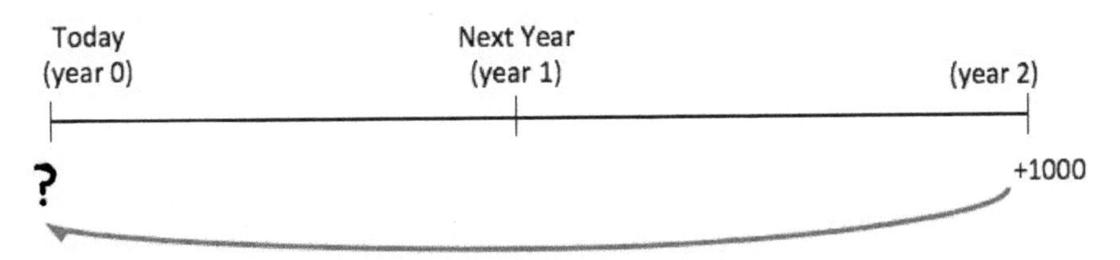

Recall that the formula for the future value of a lump sum is:

$$FV = PV(1+i)^n$$

We can re-arrange to solve for the present value, by dividing both sides by $(1+i)^n$:

$$PV = \frac{FV}{(1+i)^n}$$

The intuition behind this arithmetic is that the present value is the amount you would invest today at the interest rate *i* for *n* periods to achieve the future value.

Discounting is the process of finding a present value. You can think of discounting as the reverse of the process of compounding to find a future value. For example, what is the present value of $1,000 to be received in 2 years, if we discount at a 10% interest rate is?

$$PV = \frac{FV}{(1+i)^n} = \frac{\$1{,}000}{(1.10)^2} = \frac{\$1{,}000}{1.21} = \$826.45$$

In words, we say that $1,000 to be received in 2 years, discounted at 10% interest, has a present value of $826.45.

Another way to find a present value is to multiply the future value by the present value factor. A *present value factor* tells us how much $1 to be received in the future is worth today, given the interest rate and the number of discounting periods. The present value factor for an investment at the interest rate *i*, for the number of discounting periods, *n*, is given by:

$$PVF = \frac{1}{(1+i)^n}$$

The present value of a lump-sum amount to be received in the future is that future value multiplied by the appropriate present value factor. For example, the present value of $1,000 to be received in 2 years, discounted at 10% interest per year could also be written as:

$$PV = FV \times PVF = \$1000 \times \frac{1}{1.21}$$

$$= \$1{,}000 \times 0.82645 = \$826.45$$

Using Spreadsheets: The PV Function

The PV function implements the algebra to calculate the present value of a lump sum to be received in the future. Similar to the FV function, the PV function takes *parameters* (changeable inputs) for the rate of return, the number of periods of discounting, and the future value. To use the PV function, begin with an empty cell and type "=PV(rate, nper, pmt, fv)" (without quotation marks), and replace the parameters names *rate*, *nper*, and *fv* with the numerical values to use in the calculation. (We will discuss the *pmt* operand later in this module).

For example, you might type "=PV(8%, 5, 0, 1000)" (without quotation marks) to have your spreadsheet calculate the present value of $1,000 to be received in 5 years, discounted at a 8% annual interest rate. It is also possible to have your spreadsheet reference other cells in the spreadsheet for the inputs to the FV function. Figure 4.9 shows an example of the PV function that references other cells in the spreadsheet.

Figure 4.9 Using the PV Function to Calculate the Present Value of a Lump Sum

	A	B	C
1			
2	**PV Calculator**		
3	FV	100	
4	I	10%	
5	n	50	
6	PV	=PV(B4,B5,0,B3)	
7		PV(rate, nper, pmt, [fv], [type])	
8			

Figure 4.9 also shows the solution to the question about purchasing a cup of coffee now, and being repaid $100 in 50 years. If you could otherwise earn 10% on an alternative investment, the present value of $10 to be received in 50 years is about 85 cents today. On the other hand, if you could only earn 2% per year on your alternative investment, the present value of $100 to be received in 50 years is about $37.15 today.

Note: the PV function includes an optional parameter for "type." This is a distinction that will be explained in the section on annuities below, and we can ignore it for now (or give the type of 0).

Present Values are Additive

Cash flows that occur in the same time period may be added or subtracted from each other. However, cash flows that occur in different time periods cannot simply be added together, because of the time value of money. For example, if you receive $100 this year, and receive another $100 next year, it is not correct to say that you will have $200. These sums cannot be added together because if you had $100 right now, you could invest it and earn interest, and you would have more than $100 to add to next year's $100.

However, we can add or subtract amounts from each other so long as they are in present value. To add together amounts to be received in different years, we must find the present value of each and add the present values together.

$$PV = amount_0 + \frac{amount_1}{(1+i)^1} + \frac{amount_2}{(1+i)^2} + \cdots + \frac{amount_n}{(1+i)^n}$$

Example: Finding the Present Value of Multiple Future Amounts in Different Years

Often times, people need to save up for goals that occur in different years from one another. As an example, the payments for a college education will be due in 4 different years. Suppose your child's grandparents want to gift enough assets to pay for 4 years of college education, which will not begin for 13 more years. Assume the costs of these four years will be $100,000 for the first year; $104,000 for the second year; $108,200 for the third year; and $112,500 for the fourth year. What amount should be invested today, to fully cover the costs that will be due 13, 14, 15, and 16 years from now? Figure 4.10 shows a cash flow timeline to describe this problem visually.

Figure 4.10 Finding the Present Value of Several Future Amounts

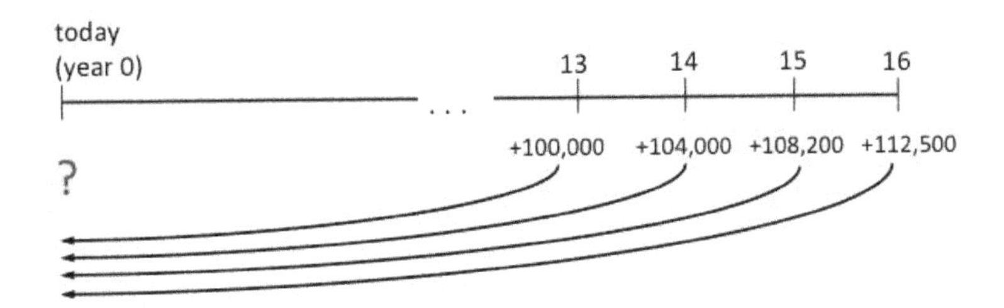

The solution to this kind of problem requires two steps: (1) find the present value corresponding to each future value, and (2) add the present values together. Suppose we could invest at 5% per year (compounded annually).

The present value of these future expenses is:

$$PV = \frac{\$100{,}000}{(1.05)^{13}} + \frac{\$104{,}000}{(1.05)^{14}} + \frac{\$108{,}200}{(1.05)^{15}} + \frac{\$112{,}500}{(1.05)^{16}}$$

$$PV = \$53{,}032.14 + \$52{,}527.07 + \$52{,}046.05 + 51{,}537.55 = \$209{,}142.80$$

Let's come back to the original question. How much should be invested today to pay for 4 years of college tuition occurring in 13, 14, 15, and 16 years? An investment of $209,142.80 today at an interest rate of 5% per year will fund these 4 future withdrawals.

Using Spreadsheets: Finding the Present Value of Several Amounts

Figure 4.11shows a simple spreadsheet solution to the same problem. For each future value (shown in column B), we can calculate the corresponding present value (in column C). Once we have amounts in present value, we can add these together to find the total present value.

Figure 4.11 The Present Value of Several Amounts

	A	B	C
1	Rate	5%	
2			
3	Years	Amount	PV(Amount)
4	13	$100,000.00	$ 53,032.14
5	14	$104,000.00	$ 52,527.07
6	15	$108,200.00	$ 52,046.05
7	16	$112,500.00	$ 51,537.55
8			
9		Total PV	$209,142.80

4.4 The Frequency of Compounding

In our discussion of interest rates and compounding, we have followed the convention of stating the interest rates as the *annual percentage rate* (APR). When interest is earned annually, there is one compounding period per year. In many cases, interest might be accrued and compounded at a different frequency. For example, many Certificates of Deposit feature monthly compounding, but state the rate of interest as an annual percentage rate. In the case of monthly compounding, there are 12 compounding periods per year. Each month, interest is earned at 1/12th of the annual interest rate. To account for monthly compounding in calculating a future value, we divide the annual interest rate by 12, and multiply the number of years by 12.

In general, we can find the future value after n years with m compounding periods per year as:

$$FV = PV \times \left(1 + \frac{i}{m}\right)^{nm}$$

For example, consider a 2-year (24-month) Certificate of Deposit with an annual interest rate of 3%, compounded monthly. The future value of this 24-month CD is:

$$FV = PV \times \left(1 + \frac{i}{m}\right)^{nm} = \$1{,}000 \times \left(1 + \frac{3\%}{12}\right)^{2 \times 12} = \$1{,}000 \times (1.0025)^{24}$$

$$= \$1000 \times 1.06175 = \$1{,}061.75$$

The effect of monthly compounding is to have more opportunities to earn interest on interest. In the example above, the future of the investment for 2 years would be $1,076.18. With annual compounding, the future value would only be $1,060.90. On account of monthly compounding, the investment earns $15.28 more over the 2 years than a similar investment with annual compounding.

It is important to always match the interest rate with the number of compounding periods per year. When we have 12 compounding periods per year, the appropriate interest rate is the monthly rate. The *effective annual rate* (EAR) is the rate of return actually earned (paid) per year, after accounting for the frequency of compounding. In general, when we know the APR and the number of compounding periods per year m, we can find the EAR as:

$$EAR = \left(1 + \frac{APR}{m}\right)^{m} - 1$$

For example, suppose a local bank offers a CD with an APR of 2.5%, with monthly compounding. The effective annual EAR would be:

$$EAR = \left(1 + \frac{.025}{12}\right)^{12} - 1 = 2.528\%$$

That is, $100 invested in this CD with monthly compounding would grow to a future value of $102.53 at the end of one year. In general, you can expect an investment like a CD to state clearly the annual percentage rate and the frequency of compounding, from which you can calculate the effective annual rate. In many cases, banks will even advertise the effective annual rate, since this is always higher than the annual percentage rate.

When Should we use the Annual Rate versus a Monthly Rate?

In general, the interest rate to use is the one that matches the frequency of compounding, the frequency of payments, or the frequency of investments. If a problem states that invested amounts are compounded annually, use the annual rate. If the problem describes payments that are made on a monthly basis (for example, with a student loan), use the monthly rate.

Continuously Compounded Interest

Increasing the frequency of compounding (i.e., the number of compounding periods per year) results in a higher effective annual rate of return. How many compounding periods can we have and continue to increase the effective annual rate of return?

Consider for example an investment that pays interest at an annual rate of 100%.

With interest paid once per year, the amount at the end of one year is $\$1 \times (1 + 100\%) = \$2$.

With interest paid semi-annually, the amount at the end of the year is $\$1 \times \left(1 + \frac{100\%}{2}\right)^2 = \$2.25$.

With interest paid quarterly, the amount at the end of the year is $\$1 \times \left(1 + \frac{100\%}{4}\right)^4 = \$2.44$.

With interest paid monthly, the amount at the end of the year is $\$1 \times \left(1 + \frac{100\%}{12}\right)^{12} = \$2.61$.

With interest paid monthly, the amount at the end of the year is $\$1 \times \left(1 + \frac{100\%}{365}\right)^{365} = \$2.71$.

The spreadsheet in Figure 4.12 illustrates the effective annual rate and future value after one year for several different compounding frequencies.

Figure 4.12 Compounding Periods and the Effective Annual Rate of Return

	A	B	C	D
1	Annual Rate	100%		
2	Principle	$1.00		
3				
4	Compounding Periods	Periodic Rate	Effective Annual Rate	Future Value
5	1	100.00%	100%	$ 2.00
6	2	50.00%	125%	$ 2.25
7	4	25.00%	144%	$ 2.44
8	12	8.33%	161%	$ 2.61
9	365	0.27%	171%	$ 2.71
10	1000	0.10%	172%	$ 2.72
11	10000	0.01%	172%	$ 2.72

As the number of compounding periods gets very large, the future value (i.e., the compounded growth of one dollar) approaches the limit of $2.72. The mathematical constant e, also known as Euler's number, is the limit of exponential growth as the number of periods approaches infinity:

$$e = \lim_{n \to \infty} \left(1 + \frac{1}{n}\right)^n \approx 2.7182818285$$

The limit of compounding frequency is called *continuous compounding*, where compounding happens instantaneously. The future value of an amount earning interest at an annual rate i for T years with continuous compounding is:

$$FV = PV e^{iT}$$

For example, if you invested $100 at 5% annual interest compounded continuously, after one year you would have:

$$FV = PV e^{rT} = \$100 \times e^{0.05 \times 1} = \$100 \times 1.050271 \approx \$105.03$$

In practical terms, banks and other financial institutions do not use continuously compounded interest in retail transactions with the non-banking public. However, continuously compounded interest is used in advanced financial modeling involving risk management contracts (i.e., insurance and derivatives).

The Rule of 72

Long before investors had access to handheld computers and spreadsheets programs, many used the *rule of 72* as a simple tool approximate present values and future values – at least under certain circumstances. The rule of 72 was used to help investors think about the future value of an investment in terms of how long it would take to double a present value. The rule of 72 also enables calculations to determine what rate of return is required to double an amount of money in a fixed number of years. While most of us now have full-fledged calculators in our pockets (e.g., smart phones), the rule of 72 is still useful for quick mental approximations.

Example: How long to double $100 at 5%? The number of periods as:

$$n = \frac{72}{rate}$$

At 5 percent interest, it would take about:

$$n = \frac{72}{5} = 14.4\, years$$

Example: What rate do I need to earn on an investment to double my money in 8 years? The interest rate (in percent) is given by:

$$rate = \frac{72}{n}$$

To double your investment in 8 years, you will need to earn a compounded annual return of:

$$\frac{72}{8\, years} = 9\, percent$$

Example: What is the present value of $1,000 to be received in 10 years at 7 percent interest? We know that:

$$\frac{72}{7\, percent} = 10\, years$$

The present value of $1000 to be received in 10 years discounted at 7% is about half that amount, or $500. The actual PV is:

$$PV = \frac{\$1{,}000}{1.07^{10}} = \$508.34$$

How it works

The rule of 72 applies to processes with exponential growth, as is the case with compound interest. The mathematical basis is in the ratio of the natural (base *e*) logarithms:

$$T = \frac{\ln(2)}{\ln(1+i)}$$

where T is the number of periods for doubling of a value given the growth rate i. The numerator $ln(2)$ represents the doubling of the initial value, but it could be changed to any amount of growth: 3 for tripling or 1.5 for 50% growth.

In fact, a rule of 70 or rule of 69.3 would provide a more accurate approximation for doubling (because $ln(2) = 0.693$). However, the number 72 is more convenient because of its many small divisors. It is easily divisible without a remainder (or fractions) by 1, 2, 3, 4, 6, 8, 9, and 12.

4.5 Inflation and Purchasing Power

The prices of goods and services can rise or fall over time. For example, in 1950 a cup of coffee cost 10 cents; in 2015, the cost of a cup of drip coffee at Starbucks approaches $2.00.

The *Consumer Price Index* (CPI) is a measure of the level prices of goods and services purchased by the typical household. Figure 4.13 shows the consumer price index for the years 1947-2014.

Figure 4.13 The Consumer Price Index (1947-2014)

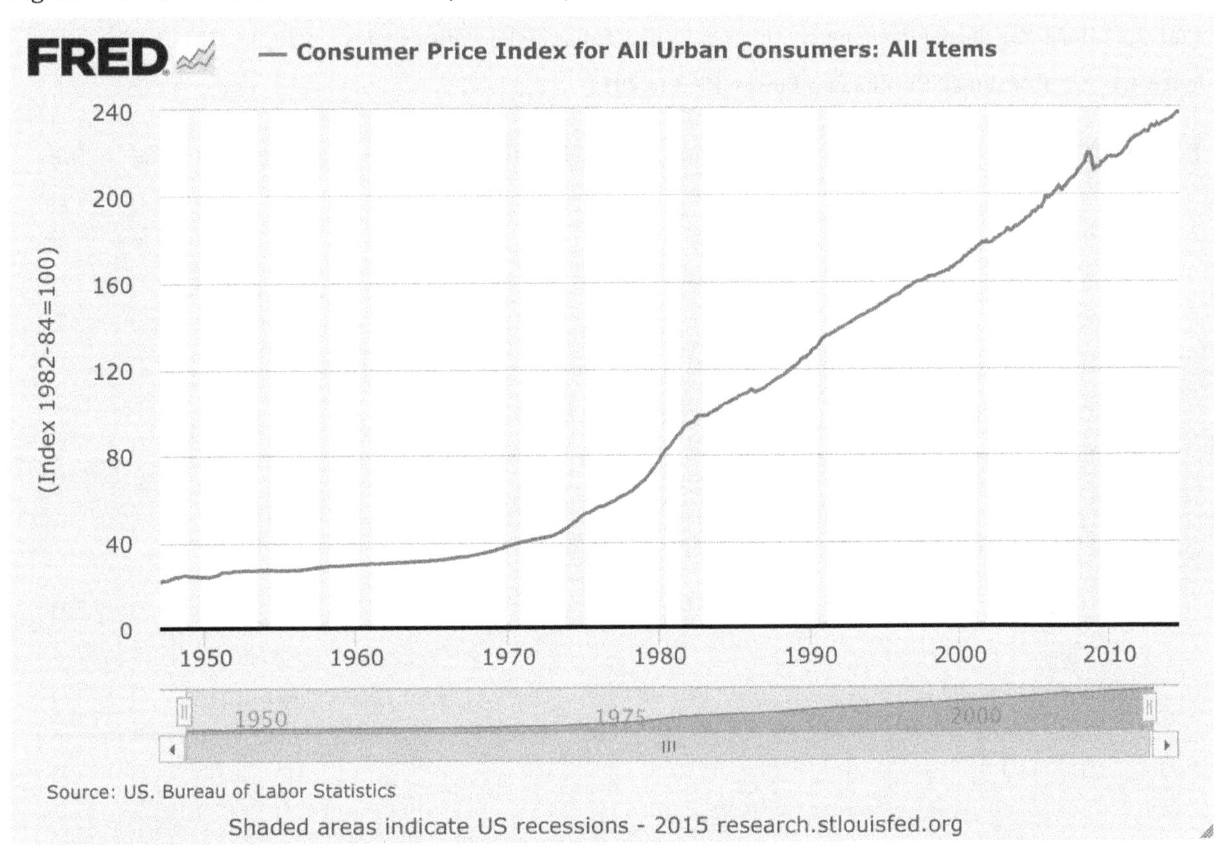

Inflation is the general increase in the level of prices in the economy. The annualized *inflation rate* measures the rate of change in prices from year to year. The effect of inflation is to reduce one's purchasing power per unit of money. There have been a few periods of *deflation* (decreasing prices), for example during the Great Recession (2007-2009), but inflation has been the general trend.

Purchasing Power

Inflation has a negative effect on your investments. Although the amount you are saving now will grow to a higher future value, there is not guarantee that those future dollars will buy the same amount of stuff as your dollars do today.

Consider Riccardo, who buys a package of M&Ms chocolate candies every day, at a cost of $1. Riccardo has $1,000 of cash today, with which he could buy 1,000 packages of M&Ms. Riccardo must choose between buying M&Ms today, or investing in a one-year certificate of deposit at a rate of 5% per year. If Riccardo invests in the CD, he knows for certain that he will have $1,050 next year. Clearly, $1,050 is certainly more than $1,000, but is Riccardo any better off for the investment? It depends on the price of what he wants to buy.

Let's assume the price of M&Ms next year is $1.05 per pack. Riccardo would have $1,050 available, which means that a year from now Riccardo would be able to buy:

$1,050 / $1.05 per package = 1,000 packages of M&Ms.

In essence, Riccardo's choice is between using his cash to buy 1,000 packs of M&Ms today, or investing his cash in a CD, and being able to buy 1,000 packs of M&Ms in a year. Despite having more dollars next year, Riccardo is actually no better off for making the investment in the CD. However, he is also no worse off for making the investment in the CD. Had he kept the $1,000 in cash and waited to purchase his M&Ms, he would only have been able to buy $1,000/$1.05 = 952 packs of M&Ms a year later.

Figure 4.14 illustrates the declining purchasing power of a dollar in the United States.

Figure 4.14 US Consumer Purchasing Power 1913 to 2014

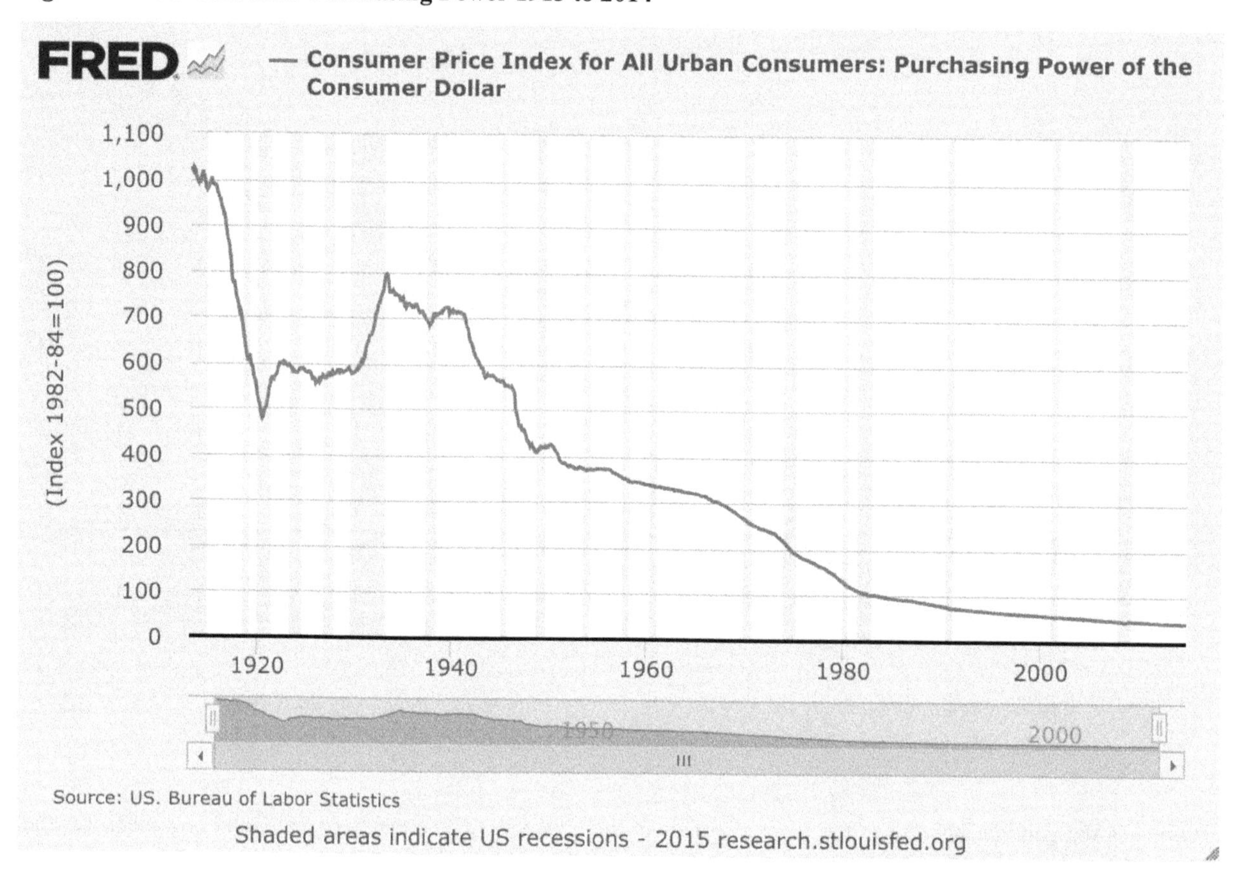

Nominal Dollars versus Real Dollars

Considering that the purchasing power of a future value might be adversely affected by inflation, it is crucial to draw a distinction between the quantity of dollars you will have in the future, and the purchasing power of those dollars.

Most non-economists speak about dollars – the money we use to make transactions – as just dollars. Economists label the dollars used in common speech as *nominal dollars* – dollars in name only. We can describe an amount of nominal dollars, but we do not know how much stuff (goods or services) those dollars will be able to purchase. For example, we know that if you save $6,000 per year, for 30 years and earn an annualized rate of return of 12%, at the end of 30 years you will have accumulated $1.2 million nominal dollars. What we don't know is how much purchasing power those dollars will have in 30 years.

Economists use the term *inflation-indexed dollars* or *real dollars* to describe the future purchasing power as a quantity of currency. An amount of real dollars describes the amount of purchasing power you would have with that quantity of dollars today. For example, suppose you can buy a car for $12,000 today and that the prices of cars keep pace with inflation. We could say that the cost of a car in 20 years will be $12,000 *real dollars*. The price of the car in *nominal dollars* might be $25,000 or $50,000– we don't know, and it will depend on the amount of inflation that occurs in the intervening years. By referring to the price in real dollars we can describe the price in terms of dollars with today's purchasing power.

Most financial investments are made in nominal dollars, rates of return are reported as nominal interest rates, and future values are stated in nominal dollars. However, only *real interest rates* and *real dollars* will be able to help us describe the actual future purchasing power of our investments. When we return to the life-cycle model, we will pay special attention to whether amounts are reported in nominal dollars or real dollars.

Great Minds in Applied Mathematics/Personal Finance: Irving Fisher

Irving Fisher (1867-1947) was a professor of Economics at Yale University, and the first recipient of a Ph.D. degree in Economics from Yale University.

As a mathematical economist, Fisher's contribution was to explain *interest* as a way of valuing investors' time preference of income. The rate of interest is a function of investors' preference for having income now (i.e., consuming now) versus having income later (i.e., consuming later).[22]

Fisher provided the definition of capital as any asset that produces an income over time. Using this definition, we can include not only financial capital but human capital as well. Fisher's definition, and the mathematics of interest that he explained in his *Theory of Interest*, provide a method of valuing an asset as the *present value* of the income to be received from that asset.

In addition to explaining interest, Fisher was the first economist to clearly distinguish between *real and nominal interest rates*. When thinking about future consumption, only the real interest rate provides a meaningful measure of future value, since the future value calculated with the real interest rate is measured in *real dollars* (i.e., dollars in today's purchasing power).

Irving Fisher died in 1947. The Sveriges Riksbank Prize in Economic Sciences in Memory of Alfred Nobel was first awarded in 1968, and the Nobel Foundation does not award the prize posthumously (an official policy since 1974). Were it not for this timing, Fisher would certainly have won the Nobel Prize in Economics.

Nominal Interest Rates and Real Interest Rates

Let's go back to Riccardo and his M&Ms. Suppose Riccardo decided to invest his $1,000 into a 5-year certificate of deposit at 5 percent annual interest. At the end of 5 years he would have:

$$FV = \$1{,}000(1.05)^5 = \$1{,}276.28$$

Now suppose that 5 years from now, the cost of M&Ms is $1.20 per pack. This means that Riccardo would be able to buy $\$1{,}276.28 \;/\; \$1.20 \;=\; 1{,}063$ packs of M&Ms. Put another way, Riccardo is better off not only in terms of nominal dollars, but also in purchasing power. The amount of increase in Riccardo's purchasing power of M&Ms is his *real return* from the investment in the certificate of deposit. Riccardo has had about a 6.3% increase (1,063 versus 1,000 M&Ms) in purchasing power over 5 years. To find Riccardo's average annual increase in purchasing power over these 5 years (i.e., his real rate of return), we solve for:

$$\sqrt[5]{1.063} - 1 \approx 1.0124 - 1 = 1.24\%$$

That is, on average Riccardo is 1.24% better off for each year that he delayed his consumption of M&Ms.

[22] Source: The Concise Encyclopedia of Economics, *Irving Fisher*, http://www.econlib.org/library/Enc/bios/Fisher.html

The Fisher Equation

The relationship between the nominal interest rate, the inflation rate and the real interest rate, is given by the Fisher Equation:

$$(1 + r) = \frac{(1 + i)}{(1 + \pi)}$$

where i = nominal interest rate, r = real interest rate, and π = inflation rate. For a one-year period, the following approximation is usually close enough: $r \approx i - \pi$.

For example, suppose that over the past one year the nominal interest rate was 8 percent and the inflation rate was 5 percent. Using the Fisher Equation, we find that the real interest rate is about 3 percent:

$$r = \frac{(1 + i)}{(1 + \pi)} - 1 = \frac{1.08}{1.05} - 1 = 0.02857 = 2.857\%$$

4.6 Summary

Savings is placed in an investment to earn interest, with the aim of maintaining or increasing future purchasing power. The future value describes the value of an asset at the end of the investment time period, and the present value describes the amount that must be set aside today to fund a goal in the future. We find a future value by compounding the present value at the interest rate, and we find a present value by discounting a future value.

A future value in nominal dollars does not guarantee the quantity of goods or services that can be purchased in the future, only the quantity and units, i.e., $1 million dollars. Inflation is the rise of prices over time, and the result of inflation is the decreasing purchasing power of nominal dollars. The primary concern of investors is maintaining or increasing their purchasing power in the future. The real rate of return is the net change in purchasing power over time, expressed as a percentage of the present value or amount invested.

4.7 Conceptual Review Questions

1. Explain these ideas: interest, future value, and compound interest. Give an example to illustrate.
2. Explain the concept of a present value. Why is a present value worth less than a future value?
3. Explain by giving an example how the Rule of 72 can help you calculate present values and future values.
4. What is inflation? What is the inflation rate? How does this relate to purchasing power?
5. What are the implications of inflation on interpreting future values?
6. Explain the difference between nominal dollars and real dollars, and the relationship between the two.
7. Explain the nominal rate of return and the real rate of return, and the relationship between the two.
8. Write and explain the Fisher Equation. Given a numeric example.
9. What are the main characteristics of Series I Bonds and TIPS?

4.8 Numerical Questions

10. What is the future value in 12 years of $500 invested at 4% interest? At 5% interest? Describe what happens to the future value as the interest rate increases.
11. What is the present value of $1,000 to be received in 8 years, assuming the prevailing interest rate is 9%? What if the interest rate is 10%? Describe what happens to the present value as the interest rate increases.
12. What is the future value of $250 invested at 7% APR for 5 years, with monthly compounding?

5 Paying on Time

Learning Objectives

- Provide a mechanism for valuing streams of payments.
- Describe the main types of loans and their features, and the mechanics of calculating loan payments.
- Introduce bonds as a major type of investment, and inflation-protected bonds as the only investment guaranteed
- Introduce the main characteristics of inflation-protected bonds.

5.1 Annuities

When we refer to multiple amounts to be received (or paid) at different points in time, we will call these a series of *payments*. An *annuity* is a series of periodic fixed payments (i.e., each payment is the same amount). Typically, annuity payments continue for many periods, often for many years, and then stop at a finite time in the future.[23]

Example: Lottery Winnings

A common example of an annuity is the typical lottery prize. For example, if you see an advertisement for the twice-weekly MegaMillions multi-state lottery, it would be typical to see the ad state a huge jackpot amount, for example \$57 million dollars. If you examine the fine print, the jackpot prize is actually 30 annual installments of \$1.9 million dollars each. The lottery authority is expecting that most people will think that 30 times \$1.9 million dollars equals \$57 million dollars. While this is true from an arithmetic perspective, it is highly misleading from the perspective of the time value of money. Figure 5.1 shows a cash flow time line depicting when the winner would receive each payment of \$1.9 million dollars.[24]

Figure 5.1 Cash Flow Timeline: Finding the Present Value of Lottery Winnings

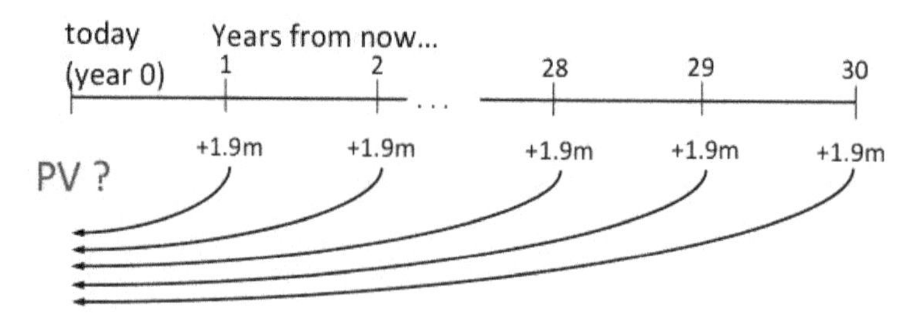

What is the present value of this annuity (the value today of the lottery prize)? Recall that present values are additive. To find the present value of the annuity, so we find the present value of each payment, and add all of them together to find the present value of the annuity as a whole. Notice that each payment is discounted for a different number of years.

Finding the Present Value of an Annuity

In general, the PV of an annuity at the interest rate i for n periods is given by:

$$PV_{annuity}\ (i, n, PMT) = \frac{PMT}{1+i} + \frac{PMT}{(1+i)^2} + \cdots + \frac{PMT}{(1+i)^n}$$

where PMT is the amount of the periodic payment, and n is the period number. In an ordinary annuity, the first payment is 1 year from now, the second payment is 2 years from now, etc., and the final payment is n years from now. The present value of all payments is the sum of the present value of each payment.

---

[23] There is also an insurance product called an annuity (formally called a single premium immediate or life annuity). In this section we are referring only to the financial arrangement with a fixed number of equal payments. We will discuss the insurance product in Module 7 (For Whom The Bell Tolls).

[24] This timeline is showing the cash flows for an ordinary annuity, but in practice lottery winnings are typically distributed as an annuity due. The distinction will be explained below.

By factoring out the PMT outside of this simplifies to:

$$PV_{annuity}\ (i, n, PMT) = PMT\left[\frac{1}{1+i} + \frac{1}{(1+i)^2} + \cdots + \frac{1}{(1+i)^n}\right]$$

A further algebraic simplification leads to the present value of an annuity as:

$$PV_{annuity}\ (i, n, PMT) = PMT\left[\frac{1-(1+i)^{-n}}{i}\right]$$

The Present Value of the Lottery Winnings

Let's return to the example of winning $57 million dollars in the MegaMillions lottery Jackpot, which is actually 30 annual payments of $1.9 million dollars each. If the appropriate discount rate is 3% per year, the present value of this annuity is equal to:

$$PV_{annuity}\ (i, n, PMT) = \$1{,}900{,}000\left[\frac{1}{1.03} + \frac{1}{(1.03)^2} + \frac{1}{(1.03)^3} + \cdots + \frac{1}{(1.03)^{30}}\right]$$

With so many terms to be added together, the present value of an annuity becomes an onerous exercise in arithmetic. Figure 5.2 presents a solution in a spreadsheet.

Figure 5.2 An Spreadsheet Calculation to Find the Present Value of the Lottery Prize

	A	B	C	D
1	Discount Rate	3%		
2				
3	Year	Payment	PV Factor	PV
4	1	$ 1,900,000	0.971	$ 1,844,660
5	2	$ 1,900,000	0.943	$ 1,790,932
6	3	$ 1,900,000	0.915	$ 1,738,769
7	4	$ 1,900,000	0.888	$ 1,688,125
8	5	$ 1,900,000	0.863	$ 1,638,957
9	6	$ 1,900,000	0.837	$ 1,591,220
10	7	$ 1,900,000	0.813	$ 1,544,874
11	8	$ 1,900,000	0.789	$ 1,499,878
12	9	$ 1,900,000	0.766	$ 1,456,192
13	10	$ 1,900,000	0.744	$ 1,413,778
14	11	$ 1,900,000	0.722	$ 1,372,600
15	12	$ 1,900,000	0.701	$ 1,332,622
16	13	$ 1,900,000	0.681	$ 1,293,808
17	14	$ 1,900,000	0.661	$ 1,256,124
18	15	$ 1,900,000	0.642	$ 1,219,538
19	16	$ 1,900,000	0.623	$ 1,184,017
20	17	$ 1,900,000	0.605	$ 1,149,531
21	18	$ 1,900,000	0.587	$ 1,116,050
22	19	$ 1,900,000	0.570	$ 1,083,543
23	20	$ 1,900,000	0.554	$ 1,051,984
24	21	$ 1,900,000	0.538	$ 1,021,344
25	22	$ 1,900,000	0.522	$ 991,596
26	23	$ 1,900,000	0.507	$ 962,714
27	24	$ 1,900,000	0.492	$ 934,674
28	25	$ 1,900,000	0.478	$ 907,451
29	26	$ 1,900,000	0.464	$ 881,020
30	27	$ 1,900,000	0.450	$ 855,359
31	28	$ 1,900,000	0.437	$ 830,446
32	29	$ 1,900,000	0.424	$ 806,258
33	30	$ 1,900,000	0.412	$ 782,775
34				
35		Total PV		$ 37,240,839

In Figure 5.2, each year's payment is represented in its own row. We find the present value of each payment by multiplying the payment amount times its present value factor. In year 1, the present-value factor is:

$$PVF = \frac{1}{1+i} = \frac{1}{1.03} = 0.971$$

To find the present value of the first year's payment, we multiple the \$1,900,000 by its present value factor:

$$PV = PMT \times PVF = \$1{,}900{,}000 \times 0.971 = \$1{,}844{,}660$$

After calculating the present value for each year, the present values can be added together to find the total present value.

Using Spreadsheets: the Present Value of an Annuity

Previously we introduced the PV function to calculate the present value of a lump sum to be received in the future. The PV function can also be used to find the present value of an annuity. The *parameters* (changeable inputs) for the PV function include the rate of return, the number of periods of discounting, amount of any periodic payments, and the future value. To use the PV function, begin with an empty cell and type "=PV(rate, nper, pmt, fv)" (without quotation marks), and replace the parameters names *rate*, *nper*, and *fv* with the numerical values to use in the calculation. The parameters for *pmt* and *fv* are both optional, and so long as either one is provided the present value will be calculated.

For example, you might type "=PV(3%, 30, 1900000)" (without quotation marks) to have the spreadsheet calculate the present value of 30 annual payments of \$1,900,000, discounted at a 3% interest rate. It is also possible to have your spreadsheet reference other cells in the spreadsheet for the inputs to the FV function. Figure 5.3 shows an example of the PV function that references other cells in the spreadsheet.

Figure 5.3 Using the PV Function to Calculate the Present Value of an Annuity

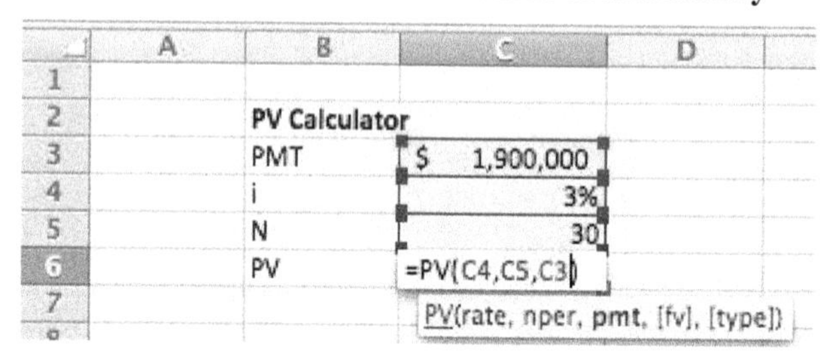

	A	B	C	D
1				
2		PV Calculator		
3		PMT	\$ 1,900,000	
4		i	3%	
5		N	30	
6		PV	=PV(C4,C5,C3)	
7			PV(rate, nper, pmt, [fv], [type])	

Ordinary Annuities and Annuities Due

The term *ordinary annuity* is applied to a stream of equal-dollar amount payments that continue for many periods, where the first payment is not received until the end of the first period. Figure 5.4 shows a cash flow diagram for an ordinary annuity.

Figure 5.4 Cash Flow Diagram: An Ordinary Annuity

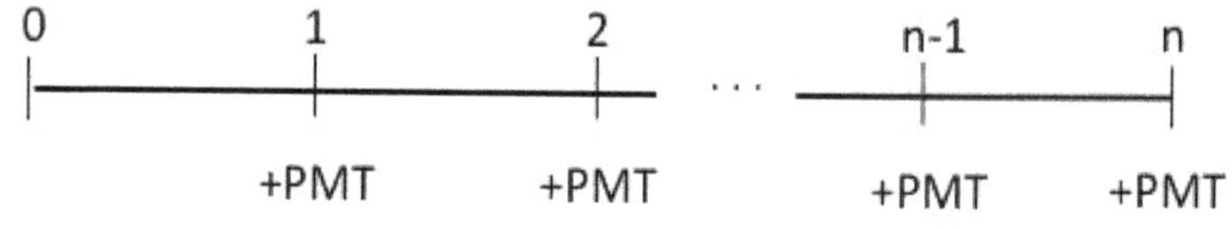

In an annuity due for *n* periods, the first payment is 1 discounted for 1 period, and the last payment is discounted for *n* periods.

As noted earlier, the typical lottery grand prize is paid out in annual installments beginning immediately, and continuing on for a number of years. An *annuity due* is a stream of equal dollar amount payments wherein the first payment is received immediately, and the last payment is received in period *n – 1* (Figure 5.5).

Figure 5.5 Cash Flow Diagram: An Annuity Due

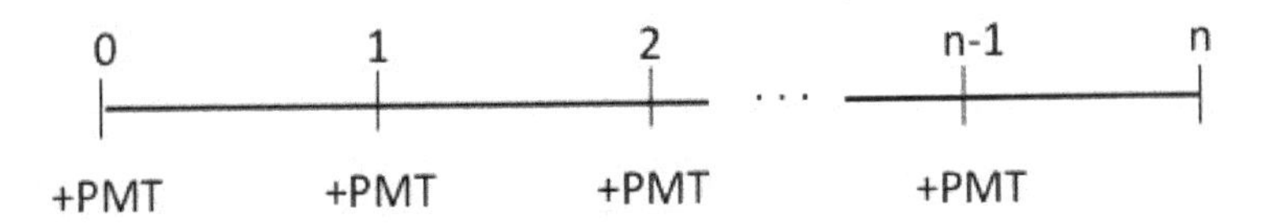

In an annuity due, the first payment is not discounted at all, and the last payment is discounted for *N-1* periods. The present value of an annuity due for *n* periods is given by:

$$PV_{annuity\ \ due}(i, n, PMT) = PMT + \frac{PMT}{1+i} + \cdots + \frac{PMT}{(1+i)^{n-1}}$$

where *PMT* is the amount of the periodic payment, and *i* is the period number. By factoring out the PMT outside of the equation simplifies to:

$$PV_{annuity\ \ due}(i, n, PMT) = PMT\left[1 + \frac{1}{1+i} + \cdots + \frac{1}{(1+i)^{n-1}}\right]$$

Notice in this equation that each payment has been brought 1 period closer to today (one fewer period of discounting). We can write the relationship between an ordinary annuity and an annuity due as:

$$PV_{annuity\ \ due}(i, n, PMT) = (1+i)\left[PV_{annuity}\ (i, n, PMT)\right]$$

In words: the PV of an annuity due is the PV of an ordinary annuity, compounded for one period since all payments are received one period sooner than an ordinary annuity.

5.2 Loans and Loan Payments

A related area of financial math involves calculating loan payments. A *loan* is a borrowing contract between a borrower (who receives money up front) and a lender (to whom the money will be repaid over time). Loans accrue interest (the rental fee for the money), which is paid by the borrower and earned by the lender. There are several common kinds of loans, broadly categorized as revolving loans and amortizing loans.

Amortizing and Revolving Loans

An *amortizing loan* is a loan in which a fixed monthly payment is made to retire the debt over a finite period of time. Common amortizing loans include student loans and car loans. In an amortizing loan, the monthly payment is the same each month (which is convenient for the borrower's cash-flow budgeting). In an amortizing loan, the borrower receives the present value at the time the loan is made, and promises to make periodic loan payments until the loan is paid off. In particular, each loan payment is the same amount in dollars. Figure 5.6 shows a cash flow diagram depicting an amortizing loan.

Figure 5.6 Cash Flow Timeline: an Amortizing Loan

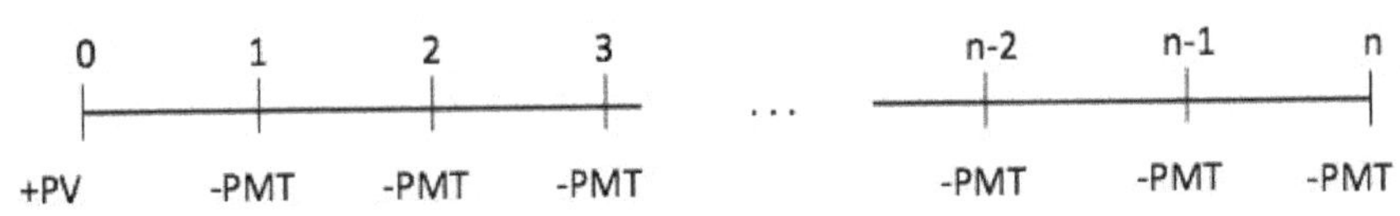

As an example of an amortizing loan, consider a loan for the purchase of an automobile. Many automobile purchasers do not have enough liquid assets to pay the entire purchase price, and must instead borrow money to finance the purchase of a car. As a result, automobile manufacturers provide vendor financing, in which a financing subsidiary of the car company (for example, General Motors Acceptance Corporation or Honda Finance Corporation) underwrites car loans to purchasers.

By contrast, in a *revolving loan,* interest accrues to the lender (usually calculated monthly), and the borrower makes minimum payments each month. Common revolving loans include credit cards and home-equity lines of credit (HELOCs for short). The minimum payment on a revolving loan could be a percentage of the amount owed (this is typical for credit card loans), or accrued interest only (in the case of HELOCs). In revolving loans, the minimum payments are not sufficient to pay off the balance.

Revolving Loans

A *credit card* is an unsecured revolving loan: the borrower does not pledge any collateral as a promise of repayment. Since credit cards are not secured by collateral, the interest rates on credit cards are typically very high, because of a substantial default risk. If borrowers have nothing to lose (but their credit rating) by late-payment or non-payment of loans, what's to stop them from simply not repaying on their loans? As a result, credit cards usually charge very high interest rates. While some credit card borrowers do not repay their loans, on average the credit card lenders are paid very handsomely for their lending services. A further discussion of interest rates is provided in the Appendix to this module.

A *home equity line of credit* (HELOC) is a revolving loan that is secured by collateral: the borrower pledges his or her house as a guarantee of repayment (a *security interest*). The collateral gives the lender some assurance of being repaid, and as a result interest rates on HELOCs tend to be relatively low.

Revolving loans are structured such that the amount you pay would decline with the outstanding balance, but would never reach zero. Some revolving loans (e.g., HELOCs) require a single large *balloon payment* at some future date to pay off the balance. In other revolving loans (e.g., credit cards) the monthly payments might go on indefinitely.

Calculating Amortizing Loan Payments

Consider the purchase of a new car, which sells for $25,000. Suppose the buyer makes a down payment of $3,000 toward the purchase, and borrows the remaining $22,000 with a new car loan. Suppose the interest rate for new car loans is 6% per year, and repayment is made over a period of 5 years.

The amortizing loan is in effect an annuity with a fixed term and equal periodic payments. The borrower makes the annuity payments, and the lender receives the payments. Recall the formula for the present value of an annuity:

$$PV_{annuity}\ (i, n, PMT) = PMT\left[\frac{1-(1+i)^{-n}}{i}\right]$$

By algebraic magic we can re-arrange to solve for the annuity's payment as a function of it's present value:

$$PMT(i, n, PV) = \frac{PV \times i}{1-(1+i)^{-n}}$$

In other words, when we know the interest rate, the number of periods over which the loan will be repaid, and the present value of the amount borrowed, solving for the amortizing loan payment is an algebraic exercise. The payment amount, when treated as an annuity and discounted, will equal the present value being borrowed.

The amortizing loan payment for the car loan of $22,000 to be repaid over 5 years (60 months) at 6% annual interest (i.e., 0.5% per month) is:

$$PMT(0.5\%, 60, \$22{,}000) = \frac{PV \times i}{1-(1+i)^{-n}} = \frac{\$22{,}000 \times 0.005}{1-(1.005)^{-60}} = \frac{\$110}{-0.2586278} = \$425.32$$

It is important to note that since the loan payments are made monthly, the appropriate interest rate to use is the monthly rate, e.g., 6% per year / 12 months = 0.5% per month, and there are 60 monthly payments in 5 years. Similar adjustments must be made for any loan payment calculation when the payments are made monthly, as is usually the case. Also, notice that the payment amount is given as a negative number, which is indicative of cash out-flow. We are interested in the absolute value of the equation's result, e.g., the payment is $425.32 per month. Figure 5.7 summarizes this amortizing loan arrangement with a cash flow diagram.

Figure 5.7 Cash Flow Timeline: a 60-Month Car Loan

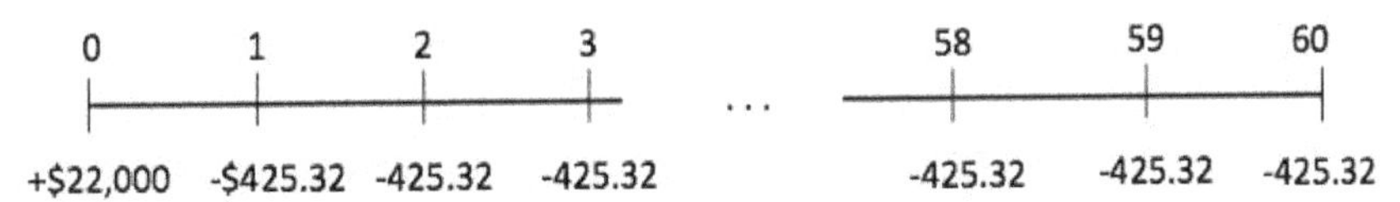

Using Spreadsheets: The PMT Function

Previously we introduced the spreadsheet PV and FV functions, which finds the present value of a lump-sum or annuity, and the future value of a lump-sum or annuity, respectively. Spreadsheets also provide the built-in PMT function for calculating an amortizing loan payment, given an interest rate, the number of periods of repayment, and the present value.

To use the PMT function, begin with an empty cell and type "=PMT(rate, nper, pv)" (without quotation marks), and replace the parameters names *rate*, *nper*, and *pv* with the numerical values to use in the calculation. Optional parameters, *FV* and *type*, allow the function to calculate the payment required to save up for a future value, or to find the payment for an annuity due.

For example, you might type "=PMT(6%/12, 5*12, 22000)" (without quotation marks) to have your spreadsheet calculate the monthly loan payments for a 5-year amortizing loan of $22,000 at 6% annual interest. It is also possible to have your spreadsheet reference other cells in the spreadsheet for the inputs to the PMT function, as shown in Figure 5.8.

Figure 5.8 Using the PMT Function to Calculate an Amortizing Loan Payment

2	**Payment Calculator**	
3		
4	i	6%
5	n	5
6	PV	-22000
7	PMT	=PMT(B4/12,B5*12,B6)
8		PMT(rate, nper, pv, [fv], [type])
9		

The Amortization Table

When a borrower enters into a loan contract for an amortizing loan (for example, a car loan or a mortgage loan on a house), the lender provides an amortization table. An *amortization table* provides proof that the required monthly payments will pay off the loan in its entirety over the desired number of periods. The amortization table shows the balance at the beginning of each period, the interest accrued in that period, the payment(s) made during that period, and the ending balance after adding the accrued interest and applying the payment. Figure 5.9 shows an amortization table for the car loan example.

Figure 5.9 An Amortization Table for a 5-year Car Loan

	A	B	C	D	E
10	Month	Beginning Balance	Interest Accrued	Payment Made	Ending Balance
11	1	$(22,000.00)	$ (110.00)	$ 425.32	$(21,684.68)
12	2	$(21,684.68)	$ (108.42)	$ 425.32	$(21,367.78)
13	3	$(21,367.78)	$ (106.84)	$ 425.32	$(21,049.30)
14	4	$(21,049.30)	$ (105.25)	$ 425.32	$(20,729.22)
15	5	$(20,729.22)	$ (103.65)	$ 425.32	$(20,407.55)
66					
67	56	$ (2,095.08)	$ (10.48)	$ 425.32	$ (1,680.23)
68	57	$ (1,680.23)	$ (8.40)	$ 425.32	$ (1,263.31)
69	58	$ (1,263.31)	$ (6.32)	$ 425.32	$ (844.31)
70	59	$ (844.31)	$ (4.22)	$ 425.32	$ (423.21)
71	60	$ (423.21)	$ (2.12)	$ 425.32	$ (0.00)

In Figure 5.9, column A shows the month number. Column B shows the beginning balance for each month. The initial balance of the car loan is $22,000. Column C shows the interest accrued for each month. The calculation of the value in column C is the beginning balance in a given month times the monthly interest rate. For example, in this spreadsheet the calculation is:

$$\textit{Interest Accrued} = \textit{Beginning Balance} \times \textit{Monthly Rate} = -\$22{,}000 \times 0.5\% = -\$110.00.$$

The values in column D are the monthly loan payment that was previous calculated using the PMT function, e.g., $425.32. Notice that the loan payment remains constant throughout the life of the loan. The values in column E are the balance of the loan at the end of each month, which is the sum of the beginning balance, the interest accrued, and the payment made. For cell E11, this works out to:

$$-\$22{,}000 + -\$110 + \$425.32 = -\$21{,}684.68.$$

At the end of the first month, the outstanding loan balance is $21,684.68. It is shown as a negative number because it is a liability. In the second month, the beginning balance is the same as the ending balance from the first month. The interest accrued will be $-\$21{,}684.68 \times 0.5\% = -\$108.42$. As the outstanding loan balance decreases each month, so does the amount of interest accrued decrease each month. The balance at the end of the second month is:

$$-\$21{,}684.68 + -\$108.42 + \$425.32 = -\$21{,}367.78.$$

The process continues month by month. In Figure 5.9, months 6 through 54 are omitted for brevity. Notice how in the last several months of this car loan, the monthly accrued interest is very little in dollar terms. In month 60, the beginning balance is –$423.21. The accrued interest on this balance is:

$$-\$423.21 \times 0.5\% = -\$2.12.$$

The loan payment is $425.32, which is exactly enough to pay off the remaining balance and accrued interest and leave the loan with a $0 balance at the end of the last month. (The amounts are off by $0.01 because the spreadsheet rounded the formatted dollar amounts, but we can safely ignore this rounding.)

Pre-Payment on an Amortizing Loan

Most amortizing loans allow the borrower the right to pay more than the required payment of the loan in any given month. The pre-payment amount is deducted from the principal balance owed, but does not waive the borrower's obligation to make regular monthly payments each month. A borrower who wishes to retire the debt sooner than required would prepay each month, and as a result she or he would pay off the loan over a shorter time period (and pay less in total interest). The amortization table helps us see the effect of prepayment in reducing the length of the repayment period.

For example, in Figure 5.10, we see an amortization table similar to the one in Figure 5.9, but with monthly payments of $500 per month.

Figure 5.10 Amortizing Table with Loan Pre-Payment

10	Month	Beginning Balance	Interest Accrued	Payment Made	Ending Balance
11	1	$ (22,000.00)	$ (110.00)	$ 500.00	$ (21,610.00)
12	2	$ (21,610.00)	$ (108.05)	$ 500.00	$ (21,218.05)
13	3	$ (21,218.05)	$ (106.09)	$ 500.00	$ (20,824.14)
14	4	$ (20,824.14)	$ (104.12)	$ 500.00	$ (20,428.26)
15	5	$ (20,428.26)	$ (102.14)	$ 500.00	$ (20,030.40)
55					
56	45	$ (2,859.28)	$ (14.30)	$ 500.00	$ (2,373.58)
57	46	$ (2,373.58)	$ (11.87)	$ 500.00	$ (1,885.44)
58	47	$ (1,885.44)	$ (9.43)	$ 500.00	$ (1,394.87)
59	48	$ (1,394.87)	$ (6.97)	$ 500.00	$ (901.85)
60	49	$ (901.85)	$ (4.51)	$ 500.00	$ (406.35)
61	50	$ (406.35)	$ (2.03)	$ 500.00	$ 91.61
62	51	$ 91.61	$ 0.46	$ 500.00	$ 592.07

In Figure 5.10, we see that the balance is repaid more quickly than in the previous amortization table (Figure 4.19), such that in after 49 months, the remaining balance is only $406.35. In practice, the borrower would make a final payment of $\$406.35 + 2.03 = \$408.38$ in month 50 to fully pay off the debt.

Amortizing a Revolving Loan

Unlike an amortizing loan, the payment due on a revolving loan, such as a credit card, is variable and changes each month. Paying the minimum required payment on a credit card loan (typically 2% or 3% of the outstanding balance) would result in payments extending infinitely into the future.

For example, suppose you have a credit card loan of $4,875 that is accruing interest at a annual rate of 18% per year or 1.5% per month. With a 2% minimum payment, the required payment is $97.50 the first month. The balance after the payment would be $\$4{,}875 - \$97.50 = \$4{,}777.50$. The interest accrued for the month would be $1.5\% \times \$4{,}777.50 = \$71.66$, bringing the new balance at the end of the month to $\$4{,}785 - \$97.50 + \$71.66 = \$4849.16$. The required payment in the second month would be 2% of $4,849.16, or $96.98. At this rate, the credit card will never be paid off!

A better technique to pay of a credit card is to find the amortizing payment required to pay off the loan in a certain amount of time. For example, to pay off the credit card in 48 months, the required fully amortizing payment would be:

$$PMT(1.5\%, 48, \$4{,}875) = \frac{PV \times i}{1 - (1 + i)^{-n}} = \frac{\$4{,}875 \times 0.015}{1 - (1 + 0.015)^{-48}} = 143.20$$

As proof that a payment of $143.20 per month will pay off the credit card in 48 months, use a spreadsheet to construct an amortization table with a beginning balance of $4,875, and interest rate of 1.5% per month, and 48 months. This is left as an exercise for the reader.

5.3 Bonds and Bond Pricing

A *bond* is a type of loan used by large borrowers (e.g., governments and corporations). The bond is a borrowing contract between the issuer (borrower) and investors. The bond contract specifies the terms of the investment, including the bond's principal, maturity, and interest rate. The bond's *principal* is the amount borrowed. The principal is sometimes referred to as the *maturity value*, *face value*, or *par value* of the bond. By convention, each bond has a principal of $1,000, and bond specialists might refer to $10,000 worth of principal as 10 bonds. In practice, bonds are often sold in larger denominations.

The bond's issue date is the date on which the investor received the bond and the issuer received the money. The *maturity* describes the length of the loan. Often, a maturity date will be listed. For example a bond issued on 1 July 2010 will state that it will mature on 30 June 2020.

Coupon Bonds

The bond contract states the annual rate at which interest is paid. While the rate is stated as an annual rate, almost all bonds pay interest in semi-annual installments (1/2 of the quoted interest amount is paid every 6 months) called *coupons*. In the days before computers, bonds were printed on special paper, with dated coupons for each interest payment. At each coupon due date, the bondholder would "clip a coupon" and turn it in to receive his or her interest payment. Today, virtually all bonds are electronically registered, but we still refer to bond interest payments as "coupons."

For example, consider a 10-year bond issued by the United States with a stated interest rate of 2.625%. Suppose an investor buys the bond directly from the Treasury and paid the "par value" (principal value) of $1,000. The investor will receive semi-annual coupons of:

$$coupon\,payment = principal \times \frac{i}{2} = \$1{,}000 \times \frac{2.625\%}{2} = \$13.125$$

The coupon payments are paid every six months for a total of 20 semi-annual coupons, and at the end of 10 years, the investor will receive his or her $1,000 back. Figure 5.11 shows a cash flow timeline to illustrate cash flows involved in this 10-year bond example.

Figure 5.11 Cash Flow Timeline for a 10-year Bond with Semi-Annual Coupon Payments

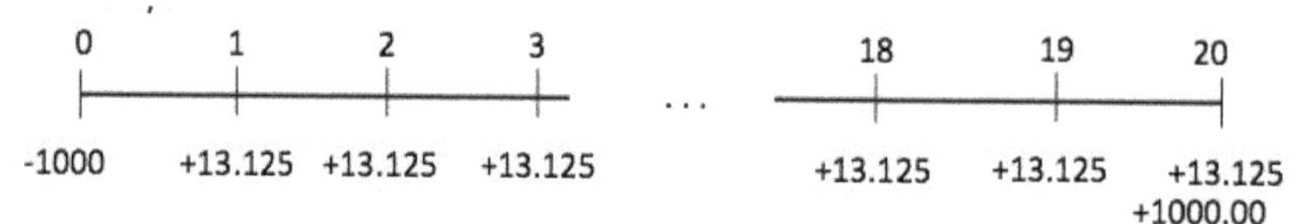

In Figure 5.11, the timeline is showing the 20 periods in which coupons are paid (e.g., half-years). At time 0, the day the bond is created, the initial cash flow (from the investor's standpoint) is a cash out-flow of $1,000. Every half-year, the investor receives the coupon amount of $13.125, which is a cash inflow. In the last period (20 half-years, or 10 years from now), the investor receives the final coupon payment of $13.125, along with the return of the $1,000 principal.

The present value of the bond's cash flows are the sum of two parts: a 20-period annuity with twice-annual payments of $13.125, and a lump sum payment of $1,000 to be received in 10 years. The present value of these payments is:

$$PV_{bond} = PMT\left[\frac{1-(1+i)^{-n}}{i}\right] + \frac{FV}{(1+i)^n}$$

$$PV_{bond} = \$13.125\left[\frac{1-(1.01325)^{-20}}{0.01325}\right] + \frac{\$1{,}000}{(1.01325)^{20}}$$

$$PV_{bond} = \$229.56 + 770.44 = \$1{,}000$$

Notice in the above calculation for the present value of the bond, the interest rate used is the half-yearly rate: $2.625\%/2 = 1.325\%$. The present value of the coupons is $229.56, and the present value of the lump-sum (the maturity value of the bond) is $770.44, and together these sum to $1,000, the par value of the bond.

Using Spreadsheets: Calculating the Bond Price Using The PV Function

Previously we introduced the spreadsheet PV function, which finds the present value of a lump-sum or an annuity. The PV function can also be used to solve for the present value of a bond. The present value of a bond is in the present value of the bond's annuity component (e.g., the coupon payments) and the present value of the lump-sum (it's maturity value). The PV function does this calculation in one step.

To use the PV function to calculate the price of a bond, begin with an empty cell and type "=PV(rate, nper, pmt, fv)" (without quotation marks), and replace the parameters names *rate*, *nper*, *pmt*, and *fv* with the numerical values to use in the calculation. An optional parameter allows the function to calculate present value for an annuity due, but coupon payments behave like ordinary annuities.

For example, you might type "=PV(2.625%/2, 10*2, 13.125, 1000)" (without quotation marks) to have your spreadsheet calculate the price of a 10-year bond with semi-annual coupon payments of $13.125 and a maturity value of $1,000, discounted at a semi-annual interest rate of 1.1325%. It is also possible to have your spreadsheet reference other cells in the spreadsheet for the inputs to the PMT function. Figure 5.12 shows an example of the PV function that references other cells in the spreadsheet.

Figure 5.12 A Spreadsheet to Calculate the Present Value of a Bond

	A	B	C
1	**Bond Price Calculator**		
2	Maturity Value	1000	
3	Coupon Rate	2.625%	
4	Coupon Amount	13.125	
5	Maturity (Years)	10	
6	Yield	3.00%	
7			
8	**Bond Price**	**=PV(B6/2,B5*2,B4,B2)**	

Discount and Premium Bonds

The above example showed that when the bonds sell at par value, the present value of the cash flows is equal to the par value price. Most newly issued bonds, and all previously issued bonds are sold in an auction market. In a bond auction, investors bid the maximum price they will pay for the bond in dollars. By paying a certain price (effectively the PV) for the bond, investors are implicitly setting the interest rate they will accept on the bonds (called the *yield to maturity* or simply the *yield*). For a given series of coupon payments and maturity value, there is exactly one interest rate that solves for the PV paid by investors. This is the interest rate earned by an investor who buys the bond at auction and holds it to maturity.

When new bonds are issued, the coupon rate is equal to the market interest rate, so that the bonds sell at or close to par value. Existing bonds must remain competitive with newly issued bonds, so the prices of existing bonds will adjust based on changes in the market interest rate. In general there is an inverse relationship between yields and bond prices. When bond yields increase, bond prices decrease. When bond yields decrease, bond prices increase.

For example, consider a 10-year bond, with a 2.625% annual coupon rate (i.e., coupon payments of $13.125 per $1,000 of principal). Suppose that the auction market requires an annual yield of 3.0% on bond. The price of a bond would be:

$$PV_{bond} = PMT\left[\frac{1-(1+i)^{-n}}{i}\right]+\frac{FV}{(1+i)^n}$$

$$PV_{bond} = \$13.125\left[\frac{1-(1.015)^{-20}}{0.015}\right]+\frac{\$1{,}000}{(1.015)^{20}}$$

$$PV_{bond} = \$225.34 + 742.47 = \$967.81$$

By discounting the bond's cash flows at 3% per year (or 1.5% per half year), the present value of the bond would be $967.81. When the yield is higher than the coupon rate, the bond would sell at a discount. A *discount bond* is one that sells for less than its par value.

On the other hand, if the yield were only 2.5%, the bond would sell for:

$$PV_{bond} = PMT\left[\frac{1-(1+i)^{-n}}{i}\right]+\frac{FV}{(1+i)^n}$$

$$PV_{bond} = \$13.125\left[\frac{1-(1.0125)^{-20}}{0.0125}\right]+\frac{\$1{,}000}{(1.0125)^{20}}$$

$$PV_{bond} = \$230.99 + 780.01 = \$1{,}011.00$$

By discounting the bond's cash flows at 2.5% per year (or 1.125% per half year), the present value of the bond would be $1,011.00. When the yield is lower than the coupon rate, the bond would sell at a premium. A *premium bond* is one that sells for more than its par value.

Zero-Coupon Bonds

Investors who buy coupon bonds are subject to reinvestment risk, i.e., they must reinvest the coupon payments as they come in and might end up reinvesting at a lower interest rate. *Zero-coupon bonds* solve this problem by not paying interest during the life of the bond, but only at maturity. Zero coupon bonds have a stated maturity value, and are sold for at a discount, i.e., the present value of the maturity value.

All short-term Treasury bonds (i.e., 3 months) are sold as zero-coupon bonds. Some longer maturity bonds (i.e., 5- or 10-year) are also sold as zero-coupon bonds, but this is less common. Since a zero-coupon bond does not pay interest until maturity, its price is more sensitive to fluctuations in interest rates than a traditional coupon bond.

Bond Investments and Purchasing Power

The bonds issued by the United States Treasury ("Treasuries") have traditionally been considered *risk-free* with respect to default (the risk of non-payment). The reason for risk-free status is that the bonds are backed by the full faith and credit of the United States, i.e., the ability of Congress to raise taxes to make all payments due to bondholders. Further, the United States has never missed a bond payment.[25]

However, traditional Treasuries promise to pay an amount in nominal dollars, and do not guarantee the investor's purchasing power (in real dollars). That is, Treasuries are not risk-free with respect to inflation risk. If inflation were to increase after you purchased a bond, the dollars you get back would have less purchasing power than the dollars you invested to buy the bond. Figure 5.13 shows the historical nominal and real return on 10-Year Treasury Bonds, calculated using historical inflation data and coupon rates.

[25] In the summer of 2011, the United States Congress delayed raising its self-imposed borrowing limit, which resulted in a default scare and a downgrade by credit-rating agency Standard and Poor's. The crisis was political rather than fiscal, and was resolved without any missed payments.

Figure 5.13 Historical Nominal and Real Rate of Return on 10-Year Treasury Bonds

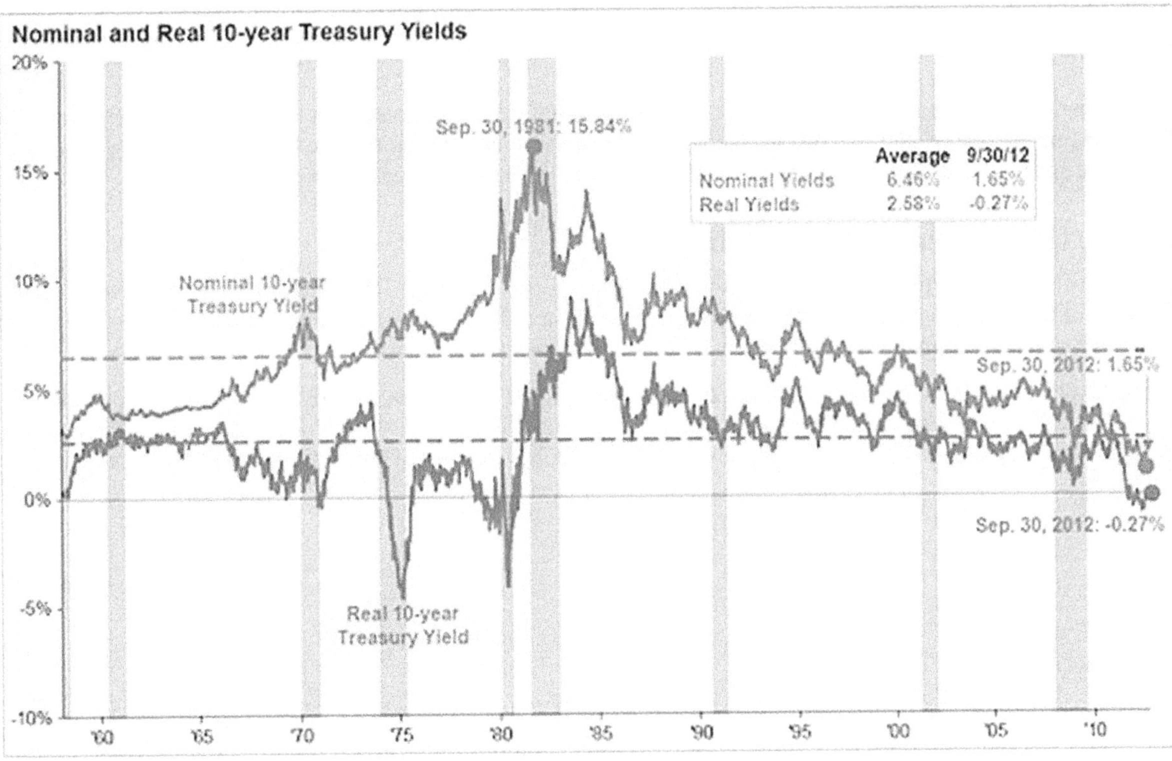

During periods of rapid inflation (for example, during 1973-1975), Treasury bonds have experienced negative real returns, i.e., the rate of interest was not sufficient even to maintain purchasing power.

5.4 Inflation Indexed Bonds

Investors should be primarily interested in maintaining their purchasing power over time. In response, many nations have begun issuing *inflation-indexed bonds*, where both the principal and coupon payments are linked to an inflation index. The United Kingdom began issuing inflation-linked bonds (called gilts) in 1981, and many countries, including the United States, Canada, Australia and Germany, have followed suit.

The two types of inflation-indexed bonds available in the United States are *Series I Savings Bonds* and *Treasury Inflation Protected Securities* or *TIPS*. Both are general obligations of the United States and share the same inflation-adjustment based on the Consumer Price Index. On account of the inflation adjustment, TIPS and I-Bonds are the only truly risk-free investments.

You can buy I-Bonds or TIPS from www.treasurydirect.gov, the website of the United States Treasury, without any fees or commissions. It's like online banking directly with the U.S. Treasury. The differences between TIPS and I-Bonds have to do with annual purchase limits, how the inflation adjustment is applied, and taxation.

Series I Savings Bonds

Series-I Savings Bonds are issued with a fixed real interest rate, which is guaranteed for 30 years. The interest rate has two components: a fixed real rate, and a variable rate that provides the inflation adjustment. The variable rate is revised twice a year to reflect changes in the Consumer Price Index. I-Bonds have a purchase limit of $10,000 per year per individual.[26]

[26] Series-I bonds are such a great investment vehicle that a purchase limit of $10,000 per year exists to encourage the use of Series-I bonds by small, individual investors, while preventing big investors from overusing this good deal.

Series-I Bonds provide guaranteed purchasing power due to the inflation adjustment. In addition, in the event of sustained deflation, the value of I-Bonds cannot decline (i.e., in nominal dollars). You can redeem Series-I Bonds anytime, but redemptions made within one year of purchase forfeit any interest earned – a small penalty given the security they provide.

Series-I Bonds come with a special tax feature: they are exempt from any state or local income tax forever, and the Federal income tax due on their interest payments may be deferred until maturity or redemption, for up to 30 years.[27] Given these special features, *I-Bonds are the perfect asset in which to invest your emergency fund.*

Treasury Inflation Protected Securities

Treasury Inflation-Protected Securities are the "big cousin" to I-Bonds. TIPS also provide protection against inflation, but are sold at auction like regular Treasury Bonds and don't have any annual purchase limits. Moreover, the interest earned and inflation-adjustment on TIPS is fully taxable each year.

The mechanics of the inflation adjustment are subtly more complex than I-Bonds. TIPS pay interest at a real interest rate, and an adjustment is made to the principal based on changed in the Consumer Price Index (CPI). The principal of a TIPS increases with inflation and decreases with deflation, but cannot fall below the original principal at issue.

[27] If the bonds are used to fund higher education (for yourself or a dependent family member), they are totally exempt from Federal taxes.

The TIPS Bond Inflation Adjustment

Every TIPS bond has an "index ratio" which is exactly correlated with the CPI. The index ratio is used to adjust the value of the TIPS when it comes time to make interest payments or to return the principal at maturity.

Suppose you bought a $10,000 10-year TIPS bond on 1/16/2007. The coupon interest rate was set to 2.375%, which is the real rate of interest. On its issue date it had an index ratio of 1.0.

On July 15, the index ratio was at 1.02773, so the principal of the bond was worth $10,277.30.

Also, semi-annual interest was paid based on the inflation-adjusted principal value:

$$Interest = Principal \times Index\,Value \times \frac{Interest\ Rate}{2}$$

$$Interest = \$10{,}000 \times 1.02773 \times \frac{2.375\%}{2}$$

$$Interest = \$10{,}000 \times 1.02773 \times 0.011875 = \$122.04$$

Suppose the index ratio on 1/15/2017 (at maturity) is 1.25. At maturity, your inflation-adjusted principal will be:

$$Adjusted\ Principal = \$10{,}000 \times 1.25 = \$12{,}500.$$

And the final coupon payment will be:

$$Interest = \$10{,}000 \times 1.25 \times \frac{2.375\%}{2}$$

$$Interest = \$10{,}000 \times 1.25 \times 0.011875 = \$148.44$$

Given the less-favorable tax treatment that TIPS receive, compared to I-Bonds, TIPS are not ideal for your emergency fund. Instead, the best way to buy TIPS is in an Individual Retirement Account that provides special tax treatment.

5.5 Summary

An annuity is a stream of payments that continue for a fixed period of time. We calculate the present value of an annuity by discounting the value of all of the payments to the present time and adding the payments.

Loans are borrowing contracts between a lender and borrower. The lender provides a lump-sum to the borrower, who makes regular payments over time to repay the loan. An amortizing loan is the inverse process of the annuity, i.e., a series of fixed payments that pays off the loan over time.

Bonds are a class of investments in which the investor lends money to a borrower in return for a promise to pay interest and return the principal at maturity.

Most bonds pay interest at a nominal rate, i.e., without guaranteeing any future purchasing power. By contrast, the US Treasury sells two types of inflation-indexed bonds, Series-I Bonds and TIPS, which provide an inflation-adjustment to protect your future purchasing power.

5.6 Conceptual Review Questions

1. What is an annuity? Draw a time line and explain. How would you calculate the present value of an annuity?
2. Explain the difference between ordinary annuity and annuity due. Draw a time-line for each to illustrate the difference.
3. Explain, giving an example, the idea of an amortizing loan. What is an amortization schedule?
4. Explain, giving and example, the idea of a revolving loan.
5. How does an amortizing loan relate to the idea of an annuity with fixed payments?
6. Explain the difference between a revolving loan (e.g., credit card) and an amortizing loan (e.g., a car loan).
7. Identify and explain the main characteristics of a bond (as an investment vehicle).
8. What are the main characteristics of Series I Bonds and TIPS?

5.7 Numerical Questions

13. Suppose you need to make 3 payments of $20,000 due in 5 years; in 6 years; and in 7 years. How much (in total) do you need to invest today at 3% to make these payments?
14. Consider a student loan of $30,000, with an APR of 6.8%. What is the monthly payment to pay off the loan over 10 years? Over 20 years?

6 Modeling The Economic Life-Cycle With Interest

Learning Objectives

- Extend the LC model by incorporating the rate of return on assets.
- Develop a spreadsheet solution for an 80-period LC model with a real return on investment.
- Evaluate the effect of the level of the real interest rate on the lifetime budget constraint, and to explain the different effect for lenders and savers.

6.1 Introduction: Incorporating Time Value of Money into the Life-Cycle Model

In Module 2, we developed both a two-period LC model and an 80-period LC model[28] for consumption smoothing. In our spreadsheet solution, we implemented a system of equations to model income, consumption, saving, and the accumulation of assets. Our simplifying assumptions included no real return on investment, and a world without taxes. In Module 4, we discussed interest, which is both the rate of return earned from an investment, and the cost of borrowing. We also introduced inflation (the change in the level of prices), which leads to a decrease in the purchasing power of nominal dollars.

In this module, we will revise the LC model to include the return on assets, i.e., a positive real interest rate. We begin by developing the conceptual framework for the life-cycle model with a positive real rate of return. Using a two-period abstraction, we explain the necessary changes to the budget constraint equations to account for the rate of return (or the cost of borrowing). Next, we develop an 80-period life-cycle model for consumption smoothing to account for return on investment.

6.2 The Two-Period Life-Cycle Model with r > 0

Recall our first approximation at a lifetime budget constraint for a two-period model with income only when young. Let's revise the two-period model to include a return on investment. Savings (the amount of income not consumed when young) is lent to another agent in the economy, and repayment is made with interest. The interest is a "rental fee" for the loan of money, paid from the borrower to the lender. (We could also think of the interest as the incentive to forgo consumption when young, e.g., payment for deferred gratification.)

As before, the amount of consumption when young is income minus savings:

$$C_y = W - S$$

Let r be the real rate of interest (i.e., net of inflation) earned as a percentage of the amount saved when young. The return on investment, rS, will be received when old. The budget constraint for old age is amended to include the return of savings and the return on savings:

$$C_o = S + rS = S(1 + r)$$

Now we combine the two periodic budget constraints and arrive at a lifetime budget constraint that incorporates savings and a real return on investment:

$$W = C_y + \frac{C_o}{1 + r}$$

The term $\frac{C_o}{1+r}$ is the present-value of consumption when old. The present value of consumption when old is the amount to be saved when young, which, after earning a return on investment at the rate r, will grow to become the amount of consumption when old.

---

[28] We develop a two-period life-cycle model to explain the conceptual framework while keeping the arithmetic as simple as possible, but the key insights and interpretation will remain the same. Later in this module, we will develop an 80-period life-cycle model.

The addition of a return on investment changes the slope of the lifetime budget constraint line. For example, consider $r = 1$, i.e., a 100% real investment return on savings when young.[29] Figure 6.1 plots the lifetime budget constraint with $r = 1$, and a numeric example follows.

Figure 6.1 A Two-Period Lifetime Budget Constraint with $r = 1$

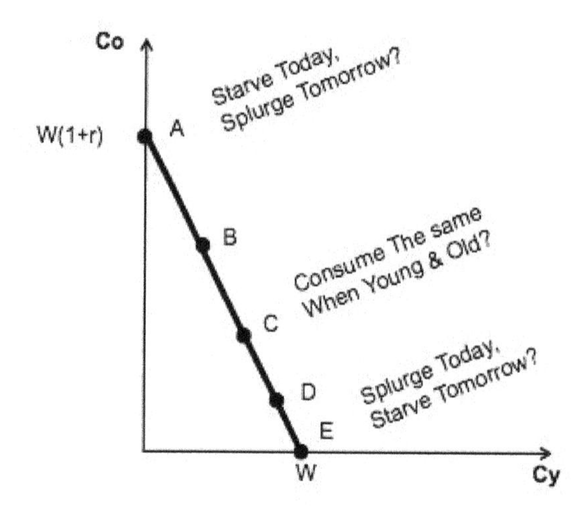

With income (W) earned only when young, notice the trade-off between consumption when young (C_y) and consumption when old (C_o). When $r = 1$, foregoing one dollar of consumption when young provides two dollars of consumption when old. Put another way, the cost when young (the present value) of consumption when old is one half of the amount of consumption when old.

Lets consider a numerical example to make the lifetime budget constraint concrete. Suppose we have a two-period life-cycle model with $r = 1$ and $W = 120$. We can solve for $\bar{C}$, the level of smooth consumption as follows:

$$W = \bar{C} + \frac{\bar{C}}{1 + r}$$

$$120 = \bar{C} + \frac{\bar{C}}{2} = 1.5\bar{C}$$

$$\bar{C} = \frac{120}{1.5} = 80$$

Given lifetime income of 120 and $r = 1$, it is possible to smoothly consume 80 in each period. To prove that this works, we can rewrite the budget constraint to solve for the required savings when young. Starting with the level of smooth consumption, we find the amount of savings required when young:

$$S = W - C_y = 120 - 80 = 40.$$

Since savings earns a 100% return on investment, the total assets available when old are equal to twice the amount saved when young:

$$C_o = S + rS = 40 + 40 = 80.$$

In this example, saving 40 when young provided resources of 80 when old, and consumption is smoothed.

[29] A 100% real return on investment is actually a very reasonable figure for a two-periodlife-cycle model.In a two-periodlife-cycle model each period might represent 40 years of adulthood. A 100% investment return would mean doubling one's real wealth over the course of a 40-year investment horizon, which is relatively conservative.

6.3 Implementing an 80-Period Life-Cycle Model with r > 0

In a previous example in Module 3, we implemented an 80-period life-cycle model as a spreadsheet (review Figure 2.8, Figure 2.9, and Figure 2.10, and the accompanying discussion). Recall the example fact pattern for Roxanne, who expects to work from age 21 to age 65, earning a real salary income of $44,000 per year. We found that Roxanne was able to smooth consumption by spending $24,750 per year. To do so, she would save $19,250 per year during her working years, and dissave $24,750 each year during retirement. We now return to this example and incorporate a non-zero real rate of return on investment.

Two main modifications are needed to incorporate the real rate of return in this model. First, we adjust the levels of accumulated wealth to account for earning compounded real investment returns on accumulated assets.

Let A_i be the assets in year i, A_{i-1} be the assets in the preceding year, and S_i be the savings in year i.
We can calculate the return earned on the previous year's assets as: $r(A_{i-1})$.
Thus, the value of A_i will be:

$$A_i = A_{i-1} + r(A_{i-1}) + S_i$$

In each year, the amount of assets will be equal to the previous year's assets, plus the return on investment at the real rate *r* for one year, plus the new savings.

Second, inasmuch as we want to keep track of the value of human capital and include it in calculating economic net worth, we need to discount human capital to present value: wages to be earned many years into the future are worth less in the present because of the time value of money. The spreadsheet in Figure 6.2 incorporates these two modifications.

Figure 6.2 A Spreadsheet Implementation of the 80-Period Life-Cycle Model with Interest

	A	B	C	D	E	F	G	H
1	Number of Periods		80					
2	Real Interest Rate		1.00%					
3	Consumption in each		24750					
4	Ending wealth		$ 509,125					
5				S = W - C			"Total Wealth"	
6	Age	Income	Consumption	Savings	Financial Assets	Human Capital	Economic Wealth	Saving or
7	(Period)	W	C	S	A	HC	ENW	Dissaving?
8	21	$44,000	$ 24,750	$ 19,250	$ 19,250	$ 1,560,040	$ 1,579,290	saving
9	22	$44,000	$ 24,750	$ 19,250	$ 38,693	$ 1,531,640	$ 1,570,333	saving
10	23	$44,000	$ 24,750	$ 19,250	$ 58,329	$ 1,502,957	$ 1,561,286	saving
11	24	$44,000	$ 24,750	$ 19,250	$ 78,163	$ 1,473,986	$ 1,552,149	saving
12	25	$44,000	$ 24,750	$ 19,250	$ 98,194	$ 1,444,726	$ 1,542,921	saving
13	26	$44,000	$ 24,750	$ 19,250	$ 118,426	$ 1,415,173	$ 1,533,600	saving

In Figure 6.2, cell C3 takes an input for the real interest rate. We reference this cell whenever we need to discount or compound using the real rate of return.[30]

Accounting for the real investment return is relatively simple. In each year, one earns an investment return on the assets held for the year (e.g., the amount of financial wealth at the end of the previous year). For example, at age 21 Roxanne saved $19,250. The next year, at age 22, she would earn asset income on her accumulated financial assets of $\$19{,}250 \times 1\% = \$192.50$.

Now suppose Roxanne saves an additional $19,250 at age 22. Roxanne's financial assets at the end of year 22 are the assets from age 21 plus interest, plus her new savings: $\$19{,}250 + \$192.50 + \$19{,}250 = \$38{,}693$.

[30] For the purpose of this example, the real rate of return on financial assets has been set to 1%, which is approximately the real rate of return on TIPS at the time of this writing.

At age 23, Roxanne earns a 1% real return of $\$38,693 \times 1\% = \$387$, and saves an additional \$19,750. Her financial assets at the end of year 23 are $\$38,693 + \$387 + \$19,250 = \$58,330$.

The second modification to the previous spreadsheet is to account for Roxanne's human capital as a present value. Recall that an amount to be received in the future must be discounted to the present. Let HC be the amount of human capital, and W be the salary, and r be the real interest rate. The value of human capital in year i is:

$$HC_i = \frac{HC_{i+1} + W_{i+1}}{1+r}$$

In effect, we are calculating the value of human capital by working backwards. In each year, the next year's salary is 1 year away, and must be discounted by $1 + r$. Figure 6.3 shows a numerical example.

Figure 6.3 Calculating the Present Value of Human Capital by Discounting

6	Age	Income	Consumption	Savings	Financial Assets	Human Capital	Economic Wealth	Saving or
7	(Period)	W	C	S	A	HC	ENW	Dissaving?
48	61	$44,000	$ 24,750	$ 19,250	$ 969,723	$ 171,686	$ 1,141,410	saving
49	62	$44,000	$ 24,750	$ 19,250	$ 998,671	$ 129,403	$ 1,128,074	saving
50	63	$44,000	$ 24,750	$ 19,250	$ 1,027,907	$ 86,697	$ 1,114,605	saving
51	64	$44,000	$ 24,750	$ 19,250	$ 1,057,436	$ 43,564	$ 1,101,001	saving
52	65	$44,000	$ 24,750	$ 19,250	$ 1,087,261	$ -	$ 1,087,261	saving

In Figure 6.3, Roxanne's human capital at age 65 is \$0 because she has future income (i.e., no salary at age 66). At age 64, Roxanne's human capital is the present value of her salary at age 65, which is:

$$HC_{64} = \frac{HC_{65} + W_{65}}{1+r} = \frac{\$0 + \$44{,}000}{1.01} = \$43{,}564$$

At age 63, Roxanne's human capital is the present value of her future salary at age 64, plus the present value of her human capital at age 64. Thus, at age 63, Roxanne's human capital, the present value of her future salary income is:

$$HC_{64} = \frac{HC_{64} + W_{64}}{1+r} = \frac{\$43{,}564 + \$44{,}000}{1.01} = \$86{,}697$$

We continue to work backwards in the same manner to calculate her human capital for each year. By working backwards, we see that Roxanne's human capital increases as we work backwards to her current age. This calculation reinforces the idea that human capital is a wasting asset that decreases with age.

Now that we have the updated annual budget constraint to account for the real rate of return in calculating the value of financial assets and human capital, we find an interesting result. Consider the graph of lifetime wealth in Figure 6.4.

Figure 6.4 Lifetime Wealth with $r = 1\%$ per Year

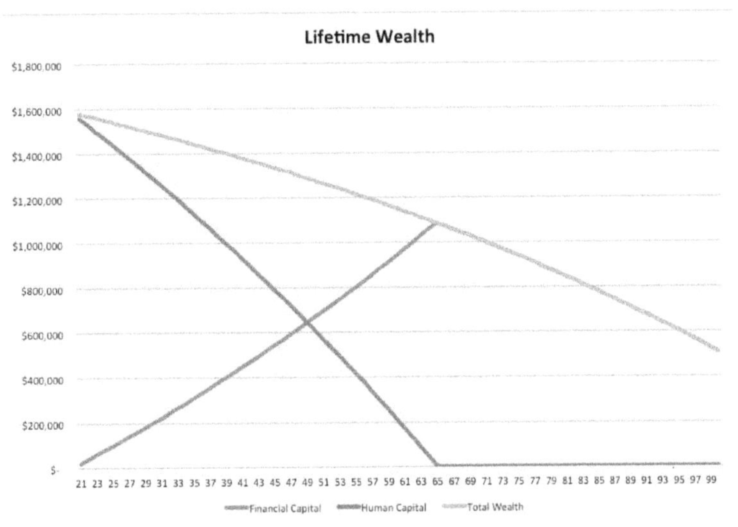

In Figure 6.4, the curves for financial capital, human capital, and total wealth are no longer linear. The graph of human capital over time is concave, the graph of financial capital is convex over the segment from age 21 to retirement, and the graph of total wealth is concave. Now the interesting part: the graph of lifetime wealth shows that her maximum age of life, Roxanne will have about $500,000 of financial assets remaining. How can we interpret this?

The $500,000 of remaining wealth at age 100 means that, by consuming $24,750 per year, Roxanne would not consume all of her resources during her lifetime. Another equally valid interpretation is that, by consuming only $24,750 per year, Roxanne would over-save during her working years. The $24,750 level of consumption was chosen when we were working under the assumption of no real investment returns (i.e., $r = 0\%$). Now that $r = 1\%$ Maria has additional resources: in addition to her accumulated savings, she also earns asset income at the real rate of return, and she can consume not only her saved assets, but also the income generated by those assets.

The rational person would not want to leave $500,000 of resources at age 100, but would prefer to consume his or her resources smoothly throughout his or her entire lifetime. With some trial-and-error (e.g., guessing), we can find the level of consumption[31] that Roxanne can afford, such that her remaining wealth at age 100 is $0. Figure 6.5 shows the updated spreadsheet with this level of consumption.

[31] It is a mathematical fact that there is a single value of consumption that will solve for all periods, such that $0 of final assets remains after age 100. A convenient way to find the correct value for consumption is to use the spreadsheet program's Goal Seek tool. We tell Goal Seek to set the assets at the end of life at $0, by changing the level of consumption. Goal Seek continues to try values for consumption until its goal is met, and this trial-and-error takes only a fraction of a second to calculate.

Figure 6.5 Consumption Smoothing with $0 of Ending Wealth at Age 100

	A	B	C	D	E	F	G	H
1	Number of Periods		80					
2	Real Interest Rate		1.00%					
3	Consumption in each		28934.42					
4	Ending wealth		$ (0)					
5				S = W - C			"Total Wealth"	
6	Age	Income	Consumption	Savings	Financial Assets	Human Capital	Economic Wealth	Saving or
7	(Period)	W	C	S	A	HC	ENW	Dissaving?
8	21	$44,000	$ 28,934	$ 15,066	$ 15,066	$ 1,560,040	$ 1,575,106	saving
9	22	$44,000	$ 28,934	$ 15,066	$ 30,282	$ 1,531,640	$ 1,561,922	saving
10	23	$44,000	$ 28,934	$ 15,066	$ 45,650	$ 1,502,957	$ 1,548,607	saving
11	24	$44,000	$ 28,934	$ 15,066	$ 61,172	$ 1,473,986	$ 1,535,159	saving
12	25	$44,000	$ 28,934	$ 15,066	$ 76,850	$ 1,444,726	$ 1,521,576	saving
13	26	$44,000	$ 28,934	$ 15,066	$ 92,684	$ 1,415,173	$ 1,507,857	saving
51								
52	64	$44,000	$ 28,934	$ 15,066	$ 827,579	$ 43,564	$ 871,143	saving
53	65	$44,000	$ 28,934	$ 15,066	$ 850,920	$ -	$ 850,920	saving
54	66	$ -	$ 28,934	$ (28,934)	$ 830,495	$ -	$ 830,495	dissaving
55	67	$ -	$ 28,934	$ (28,934)	$ 809,865	$ -	$ 809,865	dissaving
85								
86	97	$ -	$ 28,934	$ (28,934)	$ 85,096	$ -	$ 85,096	dissaving
87	98	$ -	$ 28,934	$ (28,934)	$ 57,012	$ -	$ 57,012	dissaving
88	99	$ -	$ 28,934	$ (28,934)	$ 28,648	$ -	$ 28,648	dissaving
89	100	$ -	$ 28,934	$ (28,934)	$ (0)	$ -	$ (0)	dissaving

In Figure 6.5, we see that with a real interest rate of 1%, Roxanne can afford to consume $28,934 per year, as compared to only $24,750 per year with a real interest rate of 0%. Thus, a 1% change in the real interest rate resulted in a 17% increase in Roxanne's sustainable annual consumption, for every year of her life:

$$\frac{\$28{,}934}{\$24{,}750} - 1 = 17\%$$

Holding everything else constant, a real interest rate of 2% would result in Roxanne's being able to consume $32,647 per year, or an increase of about 32% compared to real interest rates of 0%:

$$\frac{\$32{,}647}{\$24{,}750} - 1 = 32\%$$

6.4 The Level of Real Interest Rates Affects the Lifetime Budget Constraint

How much current consumption must we give up to save up for future consumption? How much future consumption must we forego to pay back the cost of borrowing for current consumption? The answer to these questions depends on the level of interest rates.

Consider the impact of low or high interest rates on saving for future consumption. Figure 6.6shows the impact of different interest rates on a two-period lifetime budget constraint with income only when young.

Figure 6.6 Two-period Lifetime Budget Constraint with Different Real Interest Rates.

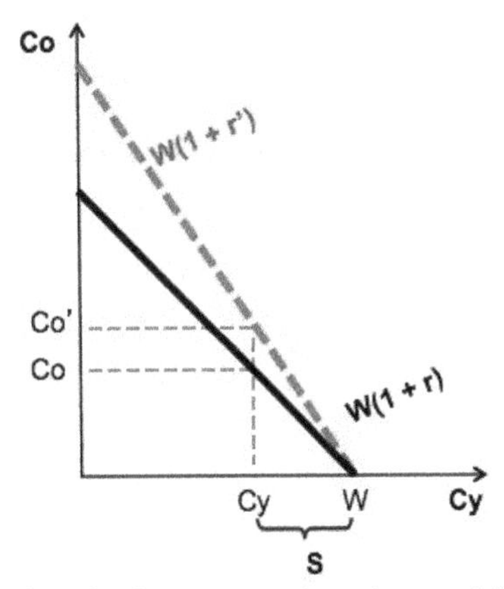

The solid line in Figure 6.6 shows the lifetime budget constraint when real interest rates are low. In order to save up for future consumption (i.e., consumption when old), we need save a large proportion of income today (i.e., when young). Since low rates mean increased saving, we can also state that low rates require us to give up a lot of current consumption to save up for future consumption.

The dashed line in Figure 6.6 shows the lifetime budget constraint when real interest rates are high. When rates are high, one can earn a large rate of return on the savingsaccumulated when young. If we save the same amount S with high interest rates, the result will be higher consumption when old ($C_0{}'$ in the graph). Since the amount saved willearn a high rate of return, we need to give up relatively little current consumption (consumption when young) to save up for future consumption (consumption when old).

The net effect of high real interest rates is to increase the feasible level of consumption in both periods. The net effect of low real interest rates is to decrease the feasible level of consumption in both periods.

6.5 Incorporating Interest Rates with Income in Both Periods

Consider a household with income in each of 2 periods. Recall the notation W_y for income when young and W_o for income when old. When real interest rates are positive ($r > 0$). Consumption when young is equal to income less savings (or income plus borrowing):

$$C_y = W_y - S$$

Consumption when old is bounded by income when old plussaving (borrowing) when young, times the investment (borrowing) rate:

$$C_o = W_o + S(1+r)$$

Recall that S (net savings) is a variable: S could be positive (saving when young) or negative (borrowing when young), and provides a common term to join the two periodic budget constraints into a single lifetime budget constraint. The lifetime budget constraint for this household is:

$$C_y + \frac{C_o}{1+r} = W_y + \frac{W_o}{1+r}$$

Figure 6.7 The Lifetime Budget Constraint with Income in Both Periods and $r > 0$

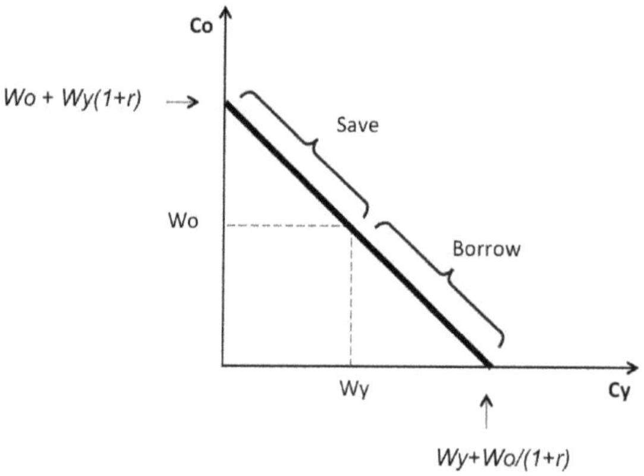

Figure 6.7 is a graphical illustration of the two-period lifetime budget constraint for a household with income in both periods and $r > 0$. On the horizontal axis (for the period of youth), notice the point labeled $W_y + W_o/(1+r)$, which is the maximum feasible consumption when young. To consume at this maximum level, one would consume all income earned when young and *borrow* the present value of income when old to supplement consumption when young. In other words, the most one can consume when young is the present value of one's income from both periods:

$$C_y = W_y + \frac{W_o}{1+r}$$

This would be consuming at the level of "splurge when young, starve when old."

Similarly, on the vertical axis of Figure 6.7, notice the point labeled $W_o + W_y(1+r)$, which is the maximum feasible consumption when old. To achieve this maximum feasible consumption when old, one would save all income when young and consume the future value of income when young during old age, in addition to income when old. This would be consuming at the level of "starve when young, splurge when old."

To maximize their economic happiness, households would like to smooth consumption. Depending on their income, this would require either saving when young and dissaving when old, or borrowing when young and repayment when old. The level of real interest rates will have an important impact on the level of smooth consumption, and the impact is different when saving or borrowing is advised.

Saving to Smooth Consumption

For a household with $W_y > W_o$, consumption smoothing requires saving (lending) when young and consuming from assets when old. Figure 6.8 shows a graphical representation of this case with real interest rates of 0 and 1, i.e., you either get back your same purchasing power (a 0% real rate of return) or you get back twice your purchasing power (a 100% real rate of return).

Figure 6.8 Lifetime Budget Constraint with $W_y > W_o$, with $r = 0$ and $r = 1$ for Comparison.

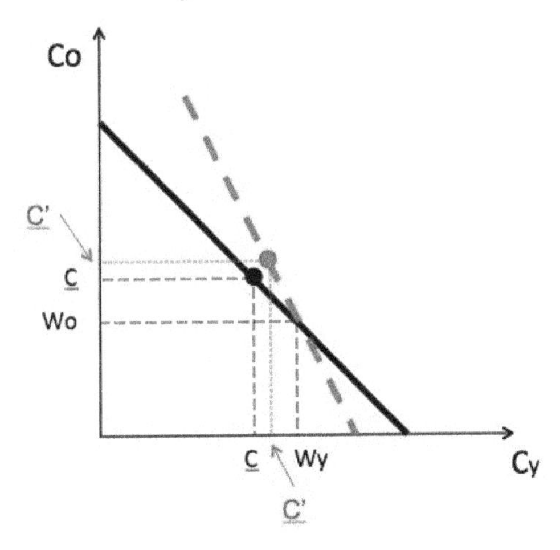

In Figure 6.8, the solid line is the lifetime budget constraint with $r = 0$, and the dashed line is the lifetime budget constraint with $r = 1$ (i.e., 100% real return on investment). When the real rate of return goes to 1, the slope of the lifetime budget constraint becomes twice as steep as compared to the slope of the budget constraint line when $r = 0$. Suppose the real rate of interest is 0. If you save \$100 when young, in old age, you would still have \$100. But if the real interest rate were 1 (i.e., a 100% real return on investment), then \$100 saved when young grows to \$200 when old. When the real interest rate is higher, future consumption requires less saving when young (i.e., you don't have to forgo as much consumption when young to pay for consumption when old).

In the graph in Figure 6.8, $\bar{C}$is the smooth level of consumption with r = 0, and $\bar{C}'$ is the smooth level of consumption when $r = 1$. When the real rate of return is higher, the smooth level of consumption is higher. For lenders, an increase in the real rate of return means that it is possible to consume more in both periods, and a decrease in the real rate of return would mean one would need to consume less in both periods.

Borrowing to Smooth Consumption

On the other hand, consider the case of a household with $W_y < W_o$: Consumption smoothing would require borrowing (dissaving) when young to consume more when young, and repaying out of future income. Figure 6.9 shows a graphical representation of this case with real interest rates of 0 and 1.

Figure 6.9 Lifetime Budget Constraint with $W_y < W_o$, with $r = 0$ and $r = 1$ for Comparison.

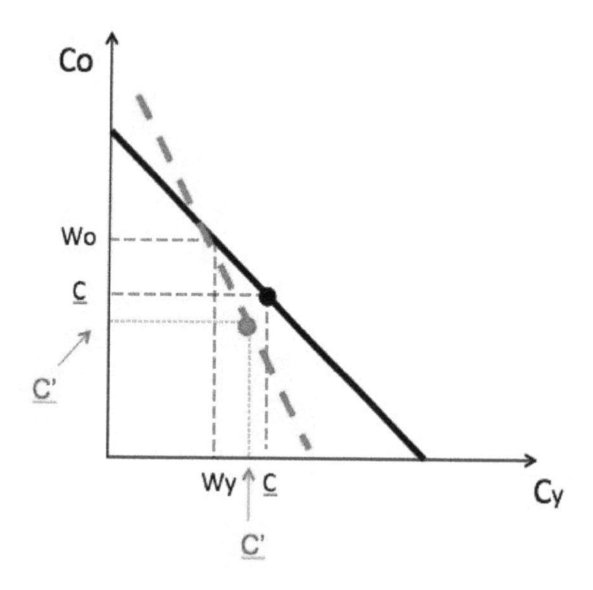

In Figure 6.9, the solid line is the lifetime budget constraint with r = 0, and the dashed line is the lifetime budget constraint with $r = 1$ (i.e., 100% for the entire period of youth). When the real rate of return goes to 1, the slope of the lifetime budget constraint becomes twice as steep as compared to the slope of the budget constraint line when $r = 0$. Suppose the real rate of interest is 0. If you borrow $100 when young, in old age, you would have to repay $100. But if the real interest rate were 1 (i.e., 100%), then $100 borrowed when young would require a repayment of $200 when old. When the real interest rate is higher, borrowing for consumption when young requires you to give up more consumption when old (i.e., you have to forgo more consumption when old to repay the borrowing used for consumption when young).

In the graph in Figure 6.9, $\bar{C}$is the smooth level of consumption with r = 0, and $\bar{C}'$ is the smooth level of consumption when $r = 1$. When the real rate of return is higher, the cost of borrowing is more expensive. For borrowers, an increase in the real rate of return means that one must consume less in both periods, and a decrease in the real rate of return would mean one could consume more in both periods.

6.6 Summary

The life-cycle model is easily adapted to account for a return on investment, without significantly changing the model. Compared to the first iteration of the life-cycle model developed in Module 3, the inclusion of a positive real return on investment means that a higher sustainable level of consumption can be achieved.

Net savers benefits substantially from a higher real rate of return. With a high real rate, savings when young will grow substantially and provide more consumption when old. In the case of an 80-period life-cycle, savings when young will have many more periods of compounding. With a high real interest rate, a saver can enjoy a higher smoothed level of consumption.

Net borrowers will suffer substantially from a higher real rate of return. With a high real rate, borrowing when young will require substantial repayment when old. With a higher real interest rate, a borrower will suffer a lower smoothed level of consumption.

6.7 Review Questions

1. Draw a graph to plot the lifetime budget constraintfor a two-period life-cycle with income earned in both periods, and a positive real rate of return.
2. Write and explain the lifetime budget constraint for a two-period life-cycle with income earned in both periods, and a positive real rate of return.
3. Consider the lifetime budget constraint for a two-period life-cycle with income earned in both periods. *Assume income when young is less than income when old.* Explain the effect of an increase in the real rate of return.
4. Consider the lifetime budget constraint for a two-period life-cycle with income earned in both periods. *Assume income when young is more than income when old.* Explain the effect of an increase in the real rate of return.

7 The Odds Are…

Introduction to Probability and Expected Value

Learning Objectives:

- Define the concepts of random events, sample spaces, and discrete probability.
- Introduce techniques for calculating the probability of events.
- Define the expected value of a random event having financial consequences.
- Introduce insurance as an application of probability and expected value calculations.

7.1 Dealing with Uncertainty

In Modules 3 and 5, we implemented the economic Life-Cycle model to illustrate modeling a system of equations. To keep things simple, we had to make some important assumptions about the future: we assumed certainty about an individual's longevity (i.e., we assumed survival to age 100), and about the rate of return one would earn on investments.

In this module, we begin to think about uncertainty: what it is, and how to factor uncertainty into mathematical and financial models. We will introduce basic concepts of probability (quantifying uncertainty) and applications related to finance.

7.2 Random Outcomes, Events, and Sample Spaces

A *random experiment* or random process is one in which the outcome is affected by randomness. Our world contains both natural random processes (e.g., the weather) and man-made random processes (e.g., the outcomes of games of chance such as the casino game of roulette). Consider a bet between you and a friend. If the subject of your bet depends on a fact (e.g., the population of Canada) then the outcome is knowable in advance, and is not random. By contrast, in a random experiment (e.g., a coin toss), it is not possible to know the result in advance.

A *trial* is a single attempt at a random experiment. An *outcome* is a possible result of a random experiment. For example, when we flip a coin, each attempt at flipping the coin is considered a trial, and the possible outcomes are heads or tails.[32]

The Sample Space

In mathematics, a *set* is a group of items (or values) that can be grouped together. The mathematical set notation uses curly braces {} to show which elements are members of a set. The *sample space* is the set all of the possible outcomes for an experiment, and we need to know the possible outcomes before we can determine how likely it is for an outcome to be achieved.

Using set notation, we can write the sample space for the coin toss experiment as:

$$S = \{h, t\}$$

What are the possible colors that can be selected randomly using a coin toss? If a sample space does not contain any possible outcomes, then we would write it as an *empty set*, $S = \{\} \, or \, S = \emptyset$. Figure 7.1 illustrates the possible outcomes for rolling a 6-sided die.

[32]Coins in the United States have two sides. One has the image of the head of a famous American statesman (called "heads"), and the other has a different image of a monument or other symbol (called "tails").

Figure 7.1 Common 6-sided Dice

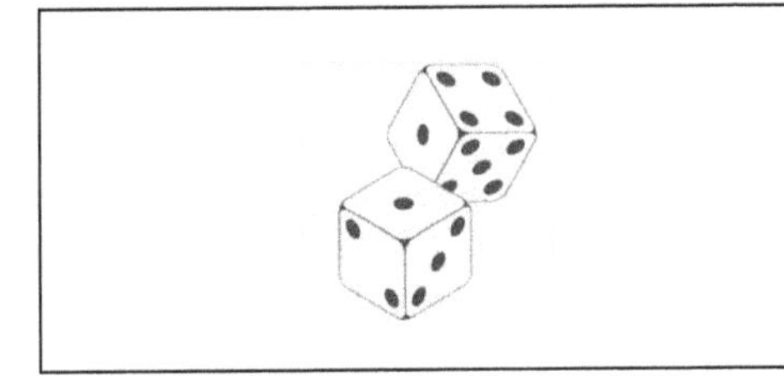

Dice are often used in games of chance, such as Backgammon or Craps. A six-sided die has 6 possible outcomes, so the sample space for rolling the die is:

$$S = \{1, 2, 3, 4, 5, 6\}$$

Using Venn Diagrams to Visualize Sets and Sample Spaces

A *Venn diagram* is a way to visualize sets, and we can use Venn diagrams to think about the probabilities of events. In a Venn diagram, we draw a rectangle to illustrate the universe of all elements in the sample space, and use circles to illustrate set(s) within that universe (for example, the sample space for a specific event).

For example, consider a standard deck of playing cards (see Figure 7.2). A standard deck of cards contains 52 cards, organized into 4 suits (Clubs, Spades, Hearts, and Diamonds). Each suit has 13 cards: the Ace, numbers 2-10, and the Jack, Queen, and King.

Figure 7.2 A Deck of Cards

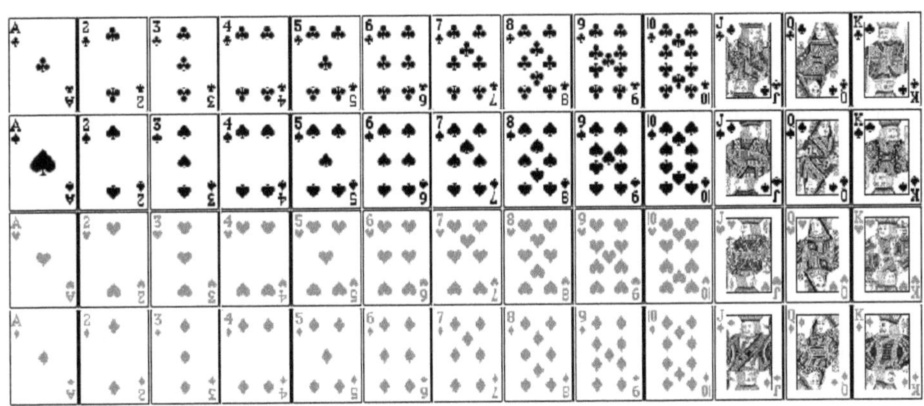

For example, Figure 7.3 shows a Venn diagram that illustrates all possible cards in the deck, grouped by color.

Figure 7.3 Venn Diagram with Cards Grouped by Color

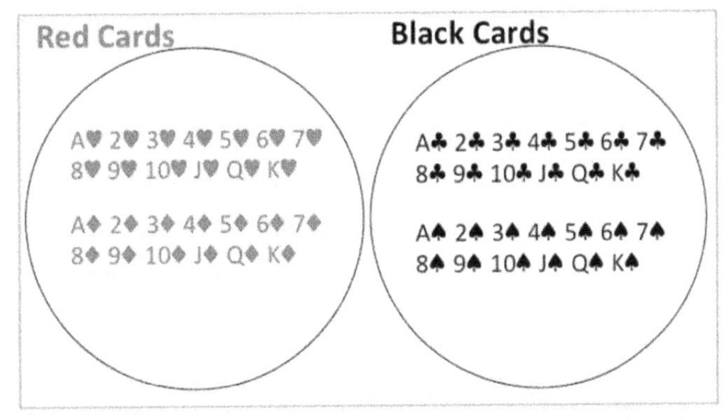

The diagram in Figure 7.3 illustrates the *universe* of all cards in a regular deck of cards. The universe is divided into 2 sets, representing the red cards and the black cards, respectively. We can count the items to find that there are 26 cards in each set, and 52 cards total. Every card is a member of only one of these sets and no the other set. We say that the set of red cards and the set of black cards are *disjoint sets*, because they have no elements in common. Finally, there are no cards in the deck that are neither red nor black.

Events

An *event* is an outcome, or a set of outcomes from a random experiment. An event is a subset of the sample space. For example, if the sample space for rolling a die is $S = \{1, 2, 3, 4, 5, 6\}$, one outcome is the event $E = \{4\}$, i.e., the event in which we roll a 4.

Other possible events include:

- The die shows an odd number: $E = \{1, 3, 5\}$
- The die shows a prime number: $E = \{2, 3, 5\}$
- The die shows a number greater than or equal to 4: $E = \{4, 5, 6\}$

An *impossible event* is one that is not part of the sample space. For example, it is not possible to roll a 7. We would write this event as the empty set, i.e., $E = \{\}\ or\ E = \emptyset$.

Independent events are events that have no bearing on one another. For example, the outcome a coin toss has no impact of effect on the outcome of rolling a die. Thus, we say that the coin toss and die roll are independent events.

7.3 Probability

The *probability* or chance describes how likely it is for a random outcome to occur. For any outcome within a sample space, the probability of that outcome is $0 \le p \le 1$. A probability of 0 means the outcome is impossible, and a probability of 1 means the outcome is a certainty (i.e., not affected by randomness).

The sum of the probabilities for all possible outcomes must equal 1. In a sample space with N outcomes,

$$\sum_{i=1}^{n} p_i = p_1 + p_2 + p_3 + \cdots + p_N = 1$$

We can describe the probability as a fraction or a percentage, i.e., "there's a 75% chance of rain tonight." Note that there cannot be a 110% chance of anything.

The term *discrete probability* is used when the outcomes in a sample space are discrete and countable. By contrast, the term *continuous probability* is used when the distribution of outcomes follows a continuous function. This module introduces examples and applications of discrete probability. Modules 7 and 10 include examples and applications of continuous probability.

Calculating the Probability with Equally Likely Outcomes

When we know all of the possible outcomes in a sample space, we can calculate the probability that an event will occur. The easiest probabilities to calculate are when each outcome is *equally likely* to occur, such as in games of chance.[33] In those cases, the probability of an event is given by:

$$p(E) = \frac{n(E)}{n(S)}$$

where $p(E)$ is the probability of event E occurring; $n(E)$ is the number of outcomes which are part of this event; and $n(S)$ is the number of outcomes in the sample space.

[33] Games of chance are games in which the outcomes depend on randomness. Popular games of chance include casino games like roulette and blackjack, as well as any game that involves picking a random number using dice, such as Backgammon or Monopoly. By contrast, games of skill such as chess do not depend on randomness. Some games, such as Scrabble, combine skill (a player's repertoire of words and creativity in finding placement) and chance (i.e., selecting each letter is a random event).

Example: A Coin Toss
Consider the probability of getting heads in a single coin toss. If the sample space is $S = \{h, t\}$ and the event in which we are interested is $E = \{h\}$, then the probability of getting heads is:

$$p(heads) = \frac{n(E)}{n(S)} = \frac{1}{2} = 0.5$$

We say there is a 50% chance of the coin landing on heads.There is also a 50% chance that the coin will land on tails, since heads and tails are the only possible outcomes in the sample space, and we know that probabilities of all events must add up to 100%. On any given flip, we expect that heads and tails are equally likely to occur. This does not mean that we will get heads exactly once out of every two flips. When a coin is flipped many times, there are likely to be runs of many heads in a row or many tails in a row – this is normal and expected. However, if we were to flip a fair over many repeated trials (e.g., a hundred or a thousand or ten thousand times), we would observe that each outcome occurs very close to half of the time.

The 50% chance is the *relative frequency probability*, since we know that with a fair coin, both possible outcomes and both are equally likely to occur.

We also know that each flip of the coin is an independent event, which means that previous results have no influence on the next flip. Even after having landed on heads 5 times in a row, the probability of the next coin toss landing on heads remains 50%.

Illustrating the Probability Using Venn Diagrams

In each of the following examples, we are trying to find the probability of selecting a card at random, i.e., from a standard deck of cards that has been shuffled.

What is the probability of getting a red card?
Look at Figure 7.3 again. There are 26 cards in the sample space of all red cards, and 52 cards in the deck as a whole. The probability of drawing a red card is:

$$p(red\ card) = \frac{26}{52} = \frac{1}{2} = 0.5$$

Figure 7.4 shows a Venn diagram that illustrates all possible cards in the deck, grouped by suit (diamonds, hearts, clubs, and spades). The suits are *subsets* of the colors, in that each card that is a member of the set of hearts is also a member of the set of red cards, etc.

Figure 7.4 Venn Diagram with Cards Grouped by Suit

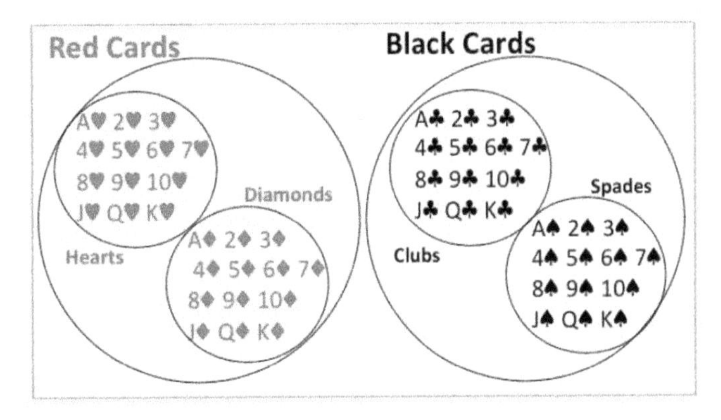

What is the probability of getting a heart?
There are 13 cards in the set of hearts, and 52 cards in the entire deck, so the probability of drawing a heart is:

$$p(heart) = \frac{13}{52} = \frac{1}{4} = 0.25$$

Within the universe of all cards, many other groupings are possible. Figure 7.5 illustrates the set of all 7s, which combines the 7 cards from each of the 4 suits.

Figure 7.5 Venn Diagram Illustrating the Set of All 7s

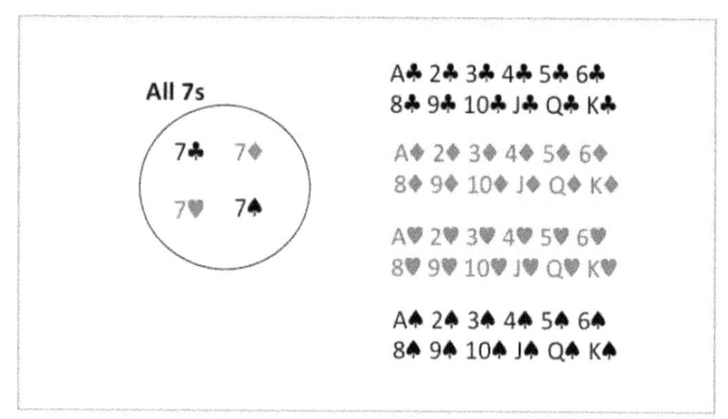

Notice that the set of 7s contains elements that are also members of other sets. For example, the 7 of hearts is also a member of the set of red cards and the set of hearts. We will return to this matter in the section on joint probability below.

What is the probability of getting a 7?
There are four 7s in the set of all sevens, and 52 cards in the entire deck, so the probability of drawing a 7 is:

$$p(7\ card) = \frac{4}{52} = \frac{1}{13} \approx 0.07692$$

The Odds Are...

Often times, the probability of success (or failure) is expressed as "the odds." Whether we talk about the probability of an event occurring, or the odds of the event occurring, we are discussing the same idea. *The odds* are a ratio used express the chance of an event occurring versus the chance against that event. To take a simple example, the probability of getting heads in a coin toss is expressed as:

$$p(heads) = \frac{n(E)}{n(S)} = \frac{1}{2}$$

The expression E' is *the complement* of E, which is the set of all outcomes that are not part of the event *E*. Since we know the number of outcomes that are part of the event *E*, and the number of outcomes in the sample space as a whole, we can find the number of outcomes in the complement of *E* as:

$$n(E') = n(S) - n(E)$$

In the example of the coin toss, the odds of obtaining heads are expressed as:

$$n(E) : n(E') = 1:1$$

We read the expression 1:1 as "one to one odds." Another way to think about this is that there is an equal chance (i.e., 50%) for and against getting heads.

As another example, consider the probability of drawing a 7 card from a full deck of 52 cards. The probability of obtaining a 7 card is:

$$p(7\ card) = \frac{4}{52} = \frac{1}{13} \approx 0.07692$$

In a deck of 52 cards, there are 4 ways to draw a 7 card, and 48 ways to draw another card that is not a 7. We would say the odds of drawing a 7 card are 4:48 or 1:12. We read this is "one to twelve odds," which means that in drawing a card there is one chance of drawing a 7 for every 12 chances of drawing a card that is not a 7.

Estimating Probability using Historical Relative Frequency

What is the probability that it will rain tomorrow? And what is the probability that your bike will be stolen in the next year? In contrast to situations where the set of possible outcomes is known and countable, there are many situations in which it is not possible to count all possible outcomes. An alternate approach is to use the *relative frequency* of past events as an estimate of probability.*Actuaries* are applied mathematicians who make their living by estimating probabilities based on historical relative frequency.

For example, consider a local fire department that is trying to plan for its staffing requirements and purchases of equipment. It would be important to know the number of emergency response calls per year. But it would also be important to know how many of those calls are real fires (i.e., requiring one or more fire trucks) and how many are false alarms. The fire department keeps records of all of the calls to which is respond. Figure 7.6 summarizes calls to a local fire department in 2012.

Figure 7.6 Fire Department Calls

Fire Department Calls	2012
Fire Surpression	97
Emergency Medical Response	2819
Hazardous Coundition	336
Service Call	412
Good Intent Call	243
False Alarms	662
Severe Weather/Natural Disaster	34
Special Type/Complaint	11
Undetermined	0
Total Calls	**4614**

Given this historical data, we could estimate the probability that a call to the fire department is a false alarm. The incidence of false alarms in 2012 was 662 out of a total of 4,614 calls. The frequency of false alarm calls was:

$$frequency(false\ alarm\ calls) = \frac{662}{4{,}614} \approx 0.1435$$

We don't know the true probability of a given call to the fire department being a false alarm. This relative frequency of 14.35% is a reasonable proxy for this probability.

Application of Historical Frequency Analysis: Weather Forecasting

Predicting the weather is an exercise in analyzing the historical frequency of events and creating a model to try to make predictions based on these events.

Meteorologists keep detailed records of the weather (i.e., temperature, rainfall, wind speed) that has occurred in the past, as well as the atmospheric conditions (i.e., temperature, wind speed and direction, cloud conditions, barometric pressure) that might have influenced the temperature or amount of precipitation on a given day. Finally, they use a statistical technique called regression to estimate the temperature, precipitation, and winds based on what has happened in the past. While the models are pretty good, the weather forecast is only an estimate.

Meteorology is one of a few fields in which applied mathematicians get paid even when their predictions are wrong!

7.4 The Multiplication Rule and Joint Probability

Repeated Trials and the Multiplication Rule

Sometimes the sample space must account for repeated trials of the same experiment. For example, consider flipping a coin three times. After flipping the first time, we will achieve an outcome of either heads or tails. Irrespective of the outcome of on the first flip, we could achieve either heads or tails on the second flip. The same applies on the third flip or any successive flips. In *independent* experiments, the outcome of one experiment has no effect on the outcome of any successive experiment. The possible outcomes are represented as a tree in Figure 7.7.

Figure 7.7 Possible Outcomes from Three Coin Flips

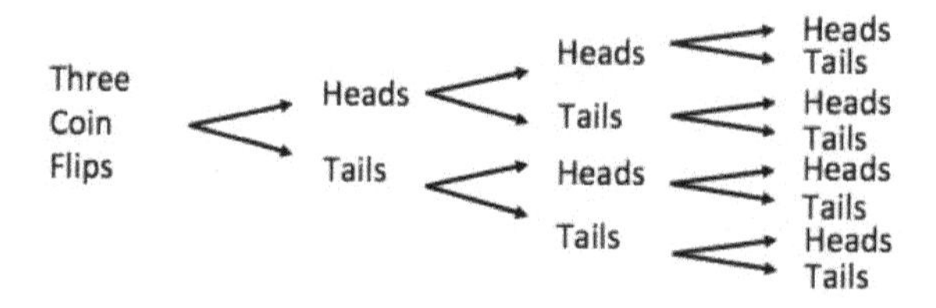

The Multiplication Rule

When an event involves multiple independent experiments (or repeated trials of the same experiment), we use the multiplication rule to create the sample space of possible outcomes. The *multiplication rule* uses multiplication to create the *product* of the sample spaces of the individual experiments. If one coin toss has a sample space of $S = \{h, t\}$, then two coin tosses will have the sample space of $S = \{h, t\} \times \{h, t\} = \{hh, ht, th, tt\}$.

Using the multiplication rule, the sample space for the event of flipping a coin three times (Figure 7.7) would include all of the possible permutations of 3 coin tosses: $S = \{hhh, hht, hth, htt, thh, tht, tth, ttt\}$.

Now, consider the sample space for the weather on any given summer's day. The weather could be sunny or rainy, and it could be cool or warm (say, over 80 degrees Fahrenheit). This sample space involves multiple domains, i.e., sky conditions and temperature. The possible outcomes include the product of the all of the outcomes from each of the two domains (see Figure 7.8).

Figure 7.8 Sample Space with Multiple Domains

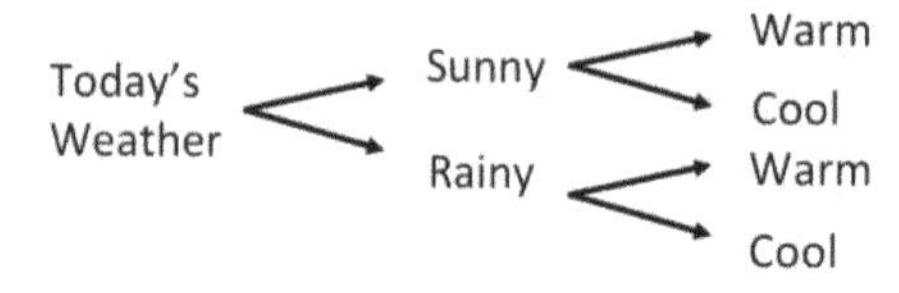

The sample space for the weather would be written as:

$$S = \{(sunny, cool), (sunny, warm), (rainy, cool), (rainy, warm)\}$$

Calculating Probability with the Multiplication Rule

Suppose we want to know the probability of obtaining two heads out of two coin tosses, i.e., $E = \{tt\}$. We can find the probability by seeing that the sample space has four possible outcomes, and only one of these contains the event for two heads:

$$p(two\ heads) = \frac{n(E)}{n(S)} = \frac{1}{4} = 0.25$$

We find also the probability of the event by multiplying the probability of an outcome satisfying that event in each trial. To find the probability of obtaining two heads out of two coin tosses, we can solve this by finding the probability of tossing heads in each trial (1/2) and multiplying these together:

$$p(two\ heads) = \frac{1}{2} \times \frac{1}{2} = \frac{1}{4} = 0.25$$

Example: Three Coin Tosses
What is the probability of obtaining at least one head out of 3 coin tosses?

The event $E = \{hhh, hht, hth, htt, thh, tht, tth\}$ describes the outcomes with at least one heads. The probability of getting at least one head is:

$$p(at\ least\ one\ head) = \frac{7}{8} = 0.875$$

Example: Craps Game
In the casino game of Craps, the shooter throws two dice, and bets are made about the sum of the numbers rolled on the dice. A simple bet is about whether the shooter rolls a 7 (win) or a 2, 11, or 12 (lose) on the first roll of the game. What is the probability that the opening roll is a winning 7?

The sample space is: *S* = {(1,1), (1,2), (1,3), (1,4), (1,5), (1,6), (2,1), (2,2), (2,3), (2,4), (2,5), (2,6), (3,1), (3,2), (3,3), (3,4), (3,5), (3,6), (4,1), (4,2), (4,3), (4,4), (4,5), (4,6), (5,1), (5,2), (5,3), (5,4), (5,5), (5,6), (6,1), (6,2), (6,3), (6,4), (6,5), (6,6)}

The outcomes which satisfy a sum of 7 are *E* = {(1,6), (2,5), (3,4), (4,3), (5,2), (6,1)}.

Therefore, the probability of winning on the opening rolls of a game of Craps is:

$$p(sum\ of\ 7) = \frac{6}{36} = \frac{1}{6} = 0.1667$$

What is the probability that the opening roll is a loser? The event E = {(1,1), (5,6), (6,5), (6,6)} describes the outcomes which are immediate losers.

Therefore, the probability of losing on the opening roll in a game of Craps is:

$$p(lose\ on\ opening\ roll) = \frac{4}{36} = \frac{1}{9} = 0.1111$$

Joint Probability

Sometimes an event combines outcomes from different sets to create a combined sample space.The ***joint probability*** represents the probability of two events occurring together.

For example, we might be interested in the probability of drawing a red 7 card of out a deck of shuffled cards. Figure 7.9 illustrates the set of all red cards, as well as the set of all 7s. Notice, of course, that two of the 7 cards are red (diamonds, hearts), but the other two are not (clubs, spades).

Figure 7.9 Venn Diagram Illustrating Overlapping Sets

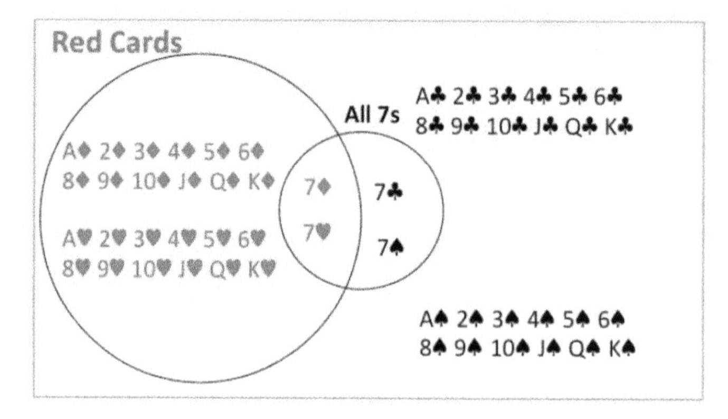

With independent events, we can use the multiplication rule to find this probability:

$$p(red\ 7\ card) = p(red\ card) \times p(7\ card) = \frac{1}{2} \times \frac{1}{13} = \frac{1}{26} \approx 0.03846$$

There is about a 3.8% chance of drawing a red 7 card.

As another example, in the game of Black Jack, all face cards (i.e., Jack, Queen, and King) have a value of 10. The set of all cards with a value of 10 includes the 10, Jack, Queen, and King of each of the 4 suits, for a total of 16 cards.

What is the probability of drawing a black card with a value of 10?

$$p(black\ card\ with\ value\ 10) = p(black\ card) \times p(value\ of\ 10) = \frac{1}{2} \times \frac{16}{52} = \frac{8}{52} \approx 0.1538$$

In words, there is about a 15% chance of drawing a black card with a value of 10.

The concept of joint probability, and the use of the multiplication rule to calculate the joint probability is an important idea to which we will return in modules 7 and 10.

7.5 Expected Values

Many real-life situations involve a financial gain or loss that is affected by a random process. When we attach a *payout* (a financial benefit or cost) to a randomly occurring event, we are interested in the expected value of this payout. The *expected value* is the probability-weighted payout from a risky process. Some applications of expected value include:

- The expected benefit from buying a $1 scratch-off lottery ticket
- Identifying which casino game is the most profitable to the casino
- Finding the cost to a retail store of offering a coupon for $10 off your purchase of $75 or more
- Calculating the cosst and benefits of insurance

Random Outcomes With Financial Consequences

We assume each event has a financial benefit or payout associated with it, and that we can calculate (or estimate) the probability of that event occurring. The expected value of such a random process is calculated as:

$$expected\ value = \sum_{e=1}^{N} p_e \times payout_e = p_1 \times payout_1 + p_2 \times payout_2 + \cdots + p_N \times payout_N$$

where N is the number of events, p_e is the probability of event e occurring, and $payout_e$ is the financial payout associated with event e. The expected value is the sum of all of the events' payouts, weighted by the probability that they would occur.

Example: Coin Toss
Consider a game between two players, in which each player bets one dollar per round. The players take turns tossing the coin, and calling the result when the coin is in the air. For example, one player flips the coin and calls "heads," expecting the coin to land heads side up. If the player guessed correctly, they take both their dollar and their opponent's dollar. What is the expected value of this game?

The probability of guessing a coin toss correctly is 50%. If you win, you will get your opponent's dollar. If you lose, you will lose your $1 bet. The expected value of each round is:

$$expected\ value = 50\% \times \$1 + 50\% \times -\$1 = \$0$$

Your expected value in playing this game is $0. Over many repeated trials you should expect to keep the same amount of money you started with. This is an example of a zero-sum game. In a *zero-sum game*, one player gains as another player loses, and vice-versa.

Example: Craps Game
Let's return to the previous example pertaining to the game of Craps. Let's make a simplification, and assume we are only concerned with the first roll of the game. Suppose the casino has the following payouts on the first roll of a game of Craps:

- If you win (roll a 7), the casino pays you 4 times the amount of your bet and returns your original bet.
- If you roll anything else, you lose your bet amount.

For example, if you bet $100 that you can roll a 7, the expected value of your bet is:

$$expected\ value = \frac{1}{6} \times \$400 + \frac{5}{6} \times -\$100 = -\$16.67$$

In words, the person placing the bet is expected to lose $16.67 per $100 bet. Sometimes the better will roll a 7 and win $400. Most of the time, the better will roll something else and lose their $100 bet.

Casino games are not zero-sum games, because the casino is in business to make money off of gamblers. In this case, the casino is expected to take 16.67% of the amount bet – not on every single game, but over the long run.

Example: Scratch-Off Lottery Tickets
Consider the "9's in a Line" scratch-off lottery game, offered by the Florida Lottery, and illustrated in Figure 7.10. This game costs $1 to purchase. The prize amounts and the chances of winning are shown on the back of the ticket and on the Florida Lottery website[34]; they are reproduced in Figure 7.10.

Figure 7.10 "9s in a Line" Scratch-off Lottery Game

Prize Amount	Probability	Prize Amount	Probability
$ 1,000	1 in 40,000	$ 20	1 in 300
$ 200	1 in 24,000	$ 15	1 in 300
$ 100	1 in 3000	$ 10	1 in 150
$ 50	1 in 4800	$ 5	1 in 150
$ 40	1 in 4000	$ 3	1 in 37.5
$ 30	1 in 2400	$ 2	1 in 15
		$ 1	1 in 10.71

The purchaser of this lottery ticket has a 1 in 40,000 chance of winning $1,000, and a 1 in 4,000 chance of winning $40. These probabilities are designed into the game, and reflect the number of winning tickets. Put differently, for every 40,000 tickets sold, one will contain a prize of $1,000.

Not every ticket is a winner, and many of the tickets that are winners will award small prizes. How much should the purchaser of a "9s in a Line" ticket expect to get when they play this game? Figure 7.11 shows the calculation of the expected value of each ticket.

Figure 7.11 The Expected Value of a Scratch-off Lottery Ticket

	A	B	C	D
1	Prize Amount	Probability	p(prize)	p(prize)*amount
2	$ 1,000	1 in 40,000	0.000025	$ 0.0250
3	$ 200	1 in 24,000	0.000042	$ 0.0083
4	$ 100	1 in 3000	0.000333	$ 0.0333
5	$ 50	1 in 4800	0.000208	$ 0.0104
6	$ 40	1 in 4000	0.000250	$ 0.0100
7	$ 30	1 in 2400	0.000417	$ 0.0125
8	$ 20	1 in 300	0.003333	$ 0.0667
9	$ 15	1 in 300	0.003333	$ 0.0500
10	$ 10	1 in 150	0.006667	$ 0.0667
11	$ 5	1 in 150	0.006667	$ 0.0333
12	$ 3	1 in 37.5	0.026667	$ 0.0800
13	$ 2	1 in 15	0.066667	$ 0.1333
14	$ 1	1 in 10.71	0.093371	$ 0.0934
15				
16		Expected Win	20.80%	
17				
18		**Expected Payout Per Ticket**		**$ 0.62**
19		**Cost of a Ticket**		$ (1.00)
20		**Net Value of Ticket**		$ (0.38)

[34]http://www.flalottery.com/scratch-offsGameDetails.do?gameNumber=1241

Figure 7.11, lists all of the events (i.e., the prize amounts) in column A, along with probability of each event in column C. Column D calculates the probability-weighted payout from this event. For example, the probability of winning $1,000 is 1 in 40,000, so the probability-weighted payout of this event is:

$$\frac{1}{40{,}000} \times \$1{,}000 = \$0.025$$

The expected value is the sum of the probability-weighted payout for all of the prizes. In this example, the expected payout per ticket is $0.62. Since it costs a dollar to play, we can summarize the net value of a ticket as the expected value of the prize money minus the cost to buy the ticket:

$$net\ value\ of\ ticket = \$0.62 - \$1.00 = -\$0.38$$

In other words, every time you play this lottery you should expect to lose 38 cents. The lottery is a game rooted in hope and greed on the part of ticket purchasers, and based on probability and expected value on the part of the lottery commission.

7.6 Application: Insurance

Every household faces different financial risks of varying magnitude or severity. An example of a risk of small magnitude would be the loss of your iPhone, whereas a risk of large magnitude would be a fire that burns down your house. Most households cannot handle the financial losses associated with a risk of large severity.

Insurance is a risk management contract between an *insured* (i.e., you) and an *insurer* (i.e., an insurance company). In an *insurance policy* (contract), the insured pays a *premium* (monetary payment) in exchange for the insurer agreeing to reimburse specific losses. Typically, an insurance contract has a limited period of coverage, such as one year at a time. Insurance is an application of probability and expected value with which most households are familiar, but few understand how insurance really works.

So Many Kinds of Insurance!

Many different kinds of insurance are available and consumers are generally familiar with how insurance works. Common types of insurance include:

- Property and casualty insurance covers losses related to personal property (i.e., houses, cars, jewelry, etc.).
- Health insurance covers the cost of medical treatments (i.e., hospitals and doctors, and in some cases prescription medication) due to an illness or accident.
- Life insurance provides benefits to the surviving dependents of the deceased.
- Disability insurance provides on-going income in the event that the insured person cannot work.

Insurance Premiums

The insurance premium depends on the probability and the severity of the loss. In an efficient insurance market, the premium would be equal to the probability-weighted expected value of the loss.

Consider an auto insurance policy that protects against the theft of your car worth $12,000 for a period of one year. There are many possible risks with having a car, including theft, collision, or liability for medical payments (i.e., if someone is injured in an accident).

Suppose that there was a 1 in 125 chance (probability) that your car would be stolen in the next year. The expected value of the loss would be:

$$expected\ value = p(loss) \times amount\ of\ loss = \frac{1}{125} \times \$12{,}000 = \$96$$

Since typical auto insurance covers many different risks, the total insurance premium is the sum of the expected values for all of these different losses. The table in Figure 7.12 shows the extended calculation for several types of risks associated with having a car, as well as hypothetical loss amounts and hypothetical probabilities associated with each risk.

Figure 7.12 Car Insurance Example

Loss Description	Loss Amount	Probability	p(loss)	p(loss)*amount
Theft of Car	$ 12,000	1 in 125	0.0080	$ 96.00
Major Collision	$ 9,600	1 in 75	0.0133	$ 128.00
Minor Collision	$ 2,400	1 in 20	0.0500	$ 120.00
Major Medical Payments	$ 100,000	1 in 750	0.0013	$ 133.33
Minor Medical Payments	$ 12,000	1 in 75	0.0133	$ 160.00
		Expected Loss		$ 477.33

In this example, the cost to insure a $12,000 car for one year would be $477.33, which is the expected value of the losses. Thus, in an efficient insurance market (i.e., not including a profit to the insurance company) the premium (paid in advance) to protect the car would be $477.33.

Insurance Deductibles

An insurance *deductible* is a form of risk-retention that requires the insured person to pay a portion of any losses. For example, if your auto insurance policy has a $500 deductible, you must pay for the first $500 of damage before the insurance company will pay the rest.

One reason for having a deductible is to reduce the small claims that would create an administrative hassle for the insurance company. Another reason is to discourage risky behavior, by adding a financial consequence to the insured person. The table in Figure 7.13 illustrates the potential losses and associated premium for the same $12,000 car from the previous example, with a $500 deductible.

Figure 7.13 Car Insurance Example (with Deductible)

Loss Description	Loss Amount	Probability	p(loss)	p(loss)*amount
Theft of Car	$ 11,500	1 in 125	0.0080	$ 92.00
Major Collision	$ 9,100	1 in 75	0.0133	$ 121.33
Minor Collision	$ 1,900	1 in 20	0.0500	$ 95.00
Major Medical Payments	$ 99,500	1 in 750	0.0013	$ 132.67
Minor Medical Payments	$ 11,500	1 in 75	0.0133	$ 153.33
		Deductible		$ 500.00
		Premium		$ 441.00

As we see in Figure 7.13, the deductible amount also reduces the expected loss to the insurance company. As a result, higher deductible insurance policies carry lower premiums.

The Business of Insurance

Insurance companies are in the business of playing the odds. Insurance companies employ teams of actuaries who apply detailed statistical models that try to gauge the risk of loss on highly specific data. For example, the actuaries who work for a car insurer will find the risk of you having a car accident, given that you are 22 years old, have six years of driving experience, drive a 2002 Honda Civic LX 4-door 5-speed sedan, and drive 7,500 miles per year. Based on these factors (and others), the actuaries estimate the likelihood and severity of you having a loss, and they set the premium at a level that will cover your expected loss (plus administrative costs).

Diversification is a risk-management strategy that involves having exposure to many small risks that are unlikely to occur at the same time (instead of one big risk).[35] Insurance companies practice diversification on a grand scale. For example, a large property and casualty insurer will write hundreds of thousands of policies on different cars. The risk of any one car being in an accident is independent from other cars being in an accident. On average, only a small fraction of the policies will suffer a large loss (i.e., a major medical liability) in a given year. For the other policies, the losses will be minimal or none. While the insurer will lose big on the policies that pay out for large losses, on most policies it will collect the premium and not pay out any losses at all.

Most insurance companies operate their underwriting at cost (i.e., zero-sum), or even at a slight loss (i.e., the premium dollars collected are more or less sufficient to pay out expected losses). An insurance company makes a profit by investing the premium dollars (called its *float*) in bonds, real estate investments, or even in the stock market. The return on investment is a profit to the insurance company. As losses occur, the insurance company can sell its investments to pay claims.

Insurance companies are regulated by the states, and must hold adequate reserves to be able to pay anticipated losses. Ratings agencies such as Fitch, Moody's, and Standard and Poors rate the financial strength of insurers based on their ability to pay claims.

Inefficiencies in the Insurance Market

Insurance markets tend to be inefficient because of moral hazard and adverse selection.

Moral hazard is a situation in which an insured person engages in risky behavior because they have insurance to cover their losses. For example, people who purchase loss damage waiver (LDW) insurance for a rental car might drive more recklessly, knowing that their losses will be paid in full by the LDW coverage. Preventing moral hazard is an additional reason for insurance deductibles, because the insured will be partially responsible for losses.

Adverse selection occurs when the individuals buying insurance know that they are likely to incur the losses for which they are buying insurance. For example, in the health insurance market, people with chronic health conditions (e.g., diabetes, hypertension, or cancer) face the highest anticipated health care costs. For this reason, the people in a health insurance pool might be sicker, on average, than the general population. Health insurers know this, and try to account for adverse selection by screening applicants for pre-existing conditions, and charging higher premiums for those cases.[36]

---

[35] Diversification will be further discussed in Module 9 (Statistics for Stock Market Investors).

[36] The Affordable Care Act of 2009 ("Obamacare") made it illegal for health insurers to practice discrimination based on pre-existing conditions. In order to prevent only the sickest people from buying health insurance, the Affordable Care Act requires everyone to buy insurance (i.e., the "individual mandate"). By requiring even healthy people to purchase health insurance, the insurers can diversify the cost of sicker individuals over a larger pool of healthy individuals.

7.7 Summary

Random events are those for which the outcome is not known in advance, and depends on chance. The sample space describes the possible outcomes from a random experiment or trial. Often, we will illustrate a sample space using a Venn Diagram or using set notation. Probability describes the likelihood of an event occurring. In discrete probability, we study the possible outcomes within a sample space, and the outcomes that define an event occurring to find the probability of an event occurring.

The expected value is a probability-weighted financial outcome. One application of expected values is to describe the rewards or payouts from games of chance (e.g., casino or lottery games). A zero-sum game is one in which the expected value is 0. Casino games and lotteries are not zero-sum games, because the casino or lottery agency retains a portion of the amount wagered.

Insurance is a risk management product that provides protection against big financial losses that might occur. Actuaries are applied mathematicians who calculate the probabilities of financial losses occurring. Insurance companies use the expected value of losses to calculate premiums, or the cost of insurance. The insured usually pays a deductible on any insurance claim, which is a form of risk-retention.

7.8 Conceptual Review Questions

1. Define these terms, and give an example of each: random experiment, trial, outcome, sample space, event.
2. What is an independent experiment? Explain and give an example.
3. Explain the concept of discrete probability. Give an example to illustrate.
4. Explain, giving an example, the multiplication rule.
5. Explain, giving an example, the concept of joint probability.
6. Explain, giving and example, the concept of expected value.
7. Are lottery tickets ever worth their cost? Make an argument using concepts of probability and expected value.
8. Explain, giving an example, the concept of insurance.
9. Explain these terms: insurance premiums and deductibles. How are insurance premiums determined?
10. Identify and explain the main inefficiencies in the insurance market.

7.9 Applied Questions

1. Consider the following game in which you roll a 6-sided die and then flip a coin.
    a. Write out the sample space for this game's outcomes.
    b. To win the opening round of this game, you must roll an even number on the die and heads on the coin toss. Write out the outcomes that would satisfy the event of winning this round, and then calculate the probability of winning a round.
    c. To win the second round of this game, you must roll a value that is at most 3, and also obtain tails from the coin toss. Write out the outcomes that would satisfy the event of winning this round, and then calculate the probability of winning a round.
2. Consider a game that involves tossing a coin 5 times.
    a. Write out the sample space for this game's outcomes.
    b. You can win this game by obtaining 4 or more of the same side (i.e., 4 heads or 4 tails). Find all outcomes that are part of the event of winning.
    c. Calculate the probability of winning this game.
3. The game in question 2 costs $2 to play, and has a prize of $5 if you win. What is the expected value of playing this game?
4. Suppose you have a car that is worth $10,000 right now. Over the next year, there is a 1% chance of having a car crash that would result in damage of $7,000, and a 15% chance of having a car crash with damage of $1,000. What is the expected value of these losses?

8 For Whom the Bell Tolls

Survival Probability and Actuarial Present Value

Learning Objectives:

- Think quantitatively about the rates of death and survival as a function of a person's age.
- Consider the *rate of change* in the survival rate as a function of a person's age.
- Use analytical techniques to model the probability of survival to a certain age.
- Use survival probabilities and discounted cash flows to develop the actuarial present value.

8.1 Death Rates and Survival Rates

In the previous module, we calculated the discrete probability of achieving a random outcome, such as the probability of achieving a certain value by rolling two dice. Recall the formula to find the probability of an event E in a sample space S:

$$p(E) = \frac{n(E)}{n(S)}$$

Estimating Death Rates

Each year, a number of individuals within a population will die. While some factors have been associated with early death (e.g., smoking, obesity, etc.), which individuals die and which ones survive in any given year is unknown and random.

Actuaries have been tabulating statistics about death rates for centuries. The *death rate* or *mortality rate* describes the fraction of a population that dies within one year, i.e., a relative frequency probability.

$$death\ rate = \frac{n(deaths)}{population\ size}$$

For example, suppose a farm has 137 cottontail rabbits born within one year.[37] At the end of the year, 42 of the rabbits have died of natural causes. We could calculate the death rate among these rabbits as:

$$death\ rate = \frac{42}{137} \approx 0.307$$

The death rate describes what has occurred in the past, but the *probability of death* describes what we expect to occur in the future. When the historical death rate is known and the population is sufficiently large, we can use the population death rate as an estimate of the probability of death in the future. Due to randomness, we don't know which individuals will die, but we can make an estimate of approximately how many will.

Now consider the population of 45-year old males in the United States. In 2009, the Social Security Administration estimated[38] that the death rate among 45-year-old men was about 0.3396%. While we don't know which individual 45-year-old men will die, we can estimate that about 1 in 300 45-year-old men will die before age 46.

[37] This example is totally made up, but according to Rexford D. Lord, Jr., "it is apparent that exceedingly few rabbits live beyond 3 years old, even though their potential lifespan is probably about 10 years."Source: "Mortality Rates of Cottontail Rabbits," Journal of Wildlife Management, January 1961.

Actuaries usually present this kind of information in a form called a life table. *A life table* shows the probability of death for people grouped by sex and age. The table in Figure 8.1 is an abbreviated life table (reproduced from the Social Security Administration[39]), showing estimated death rates for men and women of various ages in the United States.

Figure 8.1 Actuarial Life Table for the United States (abbreviated)

	Male			Female		
Exact age	Death probability ᵃ	Number of lives ᵇ	Life expectancy	Death probability ᵃ	Number of lives ᵇ	Life expectancy
0	0.006990	100,000	75.90	0.005728	100,000	80.81
1	0.000447	99,301	75.43	0.000373	99,427	80.28
2	0.000301	99,257	74.46	0.000241	99,390	79.31
3	0.000233	99,227	73.48	0.000186	99,366	78.32
4	0.000177	99,204	72.50	0.000150	99,348	77.34
25	0.001370	98,043	52.18	0.000527	98,836	56.67
26	0.001364	97,909	51.25	0.000551	98,784	55.70
27	0.001362	97,775	50.32	0.000575	98,729	54.73
28	0.001373	97,642	49.39	0.000602	98,673	53.76
29	0.001393	97,508	48.45	0.000630	98,613	52.79
80	0.061620	49,421	8.10	0.043899	62,957	9.65
81	0.068153	46,376	7.60	0.048807	60,194	9.07
82	0.075349	43,215	7.12	0.054374	57,256	8.51
83	0.083230	39,959	6.66	0.060661	54,142	7.97
84	0.091933	36,633	6.22	0.067751	50,858	7.45
95	0.264277	5,375	2.81	0.213849	11,795	3.39
96	0.284168	3,954	2.64	0.231865	9,273	3.18
97	0.303164	2,831	2.49	0.249525	7,123	2.98
98	0.320876	1,972	2.36	0.266514	5,345	2.81
99	0.336919	1,340	2.24	0.282504	3,921	2.65
100	0.353765	888	2.12	0.299455	2,813	2.49

In the table in Figure 8.1, the second and fifth columns show the probability of death for men and women respectively. After age 4, we observe that the probability of death is small for young people, and increases with age.

A *cohort* is a group of people with a common statistical characteristic, such as a group of people all born in the same year. To help interpret the death probabilities, the third and sixth columns (labeled "Number of lives") provide an example cohort that begins with 100,000 people alive at birth. For example, beginning with 100,000 male live births, about 99,301 to survive to age 1 and 98,043 to survive to age 25.

The fourth and seventh columns show the cohort's life expectancy. The *life expectancy* is an estimate of the remaining years of life for a cohort of a given age. Of course, this is not a prediction for any individual within that

[38] The Social Security Administration is a government agency in the United States that provides retirement and disability benefits to eligible workers (or retirees). The Administration uses records of deaths from the National Center for Health Statistics and estimates of the current population by sex and age from the Census Bureau.

[39] Source: http://www.ssa.gov/oact/STATS/table4c6.html

cohort, but rather the median (or middle) age of death for that cohort. The *median age of death* for an age cohort means that half of the individuals in that cohort are expected to die earlier and half are expected to die later.

Notice as well that the median age of death increases with the age of the cohort. For example, life expectancy at birth (for a male) is 75.9 years, but for a 25-year old life expectancy is 52.18 years. This means a 25-year old would expect to live to about age 77. Further, life expectancy for an 80-year old is 8.1 years, i.e., living to age 88.1. The reason that median age of death increases with age is subtle. The men who have already lived to age 80 are only a subset of their cohort at birth, but they are in the lucky half: they didn't die any earlier. The less-fortunate half of their cohort died before age 80. When we look at the median age of death among 25-year-olds, we don't know which half will die before age 80, and thus the life expectancy for the cohort as a whole is lower for 25-year-olds than for 80-year-olds.

As we observe in the life table in Figure 8.1, there is a strong relationship between age and death rates. All other things being equal, the older an individual is the greater the chance of death in a given year. Consider for example, that probability of death among 25-year-old men is 0.137%, or about 137 deaths per 100,000 25-year-olds. The probability of death among 80-year-old men is about 6.162%, or about 6,162 deaths per 100,000 octogenarians.

Survival Rates

Recall that the mortality rate is the probability that an individual will die within a year. We saw above that we can estimate the mortality rate based on historical data. The *probability of survival* or *survival probability* is the chance that an individual will *not* die during that year. Thesurvival probability is the *complement* of the probability of death, i.e.:

$$p(survival) = 1 - p(death)$$

Since everyone will eventually die, we need to consider the probability of survival over a finite period of time, e.g., one year. For example, consider a 45-year-old man in the United States. The probability of death before age 46 is about 0.003396 or about 1 in 300. Alternatively, we can describe that the probability of survival for one year (to age 46) is:

$$p(survival) = 1 - 0.003396 = 0.996604 = 99.66\%$$

Just as the probability of death increases with age, the probability of survival decreases with age.
Figure 8.2 illustrates historical and future survival rates for the United States graphically.

Figure 8.2 Survival Probabilities for Selected Calendar Years (1900, 1950, 2000, 2050, 2100)[40]

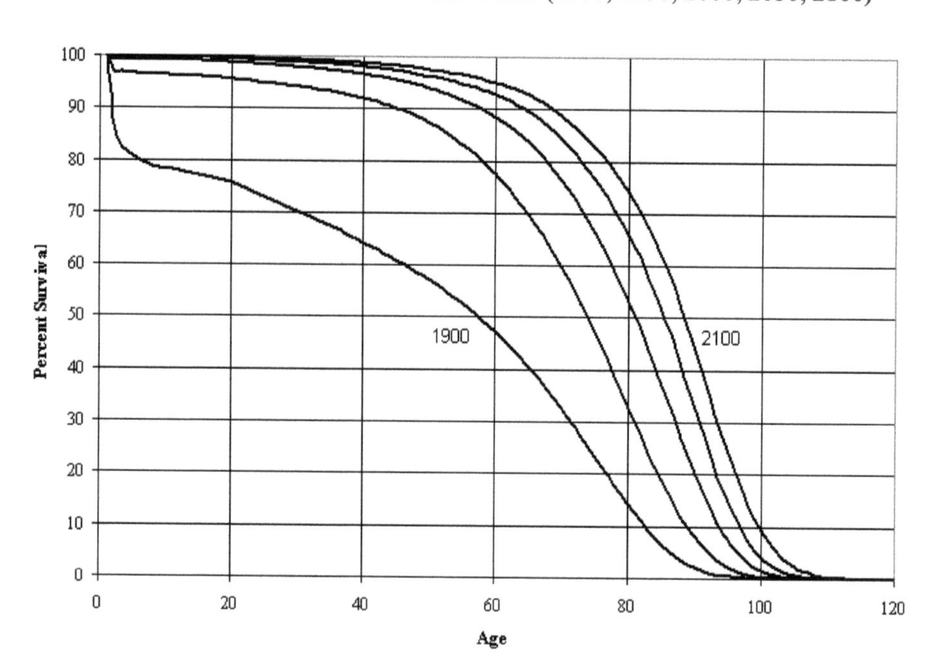

In the graph in
Figure 8.2, the left-most curve is for the year 1900, and the right-most curve is projected for the year 2100. Two important observations can be made from the data in this graph. First, survival probabilities have been increasing since 1900, and are expected to continue to increase through the year 2100. Historically, the most pronounced change in survival probabilities is among young children. Due to medical advances such as vaccinations, early childhood death rates decreased dramatically between 1900 and 1950. Second, the median age of death, i.e., the age at which only 50% of a cohort is still alive has been increasing steadily as well. In 1900, the median age of death was about 56, while in 2000 the median age of death is almost 82.

Great Minds in Applied Mathematics

Benjamin Gompertz (1779-1865) was a self-educated mathematician and actuary who lived in Great Britain. In his work as an actuary, he was intimately familiar with actuarial life tables. As chief actuary of the Guardian Insurance Office, he worked out a new series of tables of human mortality for the Royal Society. In 1825, he discovered a natural law to explain the mortality rate within a population, which is now known as the Gompertz Law of Mortality.

8.2 The Gompertz Law of Mortality

In the previous section, we observed mortality rates and survival rates, and we noted that the mortality rate increases with age. Figure 8.3 shows the mortality rates for Men aged 45-50. Benjamin Gompertz' curiosity led him to examine the *rate of change* in the mortality rate.[41]

[40] Source: http://www.ssa.gov/oact/NOTES/as120/LifeTables_Body.html

[41] You read that sentence correctly. We are curious about the rate of change in the mortality rate. In other words, we want to know the rate at which the mortality rate changes from one year to the next.

Figure 8.3 Mortality Rates for Men aged 45-50

Exact age	p(death)
45	0.003396
46	0.003703
47	0.004051
48	0.004444
49	0.004878
50	0.005347

Finding the Rate of Change in the Mortality Rate

Gompertz discovered that the mortality rate is growing exponentially, i.e., each year's mortality rate depends on the previous year's rate, compounded at an almost constant rate of growth. How much is the mortality rate changing from one year to the next? To help understand the relative magnitude of an exponentially growing value *x*, we use the natural logarithm function, i.e., *ln(x)*. For example, the mortality rate for a 45-year-old man is 0.003396 (or about 0.34%). We find the natural log of the mortality rate as:

$$\ln(p(death_{45})) = \ln(0.003396) = -5.685$$

The mortality rate for a 46-year-old man is about 0.003703 (or about 0.37%). We find the natural log of the mortality rate for a 46-year-old as:

$$\ln(p(death_{46})) = \ln(0.003703) = -5.599$$

How much did the natural logarithm of the mortality rate change from age 45 to age 46?

$$change\ in\ \ln(p(death))) = \ln(p(death_{46})) - \ln(p(death_{45})) = 0.087$$

Using Spreadsheets: Natural Logarithms

The natural logarithm function provides a way to make sense of exponentially growing values. Within a spreadsheet, the built-in LN function will calculate the natural logarithm of an expression. Look at Figure 8.4 for an example.

Figure 8.4 Using the LN Function within a Spreadsheet

	A	B	D	E
15	Exact age	p(death)		ln(p(death))
16	45	0.003396		=LN(B16)
17	46	0.003703		-5.599

To find the natural logarithm of the value in cell B16, we would use the function *=LN(B16).*
Note that natural logarithm function gives a positive result for x values greater than 1, and a negative result for x values less than 1.

Figure 8.5 shows the natural log of the mortality rate (column E) for men and the year-over-year rate of change in the natural log of the mortality rate (column F).

Figure 8.5 The Rate of Change of the Mortality Rate, ages 45-50, 65-70, and 85-90

	A	B	D	E	F
15	Exact age	p(death)		ln(p(death))	change in ln(p(death))
16	45	0.003396		-5.685	
17	46	0.003703		-5.599	0.087
18	47	0.004051		-5.509	0.090
19	48	0.004444		-5.416	0.093
20	49	0.004878		-5.323	0.093
21	50	0.005347		-5.231	0.092
22					
23	65	0.016182		-4.124	
24	66	0.017612		-4.039	0.085
25	67	0.019138		-3.956	0.083
26	68	0.020752		-3.875	0.087
27	69	0.022497		-3.794	0.087
28	70	0.024488		-3.710	0.087
29					
30	85	0.101625		-2.286	
31	86	0.112448		-2.185	0.087
32	87	0.125020		-2.079	0.087
33	88	0.137837		-1.982	0.087
34	89	0.152458		-1.881	0.087
35	90	0.168352		-1.782	0.087

Look carefully at the change in the log of the mortality rates for each of the years (column F in Figure 8.5). We can summarize Gompertz' finding by observing that *the mortality rate increases exponentially each year from adulthood until old age.* The rate of increase in the log of the mortality rate is roughly around 9% each year.

This is Gompertz big discovery: there is a function to describe the mortality rate for any given age. The slope of this function is the *rate of increase* in the mortality rate, and the intercept is the maximum age of life (i.e., where the mortality rate is 100%).

The probability of death is asymptotic to 1 (i.e., at some maximum age there would be a 100% probability of death within the next year). In turn, this means that the natural logarithm of the probability of death is asymptotic to 0 (i.e., $ln(100\%) = 0$).

8.3 Survival Probability

In financial planning, we need to think about how long our assets will last – and how long we will last. In particular, we need to think about an individual's survival probability not just for the next one year, but also for any number of years into the future. In this section, we discuss the probability of surviving for any number of years into the future. In the next section, we will use these survival probabilities in financial calculations that depend on being alive to collect future cash flows.

Calculating the Probability of Survival

What is the probability that a 45-year old will live to age 85, i.e., for *at least* 40 more years? We are not suggesting that the 45-year old will die at exactly age 85, but rather that he will survive at least to age 85 (and might live even longer).

Let's denote the probability of survival from age x for t additional years as:

$$p(survival_{x,t})$$

Note that the probability of a person age 45 surviving at least to age 85 is the *complement*(i.e., mutually exclusive opposite) of the probability of a 45-year old dying before age 85:

$$p(survival_{x,t}) = 1 - p(death_{x,t})$$

Benjamin Gompertz' law of mortality gives rise to an equation that finds the probability of survival for an individual to a certain age, based on his or her current age and two population parameters. The *Gompertz Equation* finds the *natural logarithm* of the probability of survival for t years beginning at age x:

$$\ln(p(survival_{x,t})) = \left(1 - e^{\frac{t}{b}}\right) e^{\frac{x-m}{b}}$$

The probability of survival depends on the following parameters:

x = age now in years
t = years of survival for which to find the probability
m = modal age of death within the population, in years
b = the dispersion coefficient of human life, in years

To find the probability of survival from age x for t additional years, we must undo the natural logarithm by using the exponential function:

$$p(survival_{x,t}) = e^{\ln(p(survival_{x,t}))}$$

Example 1: Probability of a 45-year-old living to 85

For example, let's find the probability of a 45 year old surviving to age 85. We need to make some assumptions about the values for m and b, which are specific to a given population (e.g., men in the United States). The values of m = 83 and b = 11 are an approximate fit for data contained in Figure 8.1.

First, we use the Gompertz equation to solve for the natural log of the probability of survival for 40 years:

$$\ln(p(survival_{45,40})) = \left(1 - e^{\frac{40}{11}}\right) e^{\frac{45-83}{11}}$$

$$\ln(p(survival_{45,40})) = (1 - e^{3.63})e^{-3.45}$$

$$\ln(p(survival_{45,40})) = (e^{-3.45} - (e^{3.63}e^{-3.45})) = e^{-3.45} - e^{0.18}$$

$$\ln(p(survival_{45,40})) = -1.16779$$

Finally, we solve for the probability of survival by using the exponential function to "undo" the natural logarithm:

$$p(survival_{45,40}) = e^{-1.16779} = 0.311$$

In words, we estimate that a 45-year-old has about a 31% chance to survive to age 85.

Example 2: Probability of a 65-year-old living to 85

As another example, let's find the probability of a 65-year-old surviving to age 85, again using the population parameters $m = 83$ and $b = 11$. First, we use the Gompertz equation to solve for the natural log of the probability of survival:

$$\ln(p(survival_{65,20})) = \left(1 - e^{\frac{20}{11}}\right) e^{\frac{65-83}{11}}$$

$$\ln(p(survival_{65,20})) = (1 - e^{1.82})e^{-1.64}$$

$$\ln(p(survival_{65,20})) = (e^{-1.64} - (e^{1.82}e^{-1.64})) = e^{-1.64} - e^{0.18}$$

$$\ln(p(survival_{65,20})) = -1.00471$$

Finally, we solve for the probability of survival, by using the exponential function to "undo" the natural logarithm:

$$p(survival_{65,20}) = e^{-1.00323} = 0.3662$$

In words, we estimate that a 65-year-old has about a 37% chance to survive to age 85.

How Age Now Impacts the Survival Probability

In general, we think that a younger person has a longer expected lifespan. However, as these examples illustrate, the 45-year-old man has a 31% chance of living to age 85, but the 65-year-old has a 37% chance of living to 85. To understand why this is, remember that not all 45-year-olds will live to age 65 (only about 85% will). Those who do make it to age 65 are the *survivors* among their cohort.

We could think of this problem in two steps. First, we find the probability of a 45-year-old surviving to age 65:

$$p(survival_{45,20}) = 0.85$$

Next, we find the probability of a 65-year-old surviving to age 85:

$$p(survival_{65,20}) = 0.3662$$

Finally, we use the multiplication rule to find the probability that a 45-year-old who survives 20 years to age 65 *also* survives another 20 years to age 85:

$$p(survival_{45,40}) = p(survival_{45,20}) \times p(survival_{65,20})$$
$$p(survival_{45,40}) = 0.85 \times 0.3662 = 0.311 = 31.1\%$$

The probability of survival from age 45 for 40 additional years is the same as the probability of survival from age 45 for 20 years multiplied by the probability of survival from age 65 for 20 years:

A Survival Probability Calculator

Using the Gompertz Equation, we can create a survival probability calculator as a spreadsheet model. The calculator requires inputs for the parameters for the values *m*, *b*,and *x*, and then calculates survival probabilities for all years based on these inputs. The graph in Figure 8.6 plots the survival probability to different ages based on these inputs.

Figure 8.6 Probability of Survival

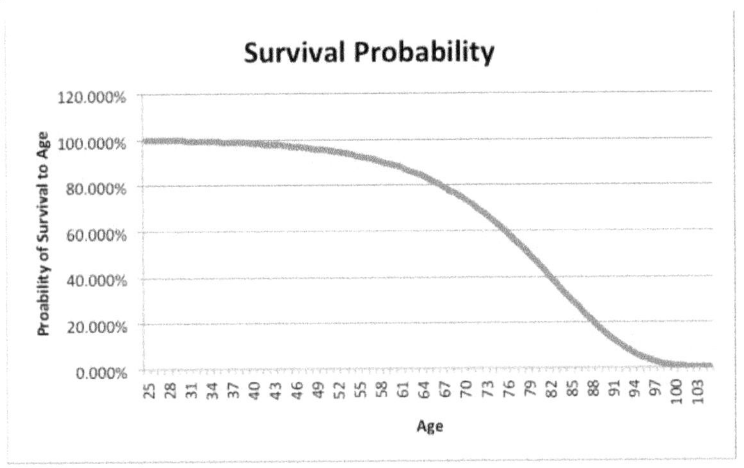

The shape of the curve plotting the survival probability should be familiar: we saw it in Figure 8.2. The details of the calculator are show in Figure 8.7 and explained below.

Figure 8.7 Survival Probability Calculator

	A	B	C	D	E
1	Parameters:				
2	m =	83			
3	b =	11			
4	x =	25	(age now)		
5					
6	Calculated Values				
7	e^((x-m)/b)	0.0051			
8					
9	Live to Age	Years (t)	(1-e^(t/b))	ln(p)	p
10	25	0	0.0000	0.0000%	100.000%
11	26	1	-0.0952	-0.0488%	99.951%
12	27	2	-0.1994	-0.1023%	99.898%
13	28	3	-0.3135	-0.1608%	99.839%
65	80	55	-147.4132	-75.6171%	46.946%
66	81	56	-161.5376	-82.8623%	43.665%
67	82	57	-177.0062	-90.7971%	40.334%
68	83	58	-193.9469	-99.4870%	36.977%
69	84	59	-212.4999	-109.0040%	33.620%
70	85	60	-232.8186	-119.4267%	30.293%
71	86	61	-255.0709	-130.8412%	27.025%
72	87	62	-279.4411	-143.3421%	23.849%
73	88	63	-306.1305	-157.0328%	20.798%
74	89	64	-335.3599	-172.0263%	17.902%
75	90	65	-367.3711	-188.4468%	15.191%

The calculator begins with inputs for the 3 parameters at the top, in cells B2, B3, and B4.

In the Gompertz equation, the value of $e^{\frac{x-m}{b}}$ depends only on the age now (x) and the population parameters (m, b). It does not depend on the variable t (the years of survival). Therefore, we can calculate the value of $e^{\frac{x-m}{b}}$ oncein cell B7 and use that value for all years.

In each row of the table that follows, we make a calculation for a specific age. We are calculating the number of years of survival to reach that age, and the probability of survival from the starting age x for that number of years.

For example, in row 10 we are calculating the probability of survival for 0 years. The probability of a 25-year-old surviving to age 25 is 100%. In row 11, we are calculating the probability of a 25-year-old surviving one year, to age 26. The natural log of the survival probability is:

$$\ln(p(survival_{25,1})) = \left(1 - e^{\frac{1}{11}}\right) e^{\frac{25-83}{11}}$$

The calculation for $\left(1 - e^{\frac{1}{11}}\right)$ is done in cell C11, to yield a value of -0.0952. We already know that the value of $e^{\frac{25-83}{11}} = 0.0051$ (as calculated in cell B7). Thus the natural log of the probability of survival (cell D11) is the product of these two cells: $-0.0952 \times 0.0051 = -0.000488 = -0.0488\%$. Finally, to isolate the survival probability, we must "undo" the natural log by taking the exponential function of this value in cell E11. The probability of a 25-year-old living to age 26 is 99.951%.

Using Spreadsheets: The Exponential Function

The exponential function is the inverse function of the natural logarithm. The relation between the two is: $x = \ln(e^x)$ or $x = e^{\ln(x)}$, for $x > 0$.

Within a spreadsheet, the built-in EXP function will calculate the exponential function of an expression. Look at Figure 8.8 for an example.

Figure 8.8 Using the EXP Function within a Spreadsheet

	A	B	C	D	E
9	Live to Age	Years (t)	(1-e^(t/b))	ln(p)	p
10	25	0	0.0000	0.0000%	100.000%
11	26	1	-0.0952	-0.0488%	=EXP(D11)
12	27	2	-0.1994	-0.1023%	EXP(number)

To find the exponential function of the value in cell D11, we would use the function *=EXP(D11)*.

8.4 Actuarial Present Value

Suppose you are 21 years old, and you receive a prize that will be worth $1,000,000 at age 80 – but will only be paid to you at age 80 and only if you are alive to receive it in person. How much is it worth now? We have seen some elements of this problem before, and you are probably already thinking about the present value of $1,000,000 to be received in 59 years (age 80 minus age 21). Recall the formula to find a present value of a lump sum:

$$Present\ Value = \frac{FV}{(1+r)^n}$$

But this problem is a bit different. Not only is the inheritance to be received in 60 years, but only on the condition that you are alive to collect it. This part requires knowing the probability of a 21 year old surviving to age 80, which we saw previously using the Gompertz survival probability:

$$p(survival_{x,t}) = e^{\left(1-e^{\frac{t}{b}}\right)e^{\frac{x-m}{b}}}$$

The *actuarial present value* is the probability-weighted expected value of the present value of an amount to be received in the future.

$$Actuarial\ Present\ Value = p(receiving\ FV) \times \frac{FV}{(1+r)^n}$$

The actuarial present value (APV) is an expected value, because the value depends on the probability of receiving the future amount, and it is also a present value, because it requires discounting an amount to be received in the future to account for the time value of money.

Example:

Let's return to the example from above, of $1,000,000 to be received at age 80 by an individual aged 21 now. Some additional facts are necessary to answer the question. We will assume the relevant interest rate is 3% per year, and the population parameters for the Gompertz Equation are $m = 83$ and $b = 11$. We will solve this problem in two parts.

First, we find the probability of survival for a 21 year old living to age 80. Recall that the Gompertz Equation gives us the natural logarithm of the survival probability, based on the current age (x), the years of survival (t), and the population parameters (m, the modal age of life, and b, the dispersal coefficient):

$$\ln\left(p(survival_{21,59})\right) = \left(1 - e^{\frac{t}{b}}\right)e^{\frac{x-m}{b}} = \left(1 - e^{\frac{59}{11}}\right)e^{\frac{21-83}{11}}$$

$$\ln\left(p(survival_{21,59})\right) = \left(1 - e^{\frac{59}{11}}\right)e^{\frac{21-83}{11}} = -212.499 \times 0.0009362 = -0.757735$$

By using the exponential function, we find the probability of a 21-year old surviving to age 80 is:

$$p(survival_{21,59}) = e^{-0.757735} = 0.4687 \approx 46.87\%$$

Next, we need to find the present value of $1,000,000 to be received in 59 years, discounted at a rate of 3% per year:

$$Present\ Value = \frac{\$1{,}000{,}000}{(1.03)^{59}} = \frac{\$1{,}000{,}000}{5.72} = \$174{,}825.08$$

Finally, we can find the actuarial present value by multiplying these two parts together:

$$Actuarial\ Present\ Value = 46.87\% \times \$174{,}825.08 = \$81{,}945.25$$

The actuarial present value of a $1,000,000 to be received at age 80 by an individual who is aged 21 now is $81,945.

An Income For Life

Suppose you are offered a magical wallet, that gives you $1,000 *each year*, as along as you remain alive to collect each year's amount. How much should you pay for this magical wallet? The ***fair value*** is whatever it would cost to reproduce the same cash flows in the same years, with the same probability of survival to receive each payment. We can find the fair value for this magical wallet by adding up the actuarial present values of each year's payment.

The actuarial present value of each payment is given by:

$$Actuarial\ Present\ Value = p(survival_{x,t}) \times \frac{PMT}{(1+r)^n}$$

where $p(survival_{x,t})$is the probability of survival given by the Gompertz Equation for survival probability from the starting age x for t additional years.

Let x be the age now and n be the maximum age of survival (i.e., the age at which the probability of survival approaches 0%). We can find total value of this magical wallet by finding the sum of the actuarial present values of each year's payment:

$$Total\ Value = \sum_{i=1}^{n}\left(p(survival_{x,i}) \times \frac{PMT_i}{(1+r)^i}\right)$$

$$Total\ Value = p(survival_{x,1}) \times \frac{PMT_1}{(1+r)^1} + p(survival_{x,2}) \times \frac{PMT_2}{(1+r)^2} + \cdots + p(survival_{x,n}) \times \frac{PMT_n}{(1+r)^n}$$

The solution to this kind of problem will require calculating the survival probability for each year until the maximum age of life, and the present value of each year's payment. The actuarial present value of each year's payment is the product of these two values, and the sum of all of these APVs is the total value of the magic wallet. The spreadsheet in Figure 8.9 calculates the total value of this magic wallet for a 21-year old.

Figure 8.9 Actuarial Present Value of a Lifetime of Payments

	A	B	C	D	E	F
1	**Gompertz Equation Inputs**					
2	m =	**87.25**				
3	b =	**9.5**				
4	x =	**21**	**(age now)**			
5						
6	**Annuity Calculation Inputs**				**Actuarial Present Value**	
7	Rate	3%				$27,315.60
8	Payment Amt	$1,000				
9						
10	**Age**	**Years**	**p(survival)**	**PV Factor**	**PV(PMT)**	**APV**
11	22	1	99.99%	0.9709	$970.87	$970.77
12	23	2	99.98%	0.9426	$942.60	$942.39
13	24	3	99.97%	0.9151	$915.14	$914.82
14	25	4	99.95%	0.8885	$888.49	$888.05
15	26	5	99.94%	0.8626	$862.61	$862.05
16	27	6	99.92%	0.8375	$837.48	$836.79
17	28	7	99.90%	0.8131	$813.09	$812.26
18	29	8	99.88%	0.7894	$789.41	$788.43
19	30	9	99.85%	0.7664	$766.42	$765.28

	A	B	C	D	E	F
60	71	50	83.54%	0.2281	$228.11	$190.56
61	72	51	81.88%	0.2215	$221.46	$181.34
62	73	52	80.08%	0.2150	$215.01	$172.17
63	74	53	78.12%	0.2088	$208.75	$163.07
64	75	54	76.00%	0.2027	$202.67	$154.02
65	76	55	73.71%	0.1968	$196.77	$145.03
66	77	56	71.25%	0.1910	$191.04	$136.11
67	78	57	68.61%	0.1855	$185.47	$127.25
68	79	58	65.79%	0.1801	$180.07	$118.47
69	80	59	62.80%	0.1748	$174.83	$109.79
	A	B	C	D	E	F
80	91	70	22.69%	0.1263	$126.30	$28.66
81	92	71	19.25%	0.1226	$122.62	$23.60
82	93	72	16.03%	0.1190	$119.05	$19.08
83	94	73	13.08%	0.1156	$115.58	$15.12
84	95	74	10.43%	0.1122	$112.21	$11.71
85	96	75	8.12%	0.1089	$108.95	$8.84
86	97	76	6.14%	0.1058	$105.77	$6.50
87	98	77	4.51%	0.1027	$102.69	$4.63
88	99	78	3.19%	0.0997	$99.70	$3.18
89	100	79	2.18%	0.0968	$96.80	$2.11
90	101	80	1.43%	0.0940	$93.98	$1.34
91	102	81	0.89%	0.0912	$91.24	$0.81
92	103	82	0.53%	0.0886	$88.58	$0.47
93	104	83	0.29%	0.0860	$86.00	$0.25
94	105	84	0.15%	0.0835	$83.50	$0.13
95	106	85	0.07%	0.0811	$81.07	$0.06
96	107	86	0.03%	0.0787	$78.70	$0.03
97	108	87	0.01%	0.0764	$76.41	$0.01
98	109	88	0.01%	0.0742	$74.19	$0.00

In the spreadsheet in Figure 8.9, there is a row for each age, which calculates the survival probability for a 21-year-old living to that age (column C). In addition, each row calculates the present value of $1,000 received that many years in the future (column E). The APV for each year's payment (column F) is the product of the survival probability and the present value. Finally, at the top of the sheet (cell F7), we have calculated the sum of each of the year's APVs, which works out to $27,315.60. This amount, $27,315.60, is the fair price for the magic wallet. If you could pay less than $27,315.60, you would be getting a good deal, and if you paid more than $27,315.60, you would be over paying.

Playing the Probabilities with Gompertz

It's important to note that an individual's actual longevity is highly idiosyncratic and an unknown random variable. Survival probabilities are most useful when thinking about a large cohort of individuals, i.e., a population. Insurance companies, governments, and pension funds are able to spread their risks across the survival probabilities of large numbers of individuals, and by doing so they can offer promises (e.g., old-age pensions) and products (e.g., immediate annuities and life insurance) which make use of the actuarial present value.

Application: Immediate Annuities

A ***lifetime annuity*** is a stream of payments that continues for the rest of the owner's life. Insurance companies sell annuities and guarantee to make the payments each month. Annuity buyers choose to purchase annuities because this product provides a guarantee: you cannot outlive the annuity payments, no matter how old you are.

For example, a 75-year old might buy an annuity contract that promises to pay $1,000 per month ($12,000 per year) for the rest of his life. The cost of the lifetime annuity is usually close to the fair value. For example, Figure 8.10 shows the calculation of the fair value of an immediate annuity for a 75-year old.

Figure 8.10 Calculating the Fair Price of an Immediate Annuity

	A	B	C	D	E	F
1	**Gompertz Equation Inputs**					
2	m =	**87.25**				
3	b =	**9.5**				
4	x =	**75**	**(age now)**			
5						
6	**Annuity Calculation Inputs**				**Actuarial Present Value**	
7	Rate	3%				$110,844.63
8	Payment Amt	$12,000				
9						
10	**Age**	**Years**	**p(survival)**	**PV Factor**	**PV(PMT)**	**APV**
11	76	1	96.99%	0.9709	$11,650.49	$11,299.70
12	77	2	93.75%	0.9426	$11,311.15	$10,604.21
13	78	3	90.28%	0.9151	$10,981.70	$9,914.09
14	79	4	86.57%	0.8885	$10,661.84	$9,230.13
15	80	5	82.63%	0.8626	$10,351.31	$8,553.46
16	81	6	78.46%	0.8375	$10,049.81	$7,885.53
17	82	7	74.08%	0.8131	$9,757.10	$7,228.11
18	83	8	69.50%	0.7894	$9,472.91	$6,583.35
19	84	9	64.74%	0.7664	$9,197.00	$5,953.74
20	85	10	59.83%	0.7441	$8,929.13	$5,342.10
21	86	11	54.81%	0.7224	$8,669.06	$4,751.52
22	87	12	49.73%	0.7014	$8,416.56	$4,185.35
23	88	13	44.63%	0.6810	$8,171.42	$3,647.02
24	89	14	39.58%	0.6611	$7,933.41	$3,140.02
25	90	15	34.63%	0.6419	$7,702.34	$2,667.70
	A	B	C	D	E	F
26	91	16	29.86%	0.6232	$7,478.00	$2,233.09
27	92	17	25.33%	0.6050	$7,260.20	$1,838.78
28	93	18	21.09%	0.5874	$7,048.74	$1,486.66
29	94	19	17.21%	0.5703	$6,843.43	$1,177.79
30	95	20	13.73%	0.5537	$6,644.11	$912.28
31	96	21	10.68%	0.5375	$6,450.59	$689.12
32	97	22	8.08%	0.5219	$6,262.71	$506.24
33	98	23	5.93%	0.5067	$6,080.30	$360.57
34	99	24	4.20%	0.4919	$5,903.20	$248.13
35	100	25	2.87%	0.4776	$5,731.27	$164.35
36	101	26	1.88%	0.4637	$5,564.34	$104.34
37	102	27	1.17%	0.4502	$5,402.27	$63.19
38	103	28	0.69%	0.4371	$5,244.92	$36.31
39	104	29	0.39%	0.4243	$5,092.16	$19.69
40	105	30	0.20%	0.4120	$4,943.84	$10.01
41	106	31	0.10%	0.4000	$4,799.85	$4.73
42	107	32	0.04%	0.3883	$4,660.04	$2.07
43	108	33	0.02%	0.3770	$4,524.31	$0.83
44	109	34	0.01%	0.3660	$4,392.54	$0.30
45	110	35	0.00%	0.3554	$4,264.60	$0.10
46	111	36	0.00%	0.3450	$4,140.39	$0.03
47	112	37	0.00%	0.3350	$4,019.80	$0.01
48	113	38	0.00%	0.3252	$3,902.71	$0.00

In Figure 8.10, we see that he 75-year-old has a 97% probability of surviving to age 76 to collect the first year's annuity payment, and a 34.6% probability of surviving to age 90 to collect that year's payments. In each case, the present value of the annuity payment is weighted by the survival probability to find the actuarial present value. Notice that there is almost a 0% chance that the 75-year-old will survive to age 110. However, were he still alive, the insurance company would continue to make payments.

8.5 Application: Life Insurance

As a young person, your human capital is by far your largest asset. A tragic accident or illness could result in losing some or all of your human capital. The premature death of the primary income earner in a household would be a devastating loss to those who depend on his or her income. While we often hear about the untimely death of a young person, disability is about ten times more likely than death. In this section, we will discuss how insurance can protect your dependents in the case of your death, and how to protect *you* in the case of a disability that prevents you from working.

Life Insurance

Life insurance provides a payoff in the event of death of the insured person during the term of coverage. In fact, the name life insurance is a bit of a misnomer, since the payoff (called a *death benefit*) only occurs in the case of death. Life insurance does not protect the insured person but rather the people who depend on the insured person's income. A more descriptive name would be "income protection for dependents." Life insurance products can be classified into two main types: permanent life insurance and term life insurance.

Permanent life insurance provides a death benefit irrespective of the date of death, so long as the policy has been paid in full. The death benefit is a known amount of money but the date at which it is received is unknown and depends on the date of death (i.e., the death benefit is a future value).

For example, consider a permanent life insurance policy with a $100,000death benefit for a 25-year old woman in excellent health. There are several choices of how to pay for this coverage. Paying annually would cost about $900 per year, every year for life. Another option is to pay $1,658 per year for 10 years, after which point the policy is fully "paid up" and no additional payments are required. Finally, a third option is a single-premium payment of $17,239, which provides lifetime coverage.[42]

Term life insurance is a pure insurance product: the insured pays a premium each year, and the policy pays a death benefit in the event of death during the year of coverage. Term life insurance policies are sold with terms of 10, 20, or even 30 years, meaning that the insured can continue to pay the same premium for each year of the term to maintain coverage.

For example, consider a term life insurance policy with a $100,000 death benefit for a 25-year old woman in excellent health. For a 10-year term of coverage, the cost would be about $120 per year; for a 20-year term, the cost would be about $163 per year; and for a 30-year term, the cost would be about $190 per year.[43]

Whole life insurance is a hybrid product that combines the death benefit of term life insurance with an investment product that accumulates a cash balance. Many life insurance agents actively sell whole life insurance products by emphasizing the investment component. However, it is almost always less expensive to buy term life insurance (if you need it) and do the saving/investing on your own.

Conditional Mortality

What is the probability of a person dying at age 75, given he is alive today at age 45? This is a case of *conditional probability*, wherein one random event (death at age 75)depends on another random event (living from age 45 to age 75).

The *joint probability* is the probability of two events both occurring. For independent events, the joint probability of both events occurring can be calculated by the multiplication of their respective probabilities.

[42] These examples reproduced from
http://www.statefarm.com/insurance/life_annuity/life/whole/whole.asp.
[43] These representative examples were generated by the MetLife online quote calculator:
https://www.metlife.com/individual/insurance/life-insurance/term-life-insurance/term-life-quote/index.html. Insurance premiums depend on both interest rates and competitive pricing.

The probability of a 45-year old surviving to age 75 is about 76.8%. The probability of a 75-year-old dying before age 76 is 3%. Thus the probability of a 45-year old dying at age 75 is:

$$p(45\ year\ old\ dying\ at\ age\ 75) = 0.03 \times 0.768 = 0.023 = 2.3\%$$

In words, we expect that of all 45-year olds living today, about 77% will still be alive at age 75. Of those who are alive at age 75, about 3% will die before age 76. But of all the 45-year olds who are alive today, only about 2.3% will die between age 75 and 76.

Life Insurance Premiums

Similar to other types of insurance, the premiums for life insurance depend on the risk of loss (i.e., the mortality rate) and the severity of the loss (the death benefit amount). However, unlike other types of insurance, it is typical for life insurance policies to have a term of more than one year, and for the annual premium to be based on that multiple-year policy period. Mortality rates depend primarily on the age of the insured. Mortality rates (and insurance premiums) are quite low for younger people, and increase rapidly with age.

Suppose that a 40-year-old man wants to buy term life insurance for one year. The one-year mortality rate for a 40-year-old is about 0.1907% (i.e., about 1 in 524). Given this risk of death, the cost to purchase a one-year, $1,000,000 term life insurance policy should be about:

$$premium = p(death) \times\ death\ benefit\ amount$$
$$premium = 0.001907 \times \$1{,}000{,}000 = \$1{,}907.$$

To calculate the price on a 10-year term life insurance policy, we would need to calculate the expected benefit in each of those 10 years, then find the present value of those expected benefits. The expected benefit in any given year is the mortality rate in that year multiplied by the death benefit amount. The tricky part is that the probability of a payout in a given year is dependent not only on the chance of death at that age, but is also conditional upon not having died earlier. For example, the conditional mortality rate for age 45 is the likelihood of death at age 45, conditional that you did not die before age 45.

The table in Figure 8.11 shows the survival rate, annual mortality rate, and conditional mortality rate for a hypothetical male aged 40. For each year, the expected benefit is the conditional mortality rate times the death benefit amount.

Figure 8.11 Ten-Year Term Life Insurance Calculation for a 40-year old, 3% discount rate.

age now	cumulative survival	annual survival	annual mortality	conditional mortality	expect benefit	present value
40	100.00%	99.81%	0.19%	0.19%	$ 1,907.16	$ 1,907.16
41	99.81%	99.79%	0.21%	0.21%	$ 2,084.49	$ 2,023.77
42	99.60%	99.77%	0.23%	0.23%	$ 2,277.87	$ 2,147.11
43	99.37%	99.75%	0.25%	0.25%	$ 2,488.68	$ 2,277.49
44	99.12%	99.73%	0.27%	0.27%	$ 2,718.38	$ 2,415.24
45	98.85%	99.70%	0.30%	0.30%	$ 2,968.53	$ 2,560.68
46	98.56%	99.67%	0.33%	0.32%	$ 3,240.82	$ 2,714.13
47	98.23%	99.64%	0.36%	0.35%	$ 3,537.02	$ 2,875.92
48	97.88%	99.61%	0.39%	0.39%	$ 3,859.03	$ 3,046.35
49	97.49%	99.57%	0.43%	0.42%	$ 4,208.83	$ 3,225.72

The present value of these 10 years of expected benefit amounts is $25,193. The premium could be pre-paid as a lump-sum amount to buy 10 years of coverage, but most often premiums were paid annually. The amortizing payment would be $2,867.43 per year (an annuity due using a 3% interest rate).

The insurance rates offered by actual life insurance companies are subject to market forces, and may under- or over-price mortality risk. When there is a discrepancy between theory and market prices, market prices should be used.

8.6 Summary

The mortality rate describes the fraction of a population that dies within a fixed period of time, such as one year. The age at which any individual will die is an unknown, random event. We can use data about the death rates within a population to make a reasonable prediction about a cohort's mortality rate, i.e., what percentage of all 80-year-olds will die within a year. The survival rate is the complement of the mortality rate, i.e., the chance of survival for one year.

Benjamin Gompertz discovered a natural law of mortality to explain the exponential growth in the mortality rate that occurs as a cohort ages. Gompertz' work gives rise to an equation to calculate an individual's survival probability.

The actuarial present value is the expected value, in the present, of uncertain future cash flows that depend on a person's survival. The actuarial present value is the present value of the future amount, multiplied by the probability of survival to that age. Annuities are guaranteed payments for life. The present value of an annuity is the actuarial present value of its payments.

The cost of life insurance, which pays a benefit upon death, also depends on the survival probability and mortality rate. The expected benefit in any year is a function is the product of the benefit amount and the conditional mortality rate at that age.

8.7 Review Questions

Conceptual Questions

1. Explain by giving an example:

    a. What is the mortality rate?

    b. What is the survival rate?

    c. What is the survival probability?

    d. What is the actuarial present value?

2. The Gompertz Equation is:

$$\ln\left(p(survival_{x,t})\right) = \left(1 - e^{\frac{t}{b}}\right) e^{\frac{x-m}{b}}$$

   Explain what the equation is used for, and identify and explain each of the parameters to the function.

3. The probability of survival for a 95-year-old man to live until 100 is higher than the probability of a 35-year-old man to live until 100. Explain why this is.

4. Identify and explain the two main types of life insurance, highlighting their key differences.

5. Write a conceptual explanation for the premium of term life insurance. Discuss the use of the survival probability, the mortality rate, and the conditional mortality rate used in this calculation.

Calculation Questions

Create a spreadsheet to solve each question.

1. Assume population parameters of m = 85 and b = 10.
    a. Find the probability of a 30-year-old living to age 90. Show your work.
    b. Find the probability of a 30-year-old living to age 60. Show your work.
    c. Find the probability of a 60-year-old living to age 90. Show your work.

2. Suppose that the population parameters for men are m = 85, and for women are m = 90, with b = 9.5 in all cases.
    a. What is the probability that a 50-year-old man will survive to age 100?
    b. What is the probability that a 50-year-old woman will survive to age 100?
    c. Explain in your own words the difference between the survival probabilities for 50-year old men and women.

3. Consider a 21-year-old, who expects to receive an inheritance of $500,000 at age 45. Calculate the actuarial present value of this inheritance at age 21. Use population parameters of m = 83, b = 11, and an interest rate of 3% per year.

4. Using population parameters of m = 87, and b = 9.5, and an interest rate of 3% per year. Consider a pension (i.e., income for life) that will pay $1,200 per month.
    a. What is the value of this pension to a 60-year-old?
    b. What is the value of this pension to a 65-year-old?

5. Using population parameters of m = 87, and b = 9.5. Reproduce the life-insurance pricing calculator that is shown in Figure 8.11. Find the price of a 20-year term insurance policy with a $500,000 death benefit for a 50-year-old male.

9 Introduction to Investing in Stocks

"What Wall Street does is package luck and sell it as skill."
–Investment Author Dan Solin

Learning Objectives

- Explain the fundamental concepts of investing in the stock market and the relationship between risk and investment return.
- Provide an overview of the historic returns from investments in various asset classes.
- Demonstrate with examples how investors make or lose money by investing in stocks.
- Identify many popular investment ideas that are not substantiated by scientific evidence.
- Provide an overview of the efficient market hypothesis and the random walk theory of stock price movements.
- Identify mutual funds as a means to achieve diversification in one's investment portfolio, and indexing as the optimal investment strategy for investing in equities.

9.1 Risk and Return

Funding long-term goals (e.g., retirement) requires investments with returns that will outpace inflation over the long run. Conventional financial wisdom holds that by investing in the stock market one can achieve high long-term investment returns to fund long-term goals (e.g., retirement or a child's education).

The approximate relationship between investment risk and expected investment return is depicted in Figure 9.1.

Figure 9.1 The Approximate Relationship Between Investment Risk and Reward[44]

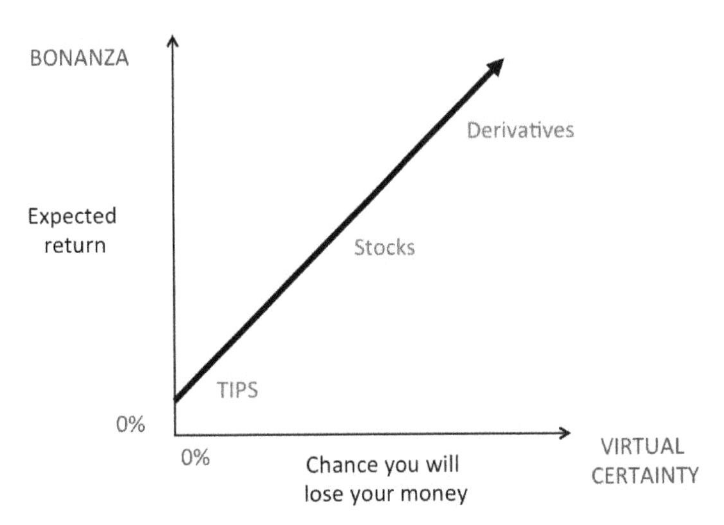

In Figure 9.1, the horizontal axis describes the chance of losing your money, and the vertical axis describes the expected investment return. The ***risk-free rate of return*** is the rate of return you would earn with a 0% chance of losing your money (or your purchasing power). The risk-free rate is achievable by investing in Treasury Inflation Protected Securities (TIPS), which were introduced in Module 4 (*Thinking in Present Value*. Investors who take on additional risk expect to be compensated by earning a ***risk premium,*** an excess return above the risk-free rate.

Stocks and the Stock Market

Stocks, or ***equities***, are a fractional ownership interest in a corporation (i.e., a for-profit business). A ***shareholder*** or ***stockholder*** is the owner of one or more shares of stock in a company. The shareholders, as a group, own the profits

[44] Source: Andrew Tobias, The Only Investment Guide You'll Ever Need.

(or losses) of the corporation and have voting rights with respect to hiring the management of the corporation. Shareholders are entitled to receive all the net profits of the business either as cash distributions called *dividends* or as *retained earnings*, which are the accumulated net worth of the corporation.

The *stock market* is where people buy or sell shares of stock. A stock market can be a physical marketplace, such as the New York Stock Exchange (NYSE), or an electronic marketplace, such as the National Association of Securities Dealers Automated Quotation (NASDAQ).

Derivative Securities

Further right along the horizontal axis of Figure 9.1 are derivatives. *Derivatives* are secondary bets on the direction of an underlying asset (e.g., a stock or a bond), and involve even greater certainty of loss. However, they offer the potential for extremely high returns.[45]

Historical Returns

Investing in stocks involves taking the risk of losing some or your invested assets, whereas certificates of deposit and TIPS bonds do not. So, why do investors take the risk of investing in the stock market? Over the eighty-year period from year-end 1925 through year-end 2012, stock returns have outpaced the returns of all other investment asset classes as well as inflation. The graph in Figure 9.2 shows the hypothetical value of $1 invested in 1926, assuming reinvestment of income and no transaction costs or taxes.

Figure 9.2 Growth of $1 Invested in Stocks, Bonds, Bills, 1926-2012.[46]

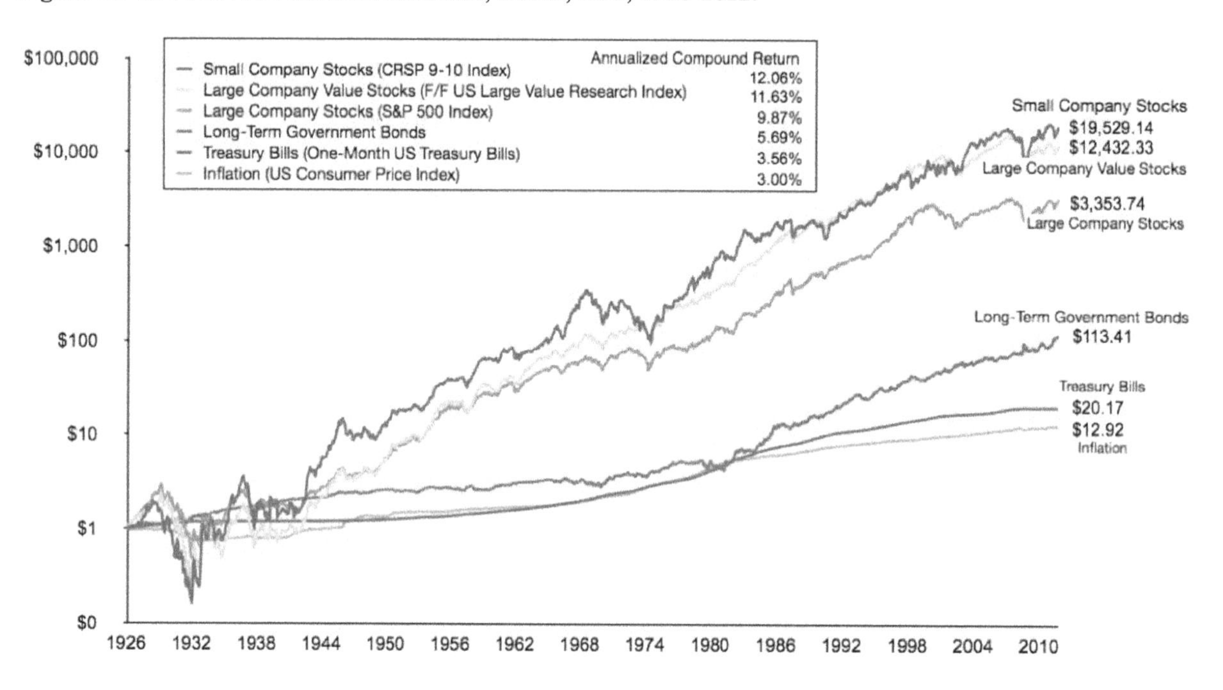

[45] Derivatives can be thought of best as side bets whose payoffs depend on the prices of the underlying asset. For example, a call option is like this bet: "I bet you that Coca-Cola stock will be worth more than 40 dollars per share on March 31. If it is, you will pay me the amount by which it exceeds $40. If the stock is below $40, you pay me nothing. I'll pay you $3 for this bet."

[46] Sources: CRSP data provided by the Center for Research in Security Prices, University of Chicago; S&P data are provided by Standard &Poor's Index Services Group; Fama/French and multifactor data provided by Fama/French; US long-term bonds, bills, and inflation data © Stocks, Bonds, Bills, and Inflation YearbookTM, Ibbotson Associates, Chicago (annually updated work by Roger G, Ibbotson and Rex A Seinquefield).

Note that the graph in Figure 9.2 has a logarithmic scale on its vertical axis. Each dotted line represents a ten-fold increase in wealth, i.e., from $1 to $10 to $100, etc. From this graph, it is clear that stock investments have outperformed all other asset classes in the eighty years from 1926 through 2012. However, note that there have also been several periods in which stock market investors lost substantial wealth. For example, from 1929-1940, 1965-1977, and during the early 2000s, the stock market experienced severe declines. Stock prices decline when there are more sellers than buyers in the market. Investors are motivated by greed and fear, and these powerful emotions result in irrational behavior (i.e., buying on greed and selling on fear).

This graph shows that an investor who invested in stocks in 1926 and remained invested in them until 2012 would have outperformed bonds and inflation. Beware: historic performance does not offer a guarantee of future results.

The Mechanics of Stock Market Investments

This section presents background information about how stocks come into existence and how investors buy and sell stocks.

The Initial Public Offering

You may have heard about a company "going public" – like Facebook – but what does this mean, exactly? The process by which a firm's stock begins to trade on the stock market is called an *initial public offering* (*IPO*). In an IPO, the firm's original owners sell some of their ownership stake to the public with the help of an investment bank. The *investment bank* is responsible for marketing the shares as an investment, including producing a *prospectus* that describes the firm's business and prospects. In the time leading up to the IPO, the investment bank places advertisements in the financial pages of newspapers, holds promotional events (called "road shows"), and does everything it can to gain attention for the event. The investment bank's goal is to sell shares at the highest possible price to maximize the amount of cash raised. The sale represents a transfer of ownership from the company's founders (or early investors) to the general public.

On the day of the IPO, the initial purchasers give their money to the investment bank, which takes a fee, and the remainder of the money goes into the firm's corporate treasury. For example, in the largest IPO in history, Facebook sold itself to the public on Friday, May 18, 2012 at $38 per share. There are approximately 2.138 billion shares outstanding (i.e., owned by investors), so the value of the firm (called the *market capitalization*) at issue was approximately: *$38 times 2.138 billion shares = $81.2 billion.* However, in the IPO, only 421 million shares were sold to the public. The company's founder, executives, employees and venture capitalists who provided initial start-up funding to the firm hold the other 1.6 billion shares.

Some companies continue to sell shares after "going public" by holding a *secondary public offering* to sell additional shares to the public.

The Secondary Market

After the IPO, all shares trade on the secondary market. The *secondary market* is the marketplace of buyers and sellers who negotiate the price at which shares trade. On the secondary market, transactions take place between the sellers and buyers of the shares. The corporation does not receive any money from this trading. The market price of the shares and the quantity of shares outstanding determine the market capitalization.

Stock purchases and sales in the secondary market are made in a *brokerage account* at a stock brokerage firm (e.g., E-Trade or Ameritrade). Investors execute trades via a *stockbroker*, whose job is to execute transactions on behalf of a customer. A stockbroker helps clients buy and sell securities and is paid a commission on each sale. Commissions on stock sales are typically around $20 per trade, but some discount brokers execute trades for as little as $8 per transaction.

For example, suppose Keith wants to buy shares in Apple Computer. He first needs to open a brokerage account with a brokerage firm and transfer money from his bank account into the brokerage account. Next, Keith instructs the broker to make the purchase for him, either by using the broker's website or by phone. For instance, he might give the order to "buy 10 shares of Apple Computer." After the trade is executed, Keith's account will be updated to show that he has 10 shares of Apple Computer, and the broker will debit the account for the cost of the shares plus his or her commission.

Making Money in Stocks

Recall that dividends are direct cash distributions from the corporate treasury to shareholders. For example, as of January 2015, the Coca-Cola Corporation distributes $1.22 in annual dividends per share. Dividends are neither required nor guaranteed: while most firms intend to keep paying dividends forever and many raise the dividend payout each year, companies that fall on hard times usually cut the dividend to preserve cash. Other companies have never paid a dividend, preferring, instead, to reinvest profits into growing their underlying business.

The *dividend yield* is the rate of return on investment due to dividends, i.e., the ratio of annual dividends to the share price:

$$dividend\ yield = \frac{dividends\ per\ share}{price\ per\ share}$$

If Coca-Cola shares trade at $42.62 per share, and dividends are $1.22 per year, the annual dividend yield for Coca-Cola is:

$$dividend\ yield = \frac{\$1.22}{\$42.62} = 0.0286 = 2.86\%$$

When the share price increases the dividend yield decreases, and vice-versa. Often, companies will raise the amount of the dividend payout annually as a way to increase investor's income over time.

Historically, approximately half of the return to shareholders has been in the form of dividends. The other source of investment returns is from capital gains. A *capital gain* is made when an asset appreciates (increases) in value. For a stock investment, the capital gain is calculated as:

$$\text{capital gain} = (\text{selling price} - \text{purchase price}) \times \text{number of shares}$$

For example, suppose you bought 100 shares of Apple Computer on January 2, 2009 at $90.75 per share. On June 15, 2010, you sold the 100 shares for $259.69 per share. Your capital gain would be:

$$(\$259.69 - \$90.75) \times 100 = \$16{,}894.$$

The actual profit is less than it appears because all stock transactions have sales commissions and any capital gains (and dividends) are taxable.[47] As of this writing, long-term capital gains (held over one year) are taxed at a flat tax rate of 15%. Short-term gains are taxed at the investor's highest marginal tax rate. For a gross capital gain of $16,894, the long-term capital gains tax would be:

$$Capital\ gains\ tax = tax\ rate \times capital\ gain$$
$$Capital\ gains\ tax = 15\% \times \$16{,}894 = \$2{,}534$$

In addition to the capital gains tax, the investment entails two trades (for buying and selling), which cost $20 each to execute. The net profit on this sale would be $16,894 (the gross capital gain), minus $2,534 in capital gains tax, minus $40 in commissions for a net return of $14,320.

Losing Money in Stocks

A *capital loss* occurs when an asset is sold for less than its purchase price. For example, if you bought 100 shares of Cisco Systems on March 1, 2000 at $77 per share and sold the shares on March 1, 2010 at $16 per share, you would have a capital loss of:

$$\text{capital loss} = (\text{selling price} - \text{purchase price}) \times \text{number of shares}$$
$$(\$16 - \$77) \times 100\ \text{shares} = -\$6{,}100$$

[47]http://www.irs.gov/taxtopics/tc409.html

Remember, in addition to the loss of $6,100 in capital, a stock sale always involves a commission. Two $20 commissions for the purchase and sale brings the net loss to $6,140.

Investing Strategies

Stock market investors can be categorized as either growth-oriented or value-oriented in their stock picking endeavors.

Growth-oriented investing is the strategy of investing in the shares of companies that you think will experience rapid growth in sales and earnings. For example, a growth investor might want to own shares in high technology companies (e.g., Apple Computer, Google, etc.) because he or she believes that the rapidly growing earnings will lead to a higher stock price.

Value-oriented investing is the strategy of buying shares in companies that you think are undervalued, i.e., selling at a low price compared to their assets or earnings. A value investor might want to own shares in mature industrial companies whose share prices are low relative to their expected earnings. For example, the *Dogs of the Dow* strategy is to invest in the stocks of the Dow Jones Industrial Average (a group of 30 of the largest stocks in the United States) with the highest dividend yield (e.g., AT&T, Verizon, Chevron, McDonalds, etc. in 2015).[48]

Holding Period

Investors' *holding period* (i.e., the length of time that an investor owns a stock) can be described by two overall strategies: buy-and-hold investing and active trading.

Buy-and-hold investors buy a stock and hold it in their investment portfolio for the long run. Warren Buffett, perhaps the most spectacular stock picker of all time, has been quoted saying his favorite holding period is "forever." Buy-and-hold investors hope to profit from the long run appreciation in stock prices and often use dividend income to fund their current needs (e.g., retirees or endowments). Additionally, by not selling stocks, buy-and-hold investors minimize the tax consequences (i.e., you do not pay a capital gains tax until you sell).

By contrast, *active traders* are stock speculators who attempt to profit by actively trading stocks with no pre-set holding period in mind. Active traders hope to take advantage of short-term stock price movements by buying low and selling high. At the extreme, *day traders* buy and sell shares on the same day (or even multiple times on the same day).

Warren Buffett and Berkshire Hathaway

Berkshire Hathaway traces its history as a textile-manufacturing firm in Rhode Island in 1839. Warren Buffett, who was already an accomplished investor, bought control of the declining manufacturing company in 1964. Under present management, Berkshire Hathaway has been transformed into an the worlds largest insurance and investment holding company, owning well-known operating subsidiaries such as Geico Insurance, Dairy Queen, Jordan's Furniture and See's Candies, as well as dozens of daily newspapers throughout the United States.[49] Through its operating subsidiaries, Berkshire Hathaway has over 288,000 employees.

Beginning in the 1960s, Berkshire Hathawayinvested in large blocks of sharesof public companies including American Express, Coca-Cola, Wells Fargo, and IBM. Some of the earliest stock holdings remain in the company's investment portfolio to-date, representing a holding period of almost 50 years. In the past 25 years, Buffett has preferred buying entire operating business to purchasing stock in public companies. Over the 50-year history of Buffet's control of the company, Berkshire Hathaway stock has experienced a compounded annual rate of return of 19.7% for the years 1965-2013. By contrast, the S&P 500 Index has had a compounded annual return of only 9.4% over the same period.

Warren Buffet has been one of the most successful value-oriented investors of all time.

[48] The Dogs of the Dow list is updated annually. See: www.dogsofthedow.com.
[49] For a complete list of operating subsidiaries, see www.berkshirehathaway.com.

9.2 Bad Ideas About Investing in Stocks

The basic rule in stock market investment is to "buy low and sell high." A lot of the marketing information coming out of brokerage firms and investment management houses encourages investors to try to pick individual stocks that will be good investments.[50] These marketing messages make investing look like a game at which the retail investor can win because of his or her intelligence and informational advantage. In practice, this is very difficult to do because the future is unpredictable.

The truth is that no one can accurately predict the future about which stocks will increase in price and which ones will decline. Nonetheless, many bad investment strategies have been offered in the attempt to select "winners." The popularity of these bad ideas is testament to the imagination and greed of their followers.

The body of economic science has disproven all of these popular investment strategies:

- *Scour the Investment Magazines*. Periodicals such as Kiplinger's Personal Finance and Money Magazine run regular features about industries and companies with exciting business opportunities. The "news" in these stories has usually been fully disseminated and is reflected in the stock prices already.

- *Follow Stock Market Newsletters*. Countless newsletters, Internet blogs, and email newsletters offer monthly, weekly, and daily advice about stocks that are poised to rise in price. Rest assured, if the authors of these newsletters knew what was going to happen in the stock market, they would invest on this information themselves rather than sell you their secrets for an annual membership fee of $99.99.

- *Follow the Pundits on TV*. The talking heads on cable news channels like CNBC and CNNfn offer daily picks of stocks that have important company news. These television shows should be treated as entertainment and marketing (for brokerage firms), not as serious investment advice.

- *Jump In On Fads*. Investment fads are industry-wide opportunities or trends that catch investor's imaginations. For example, in the late 1990s and into 2000, Internet companies were growing in popularity, and all manner of Internet-related IPOs were launched.

- *Invest in What You Know*. Many people prefer to invest in the stock of companies they do business with or brands they purchase every day, like Apple Computer or Starbucks Coffee. In reality, it is unlikely that you know something about these companies that the stock market does not already know.

- *Study the Quarterly Results*. Investment research firm Morningstar, as well as many newspapers, publish quarterly summaries of the best and worst performing stocks (and mutual funds). Unfortunately, there is no evidence that the best performing stocks will repeat their performance in the next period.

- *Act Fast on Stock Tips*. Invest based on word-of-mouth "tips" from friends and colleagues, stockbrokers, etc. For example, your stockbroker might call you and say she knows that so-and-so company is going to have a great announcement tomorrow, and you should buy the shares today. Stock tips are generally not valuable due to efficient markets (section 9.3 below).

All of these investment strategies (and many other similar ideas) share the notion that it is possible to achieve superior investment returns by having an informational advantage over other investors. An ***information advantage*** occurs when one investor has information that will affect the demand for a stock before the public (i.e., other investors) know about it. For example, if you knew that Google was about to announce its purchase of a start-up software company, you would buy shares in the start-up company, anticipating that its shares would rise on news of the Google purchase. These strategies have been disproven by the efficient markets hypothesis (section 9.3 below).

---

[50] For example, search for television commercials for E-Trade and Scottrade on YouTube.

Insider Trading

The executives and directors of a publicly traded corporation have special knowledge about new developments in the business that are not yet known to the public. The information they have (whether good news or bad news) might be good enough to make profitable stock trades.

Insider trading is buying or selling securities on material (i.e., relevant to the performance of the business), non-public information and is illegal in most jurisdictions. In the United States, the Securities Exchange Act of 1934 explicitly prohibits insider trading.

For example, suppose the Chief Operating Officer of a company knows that same-store sales declined in the most recent quarter, but this news has not been publicly released. If he or she sells shares in advance of the news release, it would be considered illegal insider trading.

Further, if an insider shares information with a non-insider who trades on the information, the latter person would still be guilty of insider trading. In 2004, television and magazine celebrity Martha Stewart was indicted of insider trading due to second-hand information she obtained from her friend Sam Waskal, who was the CEO of the ImClone Corporation. After a jury trial, Stewart was not found guilty of insider trading, but she was found guilty of four other charges and served a prison sentence. Waskal was convicted of insider trading and several other charges, and was sentenced to seven years in prison.

Other Dangerous Investment Strategies

Employee Stock Purchase Plans

An *employee stock purchase plan* (ESPP) is a mechanism for employees to purchase shares of the company they work for at a discounted price. Many corporate employers see ESPPs as a way to increase employee commitment to the company by allowing employees to gain from good corporate stock performance. Additionally, the ESPP is a way for a corporation to sell unissued shares and raise cash for the company's treasury without the publicity of a secondary public offering, which might depress the stock price.

Many employees want to invest in their employer's stock because they believe in the company (and know what it does). The risk with ESPPs is that employees are placing their financial capital in their employer. If business is bad, the employees might lose their job (income) and financial capital at exactly the same time. There are countless stories of this happening to employees in the automotive industry and in high-technology firms.

Technical Analysis

The analysis of stock charts, which is the art of looking for patterns in historical price movements, is called *technical analysis*. Technical analysts (also known as chartists) have developed countless charting techniques and clever names for patterns (such as the "head and shoulders" pattern).

Figure 9.3 shows an example of a chart produced by technical analysis, showing price support levels below which the stock price is not expected to fall, and resistance levels above which the stock price is not expected to rise.

Figure 9.3 Sample Technical Analysis Chart

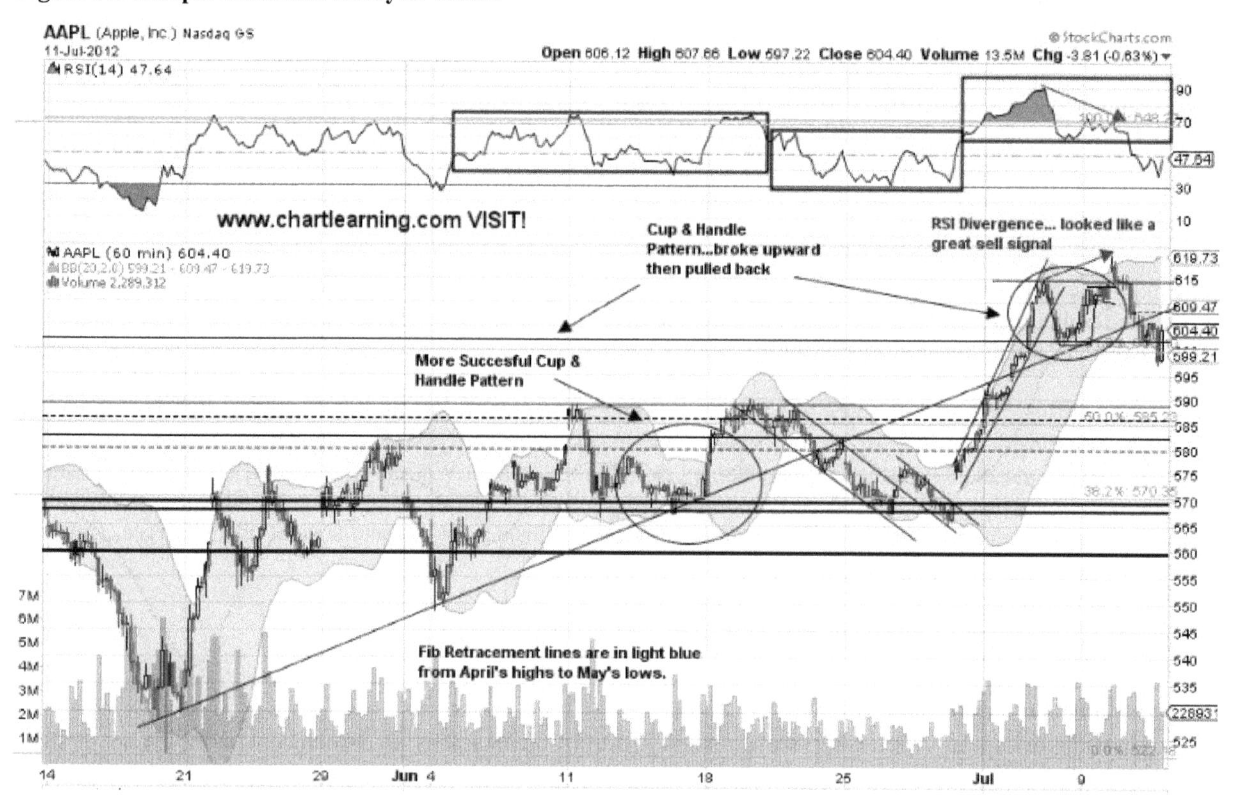

Technical analysts attempt to use historical prices or the *trading volume* (how many shares are bought and sold) to make predictions about the future. The fundamental problem with technical analysis is that stock price movements on one day are independent events from price movements on previous days. There is no evidence of serial correlation of stock prices, i.e., past prices offer no predictive power for future prices. Nonetheless, technical analysts tell *great* stories at parties.

Market Timing

"...most [stock pickers and market timers] should go out of business - take up plumbing, teach Greek..."
–Nobel Laureate Paul Samuelson

Technical analysts, among others, practice *market timing*: attempting to buy a stock before it goes up or sell it before it goes down. Market timing requires making two decisions correctly: the decision and timing of when to buy a stock and the decision and timing of when to sell it. The evidence from the flow of funds (i.e., investment dollars into and out of mutual funds) shows that, in general, retail investors are not successful at market timing. Most retail investors buy high and sell low.[51]

9.3 Good Ideas For Investing In Stocks

The best ideas about stock market investment are founded on evidence that is provable using the scientific method.

Efficient Markets

As soon as investors learn any new information about a specific company or the economy in general, they quickly incorporate this information into their opinions about whether to buy or sell any given stock, which alters the supply

[51] Read http://en.wikipedia.org/wiki/Market_timing#Evidence_against_market_timing for a more detailed discussion about market timing.

and demand for shares. The *Efficient Market Hypothesis* states that, in an efficient market, it is not possible to earn superior returns with an informational advantage because market prices fully reflect all available information.

The classic test of the efficient market hypothesis is an *event study*, which considers stock price movements leading up to a significant event, such as a corporate earnings announcement or a favorable report by an investment analyst.[52] When stock prices adjust immediately and non-randomly in response to the event, the market is said to be efficient with respect to the information of the event.

In *weak-form market efficiency*, the market responds immediately to all publicly available information. Public information is available in the news media, press releases, etc.

The stock market in the United States is considered to be at least weak-form efficient, i.e., with respect to public information. It is for this reason that information provided by the public media, magazines, etc. is not tradable information, because that information has already been incorporated into stock prices.

In *strong-form market efficiency*, the market responds immediately to all information, whether public or private. Only the insiders of a company, such as the directors or chief financial officers have access to this kind of private information. And again, insider trading is illegal.

For example, AMR Corporation, the parent of American Airlines, announced on Tuesday, November 29, 2011 that it would file for bankruptcy protection. In bankruptcy, the shareholders typically lose the value of their investment as the company's assets are transferred to creditors. We would expect that upon the announcement of its bankruptcy filing, investors would want to sell their shares at any price, thus driving down the price of shares. Indeed, the price of AMR's stock fell immediately at the time of the announcement. Figure 9.4shows the stock price for AMR Corporation for the two years leading up to the announcement of bankruptcy. This is evidence of strong-form market efficiency, i.e., that investors were incorporating news about AMR's finances into the stock price well in advance of the bankruptcy.

Figure 9.4 Stock Price for AMR Corporation, 2010-2011.

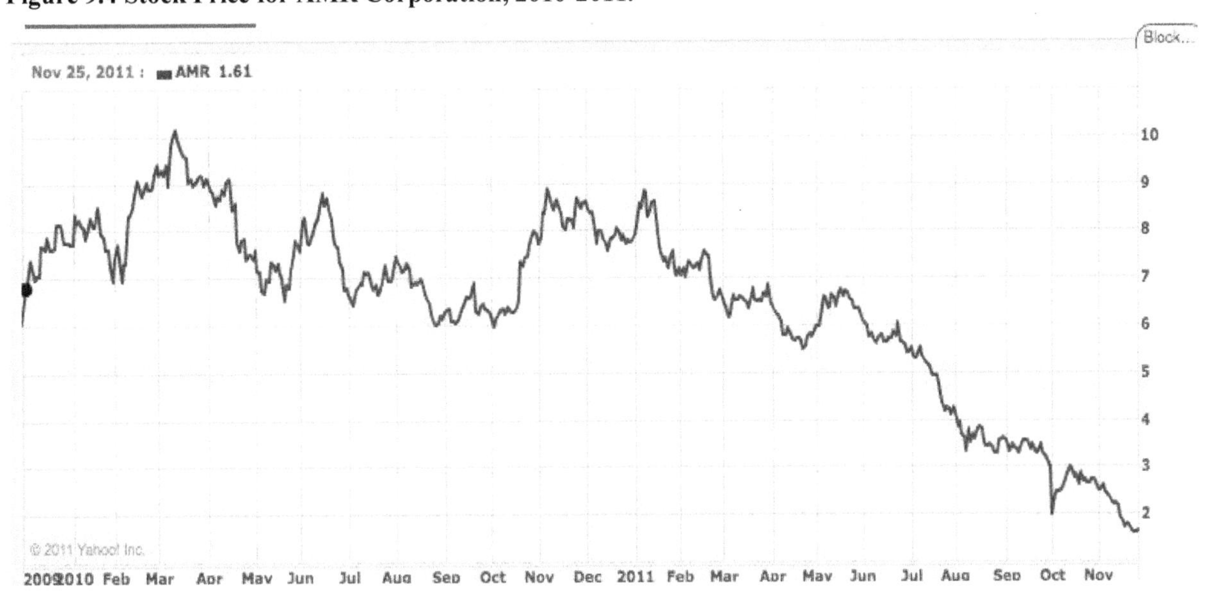

In the graph in Figure 14.4, we see that the price of AMR shares declined for most of the year leading up to its bankruptcy announcement. It seems that information was leaking out of the firm well before the announcement.

[52] Green, T. Clifton and Busse, Jeffrey A., Market Efficiency in Real-Time (May 2001). Goizueta Business School, Emory University. This video presents an event study of the market's response to a financial news show: http://www.youtube.com/watch?v=xHFVQcsJeBs.

In the most recent decade, some criticism of the efficient market hypothesis has come from behavioral economists, who argued that some degree of inefficiency in stock prices might be due to cognitive biases among investors (i.e., investors don't want to believe the news when it disagrees with their opinions).

The Random Walk Theory

A *random walk* is a path determined by a series of random incremental steps, such as flipping a coin to determine whether to turn left or right or rolling a 6-sided die to determine how many steps to take in that direction. In physics, a random walk is used to describe the movement of molecules in a liquid or gas.[53]

The *Random Walk Theory*states that the price movements of stocks follow a random walk. Decades of statistical testing by financial economists have found no convincing evidence of stock price dependencies through time.[54] The important implication of the Random Walk Theory is that stock-trading rules based on past price movements cannot outperform a simple buy-and-hold strategy.

The Random Walk Theory is consistent with the Efficient Market Hypothesis; both form the foundation for the scientific approach to investing in the stock market. Stock price movements (i.e., the percentage change in prices) appear to follow a random walk, so an important first step is to describe the statistical distribution of stock price movements.

9.4 *Quantifying Historic Stock Returns*

The first step in understanding the investment characteristics of stocks is to be able to calculate a stocks' average rate of return. Consider the year-end stock price for Coca-Cola (Figure 9.5).

Figure 9.5 Year-End Stock Price for Coca-Cola for the Years 2004-2014

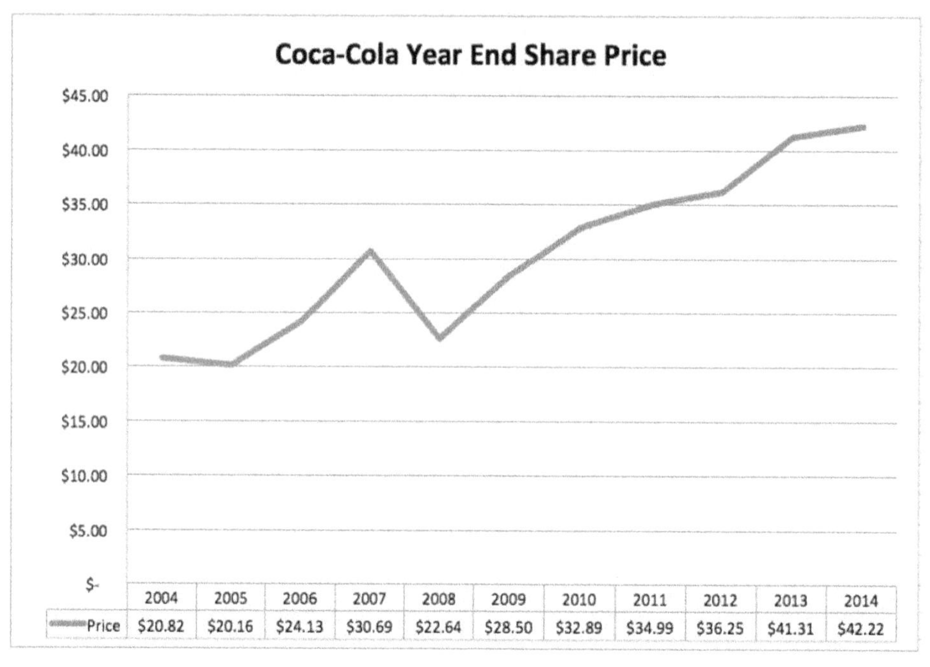

	2004	2005	2006	2007	2008	2009	2010	2011	2012	2013	2014
Price	$20.82	$20.16	$24.13	$30.69	$22.64	$28.50	$32.89	$34.99	$36.25	$41.31	$42.22

How can we describe the performance of an investment in this stock? A first approach is to discuss the annual return (i.e., the gain or loss) for each year, and the average annual return. Figure 9.6shows the year-end price ("Ending Price"), gain or loss in dollars, and the percentage gain or less for Coca Cola for the years 2005-2014.

[53] Physicists describe the movement of molecules as *Brownian motion*, in which a molecule moves in a particular direction on average with a quantifiable amount of randomness around that direction. Finance theory has borrowed from physics models to describe the movements of stock prices similarly.

[54] Jensen, Michael C., and Benington, George A., Random Walks and Technical Theories: Some Additional Evidence, The Journal of Finance, Vol. 25, No. 2.

Figure 9.6 Year-End Stock Price and Percentage Return for Coca-Cola for the Years 2005-2014

Year	Beginning Price	Ending Price	Gain or Loss	Percent Gain or Loss
2005	$ 20.82	$ 20.16	$ (0.66)	-3.19%
2006	$ 20.16	$ 24.13	$ 3.97	19.70%
2007	$ 24.13	$ 30.69	$ 6.56	27.19%
2008	$ 30.69	$ 22.64	$ (8.05)	-26.23%
2009	$ 22.64	$ 28.50	$ 5.87	25.91%
2010	$ 28.50	$ 32.89	$ 4.39	15.39%
2011	$ 32.89	$ 34.99	$ 2.10	6.39%
2012	$ 34.99	$ 36.25	$ 1.27	3.62%
2013	$ 36.25	$ 41.31	$ 5.06	13.96%
2014	$ 41.31	$ 42.22	$ 0.91	2.20%

The Rate of Return

The rate of return earned by an investor in a stock over any time period depends on the price of the stock at the start and end of the holding period. The rate of return in a given period is calculated by:

$$rate\ of\ return = \frac{new\ price - old\ price}{old\ price}$$

For example, if the price of a share of Coca-Cola at the beginning of 2014 was $41.31, and the price at the end of the year was $42.22, the rate of return for 2014 is:

$$rate\ of\ return = \frac{\$42.22 - \$41.31}{\$41.31} = \frac{\$0.91}{\$41.31} = 2.20\%$$

The Average Rate of Return

Suppose you owned shares in Coca-Cola's stock for the ten-year period from January 1, 2005 to December 31, 2014. What is your average rate of return on this investment?

The *mean* or *arithmetic average* of a series of values describes the central tendency of that value. The mean is calculated as:

$$\mu = \frac{1}{N}\sum_{i=1}^{n} x_i = \frac{1}{N}(x_1 + x_2 + \cdots + x_n)$$

The Greek letter μ (pronounced *mu*) represents the mean of a population of values. N is the number of values to be averaged, the Greek letter Σ (uppercase sigma) is the summation operator, and each value is labeled as x_i.

For example, we use the annual percentage gain or loss for each year to calculate Coca-Cola's ten-year average rate of return:

$$\mu = \frac{1}{10}[(-3.19\%) + 19.7\% + 27.19\% + (-26.23\%) + 25.91\% + 15.39\% + 6.39\% + 3.62\% + 13.96\% + 2.20\%]$$

$$\mu = \frac{84.92\%}{10} = 8.492\%$$

Using Spreadsheets: Calculating the Average

The spreadsheet AVERAGE will calculate the arithmetic average of a series of numeric values. The function takes a parameter, which is a range of cells.

Figure 9.7 Calculating the Average of a Series of Values in a spreadsheet.

53	Year	Percent Gain or Loss
56	2005	-3.19%
57	2006	19.70%
58	2007	27.19%
59	2008	-26.23%
60	2009	25.91%
61	2010	15.39%
62	2011	6.39%
63	2012	3.62%
64	2013	13.96%
65	2014	2.20%
66		
67		=AVERAGE(E56:E65)
68		AVERAGE(number1, [number2], ...)

The AVERAGE function makes it easy to calculate the average of a large series of values, e.g., monthly or annual stock returns.

We can say that the average annual rate of return on Coca-Cola stock for the ten-year period from 2005-2014 was about 8.492% per year. Now, what would have happened if you invested $1,000 in Coca-Cola stocks at the start of 2005? If you earned the average rate of return of 8.492% per year for 10 years, you would expect to have $2,259.34:

$$FV = PV(1 + r)^n = \$1{,}000 \times (1.08492)^{10} = \$2{,}259.34$$

However, as an investor owning the stock, you would *not* have earned the average rate of return. Rather, if you bought the stock at the start of 2005, and held the stock for 10 years (to the end of 2014), you would have final wealth in 2014 of $2,027.86. What explains this difference?

The average rate of return is a statistical measure of the central tendency of the annual rates of return. However, no investor actually earns the average rate of return. Rather, an investor's return depends entirely on the price at which they bought the stock, and the price at which they eventually sell the stock. The chart in Figure 9.8 shows the annual *wealth index* (amount of dollars of wealth) for an investment of $1,000 in Coca-Cola at the start of 2005.

Figure 9.8 Value of $1,000 Investment in Coca-Cola for the Years 2005-2014

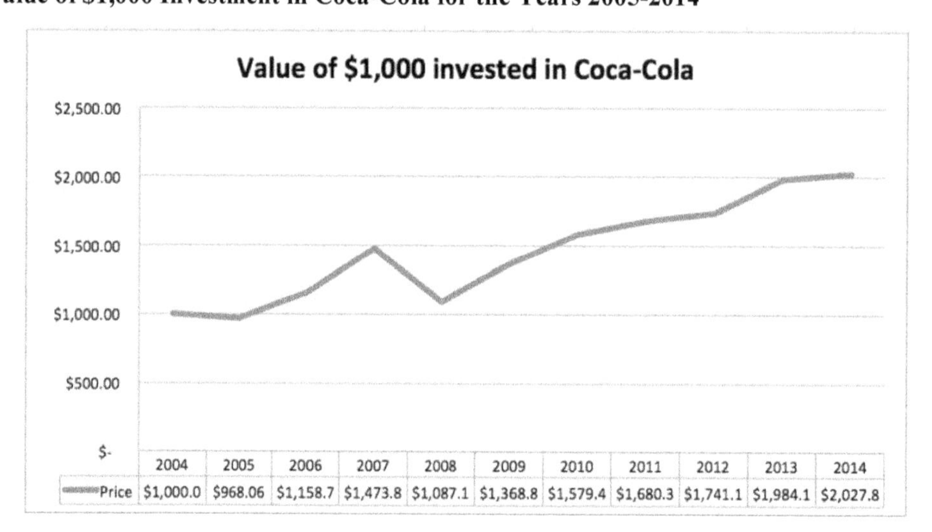

The Compounded Rate of Return

The *compounded annual rate of return* (mathematically, the *geometric mean return*) takes into consideration the ups and downs of stock movements. The compounded annual return is the effective annual rate of return that relates the final value of an asset to its starting value after more than one year of compounding. For an investment held for n years, the compounded annual return is calculated as:

$$compounded\ annual\ rate\ of\ return = \sqrt[n]{\frac{ending\ wealth}{initial\ wealth}} - 1$$

For example, if you bought Coca-Cola's stock at the beginning of 2005, you would have paid $20.82 per share. If you sold the stock at the end of 2014, you would have sold it for $42.22.The compounded annual return during the period 2005-2014 would be:

$$compounded\ annual\ rate\ of\ return = \sqrt[10]{\frac{\$42.22}{\$20.82}} - 1 = \sqrt[10]{2.02786} - 1 = 7.33\%$$

As an investor who holds shares of stock, you are really only concerned with the geometric mean, since this is the effective change in your wealth over time.Students often ask why the geometric mean is less than the arithmetic mean. The geometric mean is *always* less than the arithmetic mean, unless each year's annual return is exactly the same. This is due to the *multiplication* that is inherent to compounding returns.

Comparing Several Stocks

Consider the stocks of these 5 large U.S. companies: Coca-Cola (KO), Bank of America (BAC), Altria (MO), Southern Company (SO) and Intel (INTC). The chart in Figure 9.9 shows the amount of annual wealth an investor would have each year and after 10 years for the 5 selected stocks:

Figure 9.9 Annual Wealth Indexes for Selected Stocks

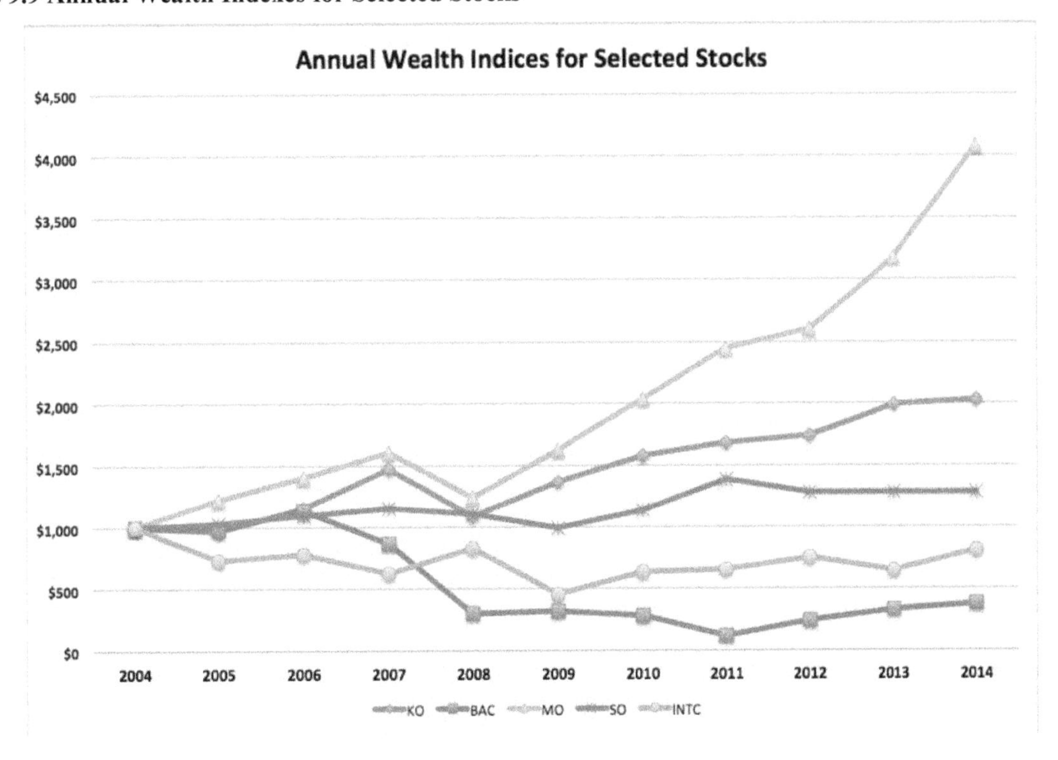

In Figure 9.9, we see that the investor experiences good years and bad years. For example, an investor in shares of Bank of America did well over the 4 years 2004 to 2006, only to see a run of bad years from 2007 to 2011, and a recovery in 2012 to 2014. Figure 9.10 compares the average return and compounded return for the 5 selected stocks.

Figure 9.10 Average Annual Return and Compounded Return for Selected Stocks for the Years 2005-2014

	KO	BAC	MO	SO	INTC
Arithemetic Mean	8.5%	2.0%	16.2%	2.9%	1.6%
Geometric Mean	7.3%	-9.2%	15.1%	2.5%	-2.1%

In each case, the average annual return exceeds the compounded annual return. In particular, notice that the average annual return for Bank of America is about 0%, whereas the compounded annual return is about -10%. The compounded annual return represents the actual change in wealth over time.

It is easy to look at the chart in Figure 9.9 and determine how you should have invested in 2005. However, to quote baseball great Yogi Bera, "It's tough to make predictions, especially about the future." We simply don't know which stocks will do well in the future. Picking individual stocks could be an exciting form of entertainment, but most small investors are better served by investing in stocks via mutual funds (see section 9.5).

9.5 Mutual Funds

From the preceding example with 5 stocks, it should start to become clear that investing in several stocks instead of just one will help to even out some of the ups and downs of individual stocks. We will discuss the benefits of diversification in greater detail in *Module 15 (Statistics for Stock Market Investors).* However, for many small investors, it would be administratively difficult and prohibitively expensive to buy shares in even a few dozen companies, let alone the hundreds of stocks that make up the entire stock market.

A *mutual fund* is an investment company that is organized to pool the resources of many individual investors. By combining the investments of many individuals, a mutual fund can hire a professional manager and invest in a diversified portfolio. At its core, a mutual fund is an investment club in which each investor chooses how much to invest. The mutual fund issues shares to its investors that represent a fractional ownership interest in the entire portfolio. Many large *mutual fund management companies* (e.g., Vanguard Group, Fidelity, TIAA-CREF, etc.) offer dozens of mutual funds from which to choose, and allow investors with as little as $25 or $100 to begin buying shares in regular or retirement accounts.

There are thousands of mutual funds from which to choose, including funds that specialize in different investment objectives and for all types of securities (stocks, bonds, money market instruments, real estate investment trusts, etc.). The Securities Act of 1933 provides the regulatory framework for mutual funds and other investments sold to the public. Under the Securities Act, a mutual fund must provide a *prospectus* to investors describing the investment objectives of the fund and disclosing important facts about the expense structure of the fund.

Buying and Selling Mutual Funds

At the close of each trading day, every mutual fund calculates its *net asset value* (NAV) per share. The NAV is the net value of all assets of the fund divided by the number of outstanding shares. New purchases of mutual fund shares and redemptions (for investors leaving the fund) on any given day are transacted at that day's NAV so that everyone pays or receives the same price on the day of the trade. Most mutual funds sell shares directly to retail investors (i.e., like you), although it is also possible to buy mutual fund shares through a brokerage account.

Mutual funds receive the dividends from their stock holdings and distribute dividends to the mutual fund investors, usually on a quarterly basis. Many investors instruct the mutual fund management company to automatically reinvest these dividends to acquire more shares in the mutual fund.

Mutual Fund Fees

Mutual funds charge fees to offset the costs of operations and staff. Some mutual funds charge a sales fee on new investments. *Loaded* mutual funds charge a sales load on new share purchases. The sales charges are in the range of 1%-6% of the amount invested. For example, investing $1,000 in a loaded mutual fund with a 5.75% sales charge would mean that only $942.50 actually gets invested into the mutual fund. *No-load* mutual funds do not charge a sales fee. In addition to sales fees, all mutual funds charge some amount of annual management expenses. Annual expenses range from as much as 2% to as low as 0.1% of assets.

There is no evidence that high fees result in higher investment returns. To the contrary, the evidence indicates that there is an inverse relationship between fees and investment returns. The more you pay in mutual fund expenses, the less of the investment return you keep. Among firms that provide a family of mutual fund investment choices, the Vanguard Group and Fidelity Investments are consistently the lowest-cost providers.

Actively Managed Mutual Funds

Many mutual funds employ a professional manager whose job it is to select securities for investment. The investment manager's goal is to purchase securities that he or she thinks are underpriced and to sell securities that he or she thinks have reached their maximum price level (i.e., buy-low, sell high).

In the earlier discussion of how not to invest in stocks, it was noted that in reality it is very difficult to consistently buy low and sell high. The record of professional investment managers is actually quite poor: evidence dating back to the 1950s shows that about 75% of actively managed mutual funds fail to beat an unmanaged stock market index.[55] The reasons for this should come as no surprise: investment managers are investing in relatively efficient markets, and cannot outperform the market due to better information;investment fund managers are not able to time the market; andinvestment managers are human, and cannot make accurate predictions about the future.

Index Mutual Funds

An *index mutual fund* invests in all of the stocks of a stock market index, in proportion to their market capitalization.[56] While it is hard to beat an unmanaged stock market index, it is fairly easy for a mutual fund to replicate the composition and returns of that index. There are a wide variety of index funds that track all the major (and many of the obscure) stock market indices, both domestically and internationally. The Vanguard Group, for example, offers almost 200 index mutual funds.

Annual expenses tend to be significantly lower for index funds than for actively managed funds, mainly because of not having to pay an investment advisory or research staff. Further, since there is relatively little trading into and out of the index, index mutual funds tend to generate much lower capital gains distributions than actively managed mutual funds (i.e., they are more tax efficient).

"Most investors, both institutional and individual, will find that the best way to own common stocks (shares) is through an index fund that charges minimal fees. Those following this path are sure to beat the net results (after fees and expenses) of the great majority of investment professionals."
–Warren Buffett[57] (perhaps America's most successful stock picker)

[55] Sharpe, William F., The Arithmetic of Active Management, The Financial Analysts' Journal Vol. 47, No. 1, January/February 1991.
[56] The first index mutual fund was invented by John Bogle, who came up for the idea as part of his undergraduate thesis at Princeton University. Bogle went on to found the Vanguard Group in 1974.
[57] Warren Buffett, 1996. Annual Report, Berkshire Hathaway.

9.6 Summary

Investing in stocks has provided fantastic investment returns over the past century, handily beating the rate of return on government bonds and inflation. Earning an investment return that exceeds the risk-free return from TIPS involves risk. The greater the expected return, the greater the risk of losing your wealth.

Investors who buy stocks earn returns from dividends and capital gains. The general rule is to buy low and sell high, earning a profit on the difference between the price at which you buy shares and the price at which you sell them.

Countless bad investment ideas have become popular over the past two centuries of stock market investing. In general, these ideas presume that you can find an informational advantage over other investors, or that you can find and exploit a trend in stock prices. The efficient market hypothesis shows that new information is quickly incorporated into stock prices, and that you have little chance of earning an excess return due to informational advantage. Further, empirical testing of the random walk theory shows that past stock price movements cannot predict future price movements.

The mean or average rate of return is an arithmetic average describing the distribution of annual returns, but investors who remain invested earn the compounded rate of return, also known as the geometric mean.

Owning several or many stocks provides a way to even out some of the ups and downs of individual stock prices. Most small investors do not have the resources to effectively build a diversified stock portfolio. Mutual funds provide a collective investment pool run by a professional manager. Actively managed mutual funds try to pick stocks that will outperform the market as a whole. Index mutual funds try to replicate the performance of the market as a whole. Historically, most actively managed mutual funds have failed to achieve investment returns in excess of the unmanaged market index.

9.7 Review Questions

1. Explain by drawing a graph the risk and return tradeoff. In particular, discuss TIPS, stocks and derivatives.
2. What rights does one obtain when he or she buys sharesof a company? How does one benefit from owning shares?
3. What is an IPO? Describe the steps involved in a conducting a successful IPO.
4. Explain how investors make money by investing in stocks.
5. List four bad investing ideas that are commonly promoted in our society. Explain the fundamental flaw in each of these bad ideas.
6. What is an information advantage? How can an investor use it to benefit from it in the stock market? Give an example to explain the concept.
7. What is the efficient market hypothesis? Compare and contrast weak form market efficiency with strong form market efficiency.
8. Describe the random walk theory and its implication for the field of technical analysis.
9. Explain the difference between the average annual return and the compounded (geometric) average return.
10. What is a mutual fund? Why is it advantageous for an individual to invest in a mutual fund vs. investing in in individual stocks? Give examples of two kinds of mutual funds.
11. What is the difference between a loaded and no-load mutual fund? Which one should one invest in, and why?
12. Explain the difference between actively managed and index funds.
13. Identify and explain three reasons why index mutual funds are more likely to yield higher net returns on investment compared to actively managed mutual funds.

9.8 Calculation Questions

14. The price of Disney Corporation stock began 2014 at $76.04 per share, and ended the year at $94.91 per share. Calculate Disney's annual rate of return for 2014.
15. The price of Southern Company stock was $33.52 at the start of 2005, and finished at $49.11 at the end of 2014. Calculate the compounded annual rate of return for this 10-year period.

10 Descriptive Statistics for Stock Market Investors

Learning Objectives

- Think about the probability distribution of random events.
- Discuss the normal distribution and the concept of confidence intervals for normally distributed data.
- Use statistical methods to describe the average rate of return on stocks and the variability of the rate of return.
- Introduce the methodology to measure the extent to which two data series move together or apart over time.
- Discuss the conventional wisdom that investing in stocks provides protection against inflation.

10.1 Introduction: Overview of Descriptive Statistics

While this is not a statistics book, many statistical concepts are important to understanding the risks and potential rewards of investing in the stock market. Descriptive statistics are used extensively to describe the rate of return and the variability of returns for individual stocks, groups of stocks held together in a portfolio, and the stock market in general.

This module provides an overview of the most important statistical concepts and terminology used to discuss investment returns.. Statistical concepts are applied to help develop investment portfolios that provide the best risk-adjusted rate of return. In Module 10 (*You Can't Handle The Truth About Stocks*), we will use this statistical information to develop simulations of stock market performance using a random number generator.

Background: The Distribution of Random Events

Consider flipping a fair coin. The probability of either outcome (heads or tails) is ½. If you flip the coin a second time, the probability of each outcome remains 50%. The distribution of possible coin toss results follows a *binomial distribution*, wherein at each point in time two branches are possible. The possible outcomes after two flips are illustrated with the following binomial tree in Figure 10.1.

Figure 10.1 Binomial Tree Showing the Possible Outcomes of Successive Coin Flips

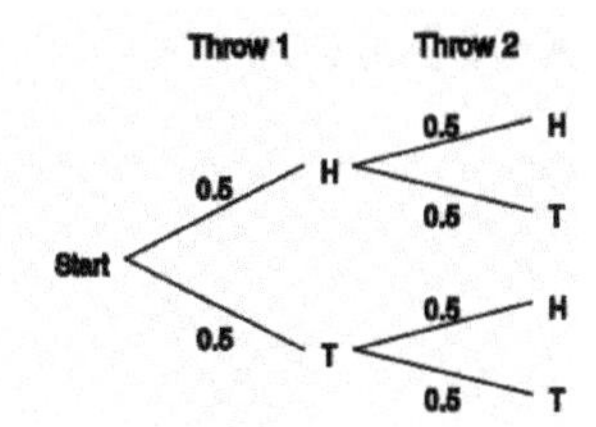

After two throws, the probability of throwing two heads is $P(HH) = 0.5 \times 0.5 = 0.25$. The probability of landing two tails is also $P(TT) = 0.5 \times 0.5 = 0.25$. What is the probability of having one head and one tail? There are two ways to achieve this result, HT and TH. The probability of heads and then tails, $P(HT)$ is 0.25, and the probability of tails and then heads $P(TH)$ is also 0.25. Thus, the probability of having one head and one tail is the sum of these probabilities, i.e., $P(HT) + P(TH) = 0.25 + 0.25 = 0.5$.

As we expand this example to more coin flips, we find an interesting pattern. If you flip a coin ten times, it is most likely that you will come up with heads four, five or six times (see Figure 10.2).

Figure 10.2 The Distribution of Heads With 10 Coin Flips

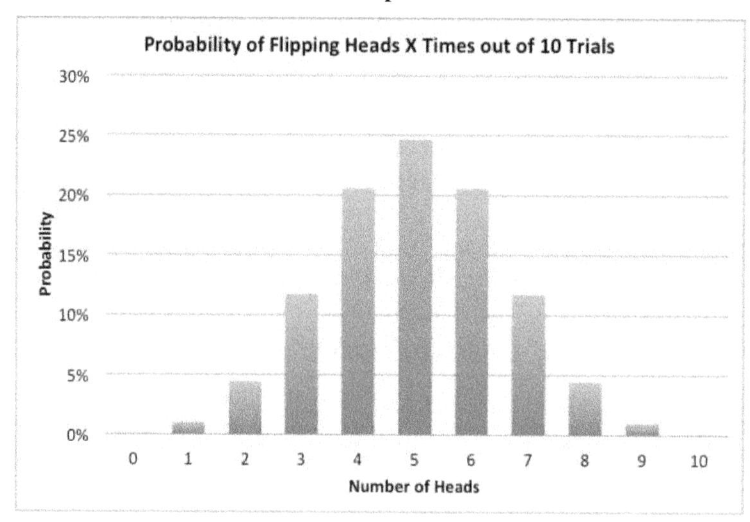

It is possible, but very unlikely that you would come up with heads 10 times in a row (a 1 in 2^{10}, or 1 in 1024 chance). It is equally unlikely that you would come up with 0 heads in 10 flips. On any given attempt, the probability of heads is 50% and we expect that most of the time the number of heads will be close to 50% of the number of flips.

As an experiment, we could run many repeated trials (say of 50 coin flips). We expect, on average, to have about 25 heads and 25 tails in each trial of 50 coin flips. While there is the possibility of an extreme result, such as fewer than 10 heads out of 50 flips, or more than 40 out of 50, these extreme results are very uncommon. It is possible, but extremely unlikely, to obtain 0 heads out of 50. Similarly, it is possible, but extremely unlikely, to get heads 50 times in a row (see Figure 10.3). Most often, the number of heads out of 50 coin flips will be somewhere in the middle, perhaps in the range of 20-30 heads out of 50 flips.

Figure 10.3 The Distribution of Heads With 50 Coin Flips

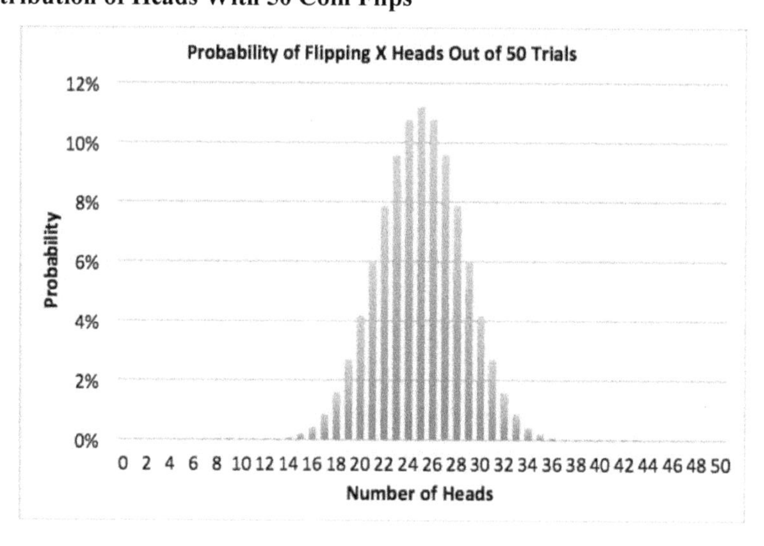

The *central tendency* is the idea of values being clustered near the middle of the distribution, and is the case in many domains within nature.

Describing the Distribution of Random Events

The *mean* is the arithmetic average of a series of numeric observations. The mean describes the central tendency of those data points. For example, the average height of an American man is 5'10", i.e., 70 inches tall. Many natural phenomena, such as the height of humans, the number of petals on a particular type of flower, or the number of bees in a hive, have values that tend to be clustered around the mean.

The *standard deviation* is a statistical measure that describes the distribution of some observations around their mean. For example, the standard deviation of the height of American men is about 3". A small standard deviation means that most observations are clustered tightly around the mean, and a large standard deviation means that observations are dispersed further from the mean. We address how to calculate the standard deviation below.

The *normal distribution* describes many phenomena where data observations are clustered around their mean. The graph in Figure 10.4 depicts the normal distribution's probability density function, also known colloquially as the *bell-shaped curve*.

Figure 10.4 The Normal Distribution

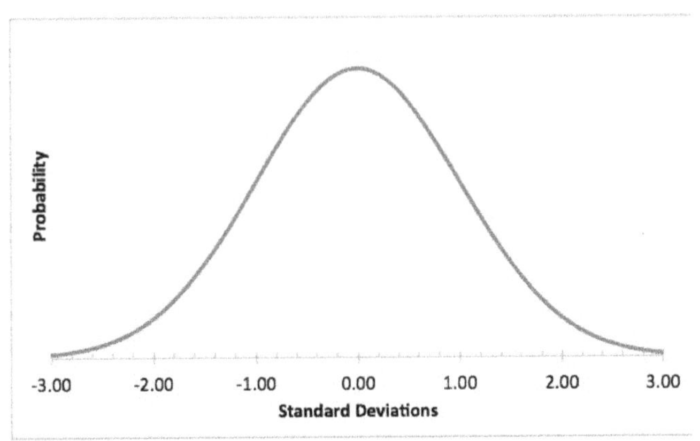

In the graph of the normal distribution (Figure 10.4), the horizontal axis shows the observed value (by its distance above or below the mean) and the vertical axis shows the *frequency* or *probability* of observations. In normally distributed data, we see that most observations occur close to the mean, and few observations occur off in the tails of the distribution.

The probability distribution of many events that have a discrete probability distribution (e.g., the number of heads achieved when flipping a coin n times) can be approximated using the normal distribution. As the number of trials becomes larger, the binomial distribution (Figure 10.2 and Figure 10.3) begins to look like the normal distribution (Figure 10.4) with most of the outcomes clustered near the middle, and few occurring in the tails (i.e., outliers).

Descriptive Statistics for a Population Versus a Sample

In the introduction to the standard deviation presented above, we glossed over a subtle issue with respect to describing a population versus describing a sample. A *population* is a complete set of observations, for example all American men. Often, it is either difficult or not possible to measure every observation for a population.

A *sample* is a subset of a population, such as a group of 30 American men selected randomly. An unbiased, randomly selected sample is useful for describing the population as a whole. For example, we could measure the height of a sample of 30 men, and we expect the *sample mean* to approximate the *population mean*. The important thing to note is that we expect (based on probability theory) that the sample mean will approximate the population mean, but we are not certain.

In this section and the sections that follow, the formulas and examples refer to the *population variance* and *population standard deviation*. Many statistics books discuss the *sample variance* and *sample standard deviation*, which are very similar metrics.

The difference between the descriptive statistics (i.e., the mean and standard deviation) for a population and descriptive statistics for a sample are subtle. These differences matter most when sample size is small, i.e., less than 30. As the sample size grows larger, the differences become less important. When we have a sufficiently large sample, the sample statistics and population statistics will be the virtually the same. At the limit, a sample that includes all elements of the population will have the same descriptive statistics as the population.

The 68%-95%-99.7% Rule

With normally distributed data, most values tend to fairly close to the mean. We can make certain statements about the percentage of all observations that occur within a certain range of values above or below the mean. We observe that about 68.3% of all observations will fall within 1 standard deviation above or below the mean, 95.4% of observations within 2 standard deviations of the mean, and 99.7% of observations fall within 3 standard deviations of the mean.

For example, Figure 10.5 shows the frequency distribution of the height of American men, with a mean of 70 inches and a standard deviation of 3 inches. The shaded area illustrates the bounds of 1 standard deviation above or below the mean, i.e., from 67 inches to 73 inches of height.

Figure 10.5 Normal Distribution of the Height of American Men with 1 Standard Deviation Shaded

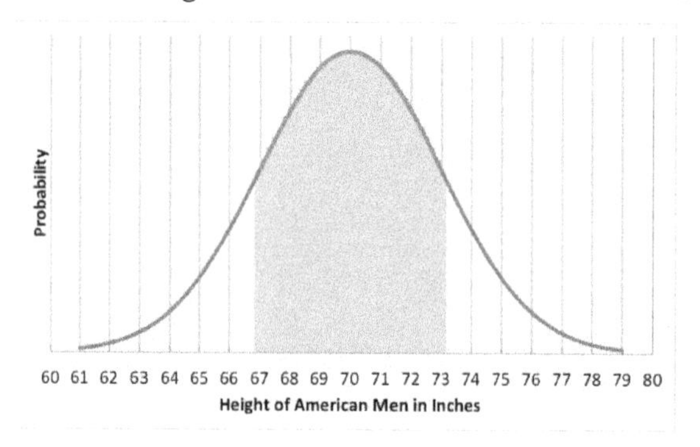

In Figure 10.5, we observe that about 68.5% of American men are between 67 inches (5'7") and 73 inches (6'1") tall.

We can further observe that about 95.4% have heights within two standard deviations of the mean, i.e., they are between 5'4" and 6'4" tall (Figure 10.6).

Figure 10.6 Normal Distribution of the Height of American Men with 2 Standard Deviations Shaded

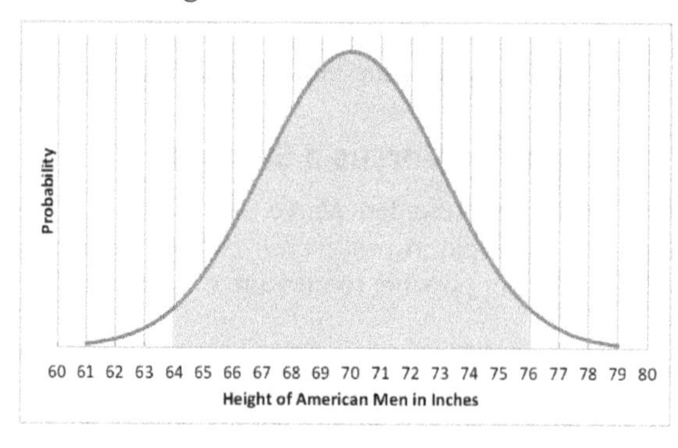

Finally, we observe that about 99.7% of American have heights within 3 standard deviations of the mean, i.e., they are between 5'1" and 6.9" tall (Figure 10.7).

Figure 10.7 Normal Distribution of the Height of American men with 3 Standard Deviations Shaded

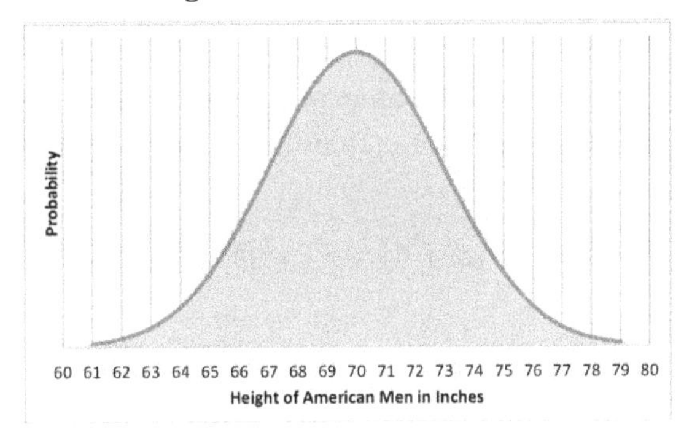

Confidence Intervals

When a statistician makes an estimate about the possible values that a random variable might achieve, the estimate is expressed with a certain degree of confidence that the actual observations will occur within a range of possible values. A *confidence interval* is constructed as a range of values centered around the mean, and provides some guidance for the range of possible values of an estimate. The upper and lower bounds of this range are given by:

$$\mu \pm Z\sigma = (\mu - Z\sigma, \mu + Z\sigma)$$

where μ represents the mean, Z is the number of standard deviations above and below the mean, and σ (the Greek letter lowercase sigma) is the standard deviation.[58]

For example, a statistician would make a statement that about 95% of the time, the height a randomly selected American man is within the range of:

$$\mu \pm Z\sigma = 70\ inches \pm 2\ \times\ 3\ inches = (64\ inches, 76\ inches)$$

This range of values between 64 and 76 inches (i.e., between 5'4" and 6'4" tall) represent a *95% confidence interval*. Another way to think about the confidence interval is to state that about 19 out of 20 American men have a height between 5'4" and 6'4" tall.

If you wanted to have a higher degree of confidence of including most men in your estimate, you would need to include a higher range of possible values. For example, we could estimate that about 98% of American men have a height between about 63 and 77 inches tall. It is possible to make statements with other confidence intervals, but the 95% confidence interval is most common.

Calculating the Population Standard Deviation

An intermediate step toward calculating the standard deviation of a data series is to calculate the variance. The *variance*, represented by σ^2 (the Greek lowercase letter sigma, squared), is the average squared deviation from the mean (μ), defined by:

$$\sigma^2 = \frac{1}{N}\sum_{i=1}^{N}(x_i - \mu)^2$$

$$\sigma^2 = \frac{1}{N}[(x_1 - \mu)^2 + (x_2 - \mu)^2 + \cdots + (x_N - \mu)^2]$$

The uppercase Greek letter Σ (sigma) is the summation operator and indicates a repeated operation that will take place N times (where N is the number of observations). The subscript *i* takes on each successive value from 1 to N,

[58] One area in which the differences between descriptive statistics for a sample and descriptive statistics for a population matter is in constructing confidence intervals. When dealing with samples, we need to account for the size of the sample when selecting the number of standard deviations to include for the confidence interval bounds. In our discussion in this module, we will use the population statistics and the standard normal distribution (i.e., Z-scores).

so that we can apply the same operation to each observation. The summation indicates that we are adding each term of $(x_i - \mu)^2$, i.e., the squared deviation of one observation x_i from its mean value.

Example: Calculating the Variance and Standard Deviation of a Series of Quiz Grades

Suppose the grades attained on an in-class quiz (out of 100) are as follows: 71, 94, 80, 91, 79, and 83. First we calculate the mean:

$$\mu = \frac{1}{N}\sum_{i=1}^{N} x_i = \frac{1}{6}(71 + 94 + 80 + 91 + 79 + 83) = \frac{498}{6} = 83$$

Next, we can calculate the variance, i.e., the average squared deviation from the mean:

$$\sigma^2 = \frac{1}{N}\sum_{i=1}^{N}(x_i - \mu)^2$$

$$\sigma^2 = \frac{1}{6}[(71 - 83)^2 + (94 - 83)^2 + (80 - 83)^2 + (91 - 83)^2 + (79 - 83)^2 + (83 - 83)^2]$$

$$\sigma^2 = \frac{1}{6}[(-12)^2 + (11)^2 + (-3)^2 + (8)^2 + (-4)^2 + (0)^2]$$

$$\sigma^2 = \frac{1}{6}[144 + 121 + 9 + 64 + 16 + 0] = \frac{354}{6} = 59$$

The variance of these quiz grades is 59. Note that the variance is measured in *square units* and thus hard to interpret. The variance of 59 is not 59 quiz points, but rather 59 *quiz points squared.*

The standard deviation, which is measured in the same units as the original data series is easier to interpret. The standard deviation of a series of observations is the square root of the variance, i.e.:

$$\sigma = \sqrt{\sigma^2} = \sqrt{\frac{1}{N}\sum_{i=1}^{N}(x_i - \mu)^2}$$

We calculate the standard deviation as the square of the variance:

$$\sigma = \sqrt{\sigma^2} = \sqrt{59} = 7.68$$

Thus, we can say that the distribution of quiz grades has a mean of 83, and a standard deviation of about 7.7.

Using Spreadsheets: Calculating the Population Standard Deviation

Spreadsheet programs provide a built-in function to calculate the population standard deviation of a series of values. The built-in STDEVP function takes a parameter that is a series of values. (see Figure 10.8).

Figure 10.8 Calculating the Standard Deviation of a Series of Values in a spreadsheet.

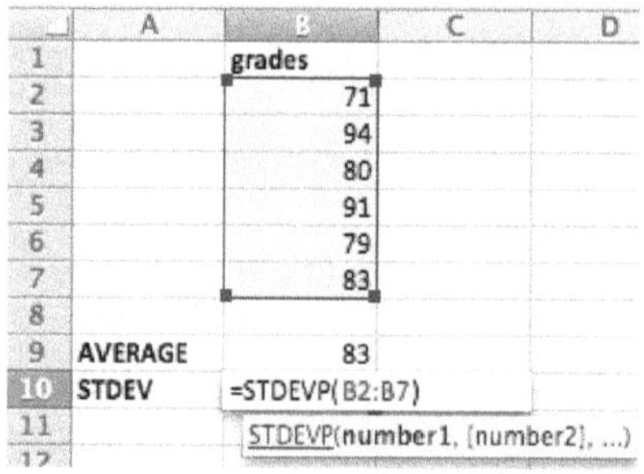

	A	B	C	D
1		grades		
2		71		
3		94		
4		80		
5		91		
6		79		
7		83		
8				
9	AVERAGE	83		
10	STDEV	=STDEVP(B2:B7)		
11		STDEVP(number1, [number2], ...)		
12				

In the examples the follow, we will use the STDEVP function to calculate the standard deviation of a large series of values, e.g., monthly or annual stock returns.

Note: there is also a function called STDEV, that calculates the *sample standard deviation.* The difference between the *population standard deviation* and the *sample standard deviation* is subtle, and really only matters when the sample size is small, i.e., less than 30.

10.2 The Statistical Distribution of Stock Returns

The historical returns earned by investing in stocks provide some guidance about what we might expect in the future. There is no guarantee that stock returns in the future will behave as they did in the past, but we can at least use the past behavior to help us understand the riskiness of investment allocations. We describe the historic performance of a stock by its mean or average annual return, and we measure the variability of a stock's return around its mean return by its standard deviation.Financial economists calculate the standard deviation of stock returns (not prices), since the stock returns measure the percentage change in price.

The distribution of historical stock returns follows an approximately normal distribution. For example, during the period 1965-2014, the *monthly* returns for Coca-Cola were approximately normally distributed with a mean of 1.39% and a standard deviation of 7.89% per month. The graph in Figure 10.9 shows the nominal monthly stock returns for Coca-Cola Stock from 1965-2014, with the normal distribution curve overlaid for comparison.

Figure 10.9 Frequency Distribution for the Monthly Percentage Rate of Return

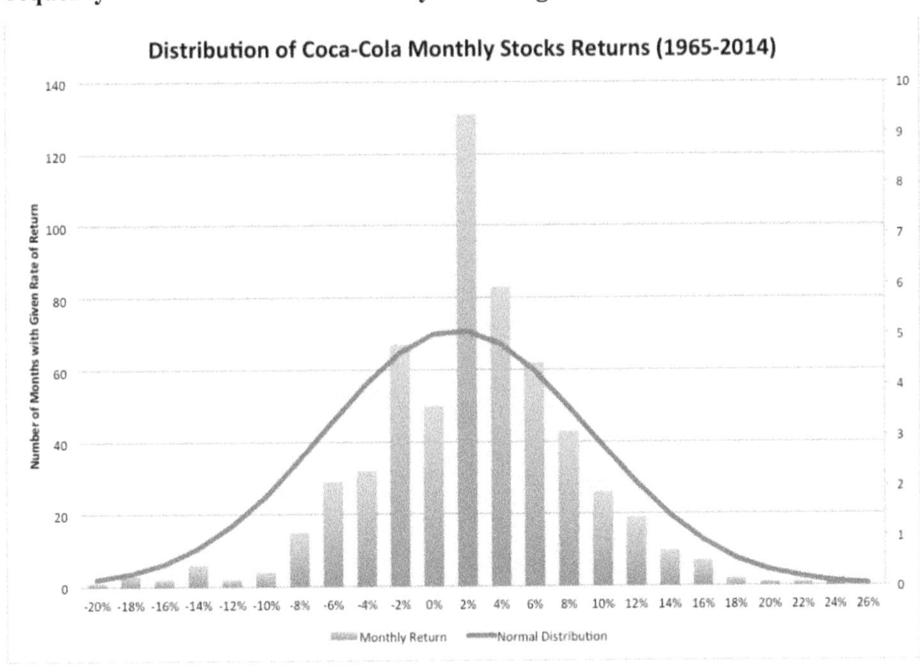

Strictly speaking, the distribution of stock returns does not follow a normal distribution. In the Coca-Cola example in Figure 10.9, there are too many values close to the mean, and too many extreme observations far below or above the mean (i.e., "fat tails"). Nonetheless, we can use the assumption of normally distributed returns to make observations about the risk of investing and the construction of portfolios.

It is common to use the historical standard deviation of returns over long periods of time to estimate the riskiness of a stock's return, i.e., how spread out are its returns from its average return. The chart in Figure 10.10 shows the average annual rate of return and the annual standard deviation for Coca-Cola (KO), Bank of America (BAC), Altria (MO), Southern Company (SO), and Intel (INTC) over the 30 years 1985-2014.

Figure 10.10 Historical Annual Returns and Standard Deviation, Selected Stocks for the Years 1985-2014

	KO	BAC	MO	SO	INTC
Mean	14.6%	12.1%	19.3%	6.7%	20.0%
Std Deviation	23.4%	38.2%	28.5%	14.7%	44.5%

A higher standard deviation means that the stock's return had a higher degree of variation about its mean, i.e., monthly returns were very spread out from the mean return. By contrast, a low standard deviation means that the stock's monthly returns were clustered closely around the mean return. We can visualize this relative amount of variability of each stock's returns by plotting a normal distribution for each stock's mean and standard deviation (see Figure 10.11).

Figure 10.11 Normal Distribution of Annual Stock Returns for Selected Stocks, 1985-2014

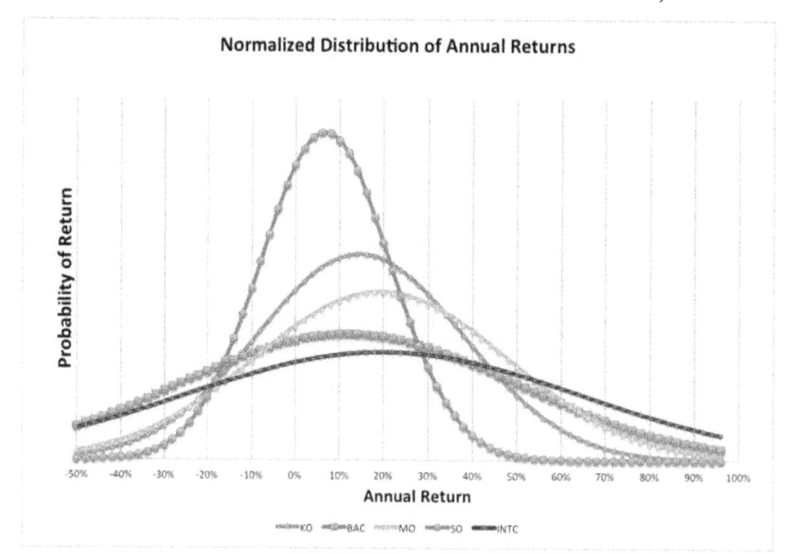

For example, we can make the statement that during the 30-year period from 1985-2014, the stock of Bank of America earned an average annual return of 12.1%, whereas the stock of Southern Company earned an average return of only 6.7%. Bank of America's annual standard deviation of 38.2% indicates that its annual returns were more spread out away from its average annual return of 12.1%. By contrast, Southern Company's annual standard deviation of 14.7% indicates that its annual returns were more tightly clustered around its average annual return of 6.7%.

10.3 Diversification and Correlation

Managing Risk in Stock Market Investments

Picking individual stocks in which to invest is effectively a game of chance. In any given year, some stocks will rise in price and others will fall. It is not possible to know in advance which ones will have gains and which ones will have losses, just like it is not possible to know the outcome of a coin toss before in advance.

Diversification is an important risk management strategy in which exposure to a single large risk is exchanged for exposure to many smaller and uncorrelated or negatively correlated risks. Diversification is the fundamental idea behind the old adage, "don't put all your eggs in one basket."

Correlation is the Secret to Effective Diversification

Correlation describes the extent to which two series tend to vary together, i.e., sharing the pattern of ups and downs. For example, there might be a relationship between sunny weather and ice cream sales. When the sun shines, more people choose to buy ice cream. When it's raining, fewer people buy ice cream.

Negative correlation describes the extent to which two series tend to move opposite of each other. For example, there might be a negative correlation between sunny weather and umbrella sales. Few people buy umbrellas when it is sunny outside, but many people buy umbrellas when caught in a rainstorm.

When there is no systematic relationship between two data series, we say that the data series are *uncorrelated*. For example, there is no relationship between the price of gasoline and the number of runs scored by the Red Sox on any given day.

Quantifying the Relationship Between Data Series

When we have more than one data series, we might want to describe the relationship between these observations. Do two (or more) values go up together, down together, or in opposite directions? For example, we could measure the relationship between height and shoe size. Do taller people wear bigger shoes? The *covariance* statistic measures the extent to which observations from the two data series vary from their respective means.

A high covariance indicates that the data series tend to vary away from their respective means at the same time, and vice versa. For example, if taller people wear bigger shoes, and shorter people wear smaller shoes, we would see a high covariance. Unfortunately, the covariance is not a standardized measure, so it's values are hard to interpret.[59] We are more interested in the correlation coefficient, whose values are easily interpreted.

The *correlation coefficient*, denoted by the Greek letter ρ (rho), is the statistical measure of how two data series co-vary with each other and runs on a standardized scale from –1.0 to +1.0. (see Figure 10.12).

Figure 10.12 Interpreting the Correlation Coefficient

Type of Correlation	Correlation Coefficient	Interpretation
Perfect Positive Correlation	$\rho = 1$	The two data series always go up or down together. When one goes up, the other always goes up as well.
Positive Correlation	$\rho > 0.5$	The two data series tend to move together. When one goes up, the other tends to go up.
No Correlation	$\rho = 0$	There is not systematic relationship between the two data series.
Negative Correlation	$\rho < -0.5$	The two data series tend to move opposite to each other. When one goes up, the other tends to go down.
Perfect Negative Correlation	$\rho = -1$	The two data series always move in opposite directions. When one goes up, the other always goes down.

For example, the retail prices of gasoline and propane have a strong positive correlation (about 0.8), which is due to the fact that both are derived from crude oil. By contrast, the correlation coefficient between annual returns in the stock market and the rate of inflation as measured by the CPI is close to 0.

Diversification in Practice

Diversification does not eliminate randomness, but it helps to minimize the negative effects of randomness. Diversification works best when you can find risks that are negatively correlated with each other: when one goes up, the other goes down.

For example, you could diversify the risk related to the weather by opening two businesses that share the same street corner. When it's sunny, sell ice cream and when it's raining, sell umbrellas (and hot chocolate). This provides diversification because when one business does poorly (because of random weather), the other business does well (because of the same random weather). However, it is possible to get some benefits of diversification even with events that are positively correlated, so long as they are not *perfectly* correlated.

[59] The process to calculate the covariance is a bit complex, and the resulting numbers are hard to interpret due to non-standardized units (e.g., the covariance between people's heights in inches and shoe size are in units that are neither inches nor in shoe sizes, but a product of inches and shoe size).

Using Spreadsheets: Measuring Correlation

The spreadsheet CORREL function calculates the correlation coefficient for two data series. The function takes two parameters, each of which is a range of cells. The ranges of cells do not need to have the same type of data, but do need to be of the same length, i.e., 10 items each.

Figure 10.13 Calculating the Correlation Coefficient Between Two Data Series

	Annual Returns (net of dividends)		
4	Year	KO	BAC
5	2004		
6	2005	-3.19%	-1.79%
7	2006	19.70%	15.69%
8	2007	27.19%	-23.94%
9	2008	-26.23%	-65.33%
10	2009	25.91%	6.96%
11	2010	15.39%	-11.42%
12	2011	6.39%	-58.32%
13	2012	3.62%	108.81%
14	2013	13.96%	34.11%
15	2014	2.20%	14.90%
16			
17			
18	CORRELATION	=CORREL(C6:C15,D6:D15)	

When two data series are related, they have a non-zero correlation coefficient. In the example in Figure 10.13, the correlation coefficient for the annual return on Coca-Cola and Bank of America is 0.24, i.e., a weak positive correlation.

10.4 Conventional Wisdom: Stocks Provide Protection Against Inflation

The Verdict of History[60]

Investment in common stocks has been extremely favorable to investors. Over the long run (i.e., 1925-2012), the compounded rate of return from common stocks has outpaced all other asset classes, and handily outpaced inflation – despite periods of loss (i.e., negative returns) in both real and inflation-adjusted terms.

Figure 10.14 Growth of $1 invested in Stocks, Bonds, Bills, 1926-2012.[61]

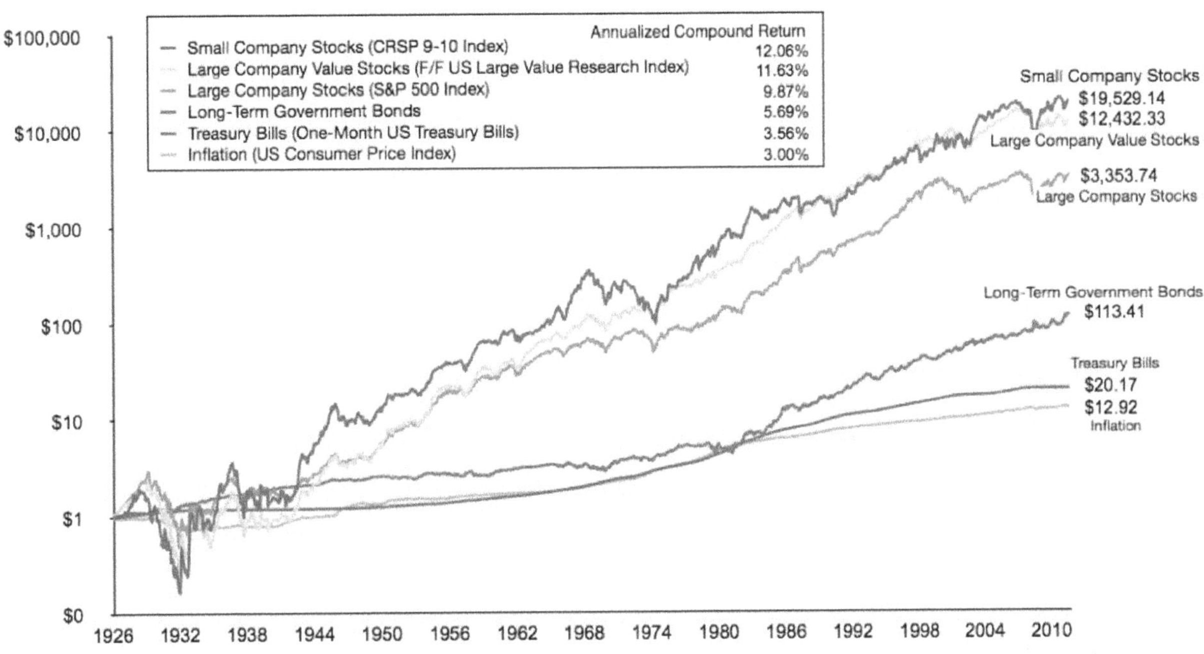

The historical evidence is not in dispute: investment in common stocks has outperformed all other asset classes over the period from 1925-2012. After reviewing the chart in **Ошибка! Источник ссылки не найден.**, many financial advisors (and academics) have concluded that it is important to invest in stocks as a way to protect against future price increases, i.e., inflation. Over the long run, stocks have had a higher rate of return than Treasury bonds, and stock returns that have exceeded the long-run rate of inflation.

Do Stocks Really Protect Against Inflation?

Stock market returns have outpaced inflation over the long run (refer back to **Ошибка! Источник ссылки не найден.**). Based on the historical record of stock market returns and inflation, financial advisors, academics, and even regulators at the Securities and Exchange Commission have recommended that investing in stocks is important to provide protection against inflation.

To determine whether stocks provide good protection against inflation, we need to observe whether or not there is a statistically significant relationship between the annual rate of return on stocks and the annual rate of inflation. From the graph in **Ошибка! Источник ссылки не найден.**, we would expect that when the annual rate of inflation is high, stocks would have higher annual returns to compensate investors for the decrease in their purchasing power.

---

[60] For further reading, see Siegel, Jeremy, *Stocks for the Long Run*, which discusses the historic return from stocks.

[61] Sources: CRSP data provided by the Center for Research in Security Prices, University of Chicago; S&P data are provided by Standard &Poor's Index Services Group; Fama/French and multifactor data provided by Fama/French; US long-term bonds, bills, and inflation data © Stocks, Bonds, Bills, and Inflation YearbookTM, Ibbotson Associates, Chicago (annually updated work by Roger G, Ibbotson and Rex A Seinquefield).

Ошибка! Источник ссылки не найден. plots the annual inflation rate and the annual rate of return on common stocks.

Figure 10.15 Stock Market Returns and Inflation Rate

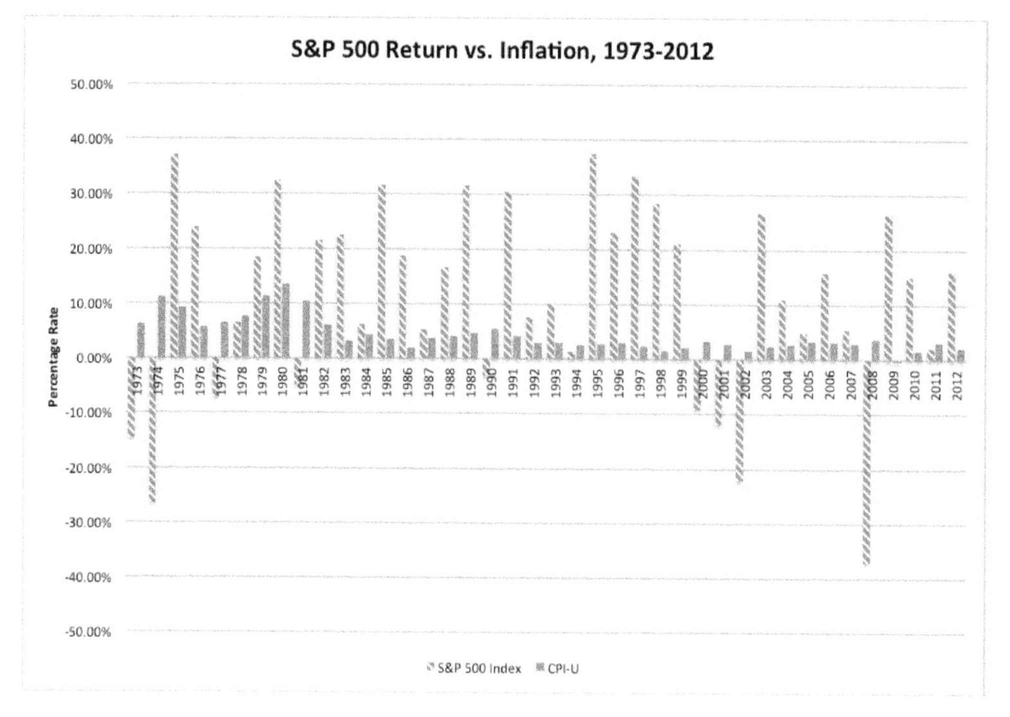

Recall thatcorrelation is the statistical measure of the relationship between two data series. A positive correlation would indicate a good hedge (i.e., when inflation raises the price level, stocks go up in value in the same year). While correlation provides evidence about the co-variation of data series, it does not tell us about the cause of this co-variation. Did stocks go up because of increases in the general price level, or did prices rise due to increased stock market wealth? It's not clear.

The correlation coefficient between the S&P 500 stock market index returns and the Consumer Price Index for the years 1926-2012 is about -0.02, or approximately 0. The statistical evidence is that there is *no correlation* between stocks and inflation. To the extent that stock returns have outpaced inflation, the strongest statement we can make is that it has been a *coincidence*. Moreover, investments in the stock market have come at the cost of requiring investors to bear stock market risk.

By contrast, TIPS bonds are indexed for inflation: the principal and interest on TIPS bonds are adjusted based on changes in the Consumer Price Index. The rate of return on TIPS held to maturity and the inflation rate have almost perfect positive correlation (i.e., close to 1.0). TIPS provide an excellent hedge against inflation. As per the definition of hedging, the investor must give up the potential returns (and losses) from stocks in order to be protected against inflation.

10.5 Summary

A scientific approach to investing in the stock market begins with a statistical description of the historical returns earned by stock investors. We consider stock returns in a given month or year to be randomly distributed, and we assume that they follow an approximately normal distribution. The mean annual return and standard deviation provide guidance about the distribution of historical returns. A higher standard deviation means that the stock's return had a higher degree of variation about its mean over time, and a lower standard deviation means that the stock's rate of return had a lower degree of variation over time.

A confidence interval provides some guidance for the range of possible values in which a random event might occur, for example the distribution of stock returns. While there is no guarantee that values will fall within the confidence interval it is still useful to set expectations about what is likely to occur.

Correlation measures the extent to which two data series tend to move together over time. A positive correlation between the returns of two stocks indicates that both stocks moved up together and moved down together. Negative correlation between the returns of two stocks indicates that the returns moved in opposite directions from each other. When there is no systematic pattern between the data series, we say there is no correlation.

Conventional wisdom holds that investing in stocks is required to outpace inflation. Correlation is the statistical measure of how two data series vary together over time. There is no evidence that annual stock returns are correlated with inflation. While compounded stock returns have exceeded the rate of inflation over the history of the 20th century, there is no guarantee that future stock returns will outpace inflation.

10.6 Review Questions

1. Explain why the distribution of random events has many values clustered near the central value.
2. What does the standard deviation measure?
3. What is a confidence interval? How does this relate to the 68-95-99.7 rule?
4. What is correlation? Explain the meaning of the correlation coefficient.
5. Over the past century, investing in stocks has provided about a 9% real rate of return. The conventional wisdom is that stocks provide a good hedge for inflation. Explain why this is wrong.

10.7 Exercises

Look up historical monthly (annual) stock prices for the past 20 years on Yahoo Finance or a similar website for any 5 stocks of your choosing. Use Excel to do the following exercises:

1. Calculate the monthly (annual) rate of return based on the changes in price for each stock.
2. Calculate the monthly (annual) standard deviation of returns for each stock.
3. Construct equally weighted portfolios of 1, 2, 3, 4, and 5 stocks. Assume annual rebalancing, so that the annual return is the equally weighted return of each of the several stocks.
4. Calculate the average rate of return and the standard deviation of returns for the portfolio as a whole. What happens to the standard deviation of the portfolio as you increase the number of stocks in the portfolio?

11 Diversification and Portfolio Construction

Learning Objectives

- Introduce diversification as a risk-management technique for investing in stocks.
- Quantify the risk-reduction benefits of investing in a portfolio of stocks.
- Provide a conceptual overview of modern portfolio theory and the efficient frontier.
- Quantify the historical performance of the stock market in statistical terms and the range of possible future stock market returns.

11.1 Diversification and Portfolios

Recall that *diversification* is a risk-management technique in which a large exposure to a single risk is replaced with smaller exposures to several risks. Diversification plays an important role in stock market investing. Instead of investing in just one stock and hoping for spectacular gains, investors who diversify are investing in many different stocks instead of just one, and hope for these investments to do well *on average*. In order for diversification to work, it is crucial that the many stocks' risks not be perfectly correlated.

A *portfolio* is a basket (i.e., a group) of multiple securities (e.g., stocks) held together by one investor. For example, we could make a portfolio consisting of the stocks of the five companies previously shown in Figure 10.10 and Figure 10.11– Coca-Cola (KO), Bank of America (BAC), Altria (MO), Southern Company (SO), and Intel Corporation (INTC) – held in equal proportion (i.e., 20% invested in each).

Calculating the Rate of Return on a Portfolio

When multiple stocks are held together in a portfolio, the return on the portfolio is the weighted average return of the stocks in the portfolio. We calculate $r_{portfolio}$, the rate of return on a portfolio, as:

$$r_{portfolio} = \sum_{i=1}^{N} w_i r_i = w_1 r_1 + w_2 r_2 + \cdots + w_N r_N$$

That is, for a portfolio of N stocks, the return of the portfolio is the sum of the return of each individual stock (i.e., r_i) times the proportion or weighting of that stock within the portfolio ($w_{i\,i}$). That is, a stock representing 1/5 of your portfolio would be responsible for 1/5 of portfolio return.

Example: The 5-stock Portfolio

Consider the portfolio of 5 stocks – Coca-Cola (KO), Bank of America (BAC), Altria (MO), Southern Company (SO), and Intel Corporation (INTC) – held in equal proportion (i.e., 20% in each stock). Figure 11.1 shows the annual returns for each stock in the portfolio, and the portfolio as a whole in 2014.

Figure 11.1 Annual Returns for Selected Stocks in 2014

Annual Returns	KO	BAC	MO	SO	INTC		Portfolio
2014	2.2%	14.9%	28.3%	19.5%	25.9%		18.2%

For example, the stock of Coca Cola had a meager return of 2.2% in 2014. By contrast, Altria (MO) had a good year in 2014, and if you had invested in it, you would have earned a very favorable 28.3% return on your investment.

The result of holding a portfolio of stocks is that the return on the portfolio evens out some of the ups and downs of the individual stocks. The portfolio return on an equally-weighted portfolio invested in these 5 stocks for the year 2014 was:

$$r_{portfolio} = \sum_{i=1}^{N} w_i r_i = 0.2 \times 2.2\% + 0.2 \times 14.9\% + 0.2 \times 28.3\% + 0.2 \times 19.5\% + 0.2 \times 25.9\%$$

$$r_{portfolio} = 0.44\% + 2.98\% + 5.66\% + 3.90\% + 5.18\% = 18.2\%$$

By investing in this portfolio, you would have earned the weighted-average annual return (gain) of 18.20%. The result of investing in a portfolio of stocks is to "average-out" the best and worst performing stocks. 2014 was a good year for stocks in general, and none of these 5 stocks suffered a loss. However, if one or several of these stocks had incurred a loss during the year, the loss would be at least partially offset by the other stocks in the portfolio.

Maintaining Portfolio Allocation with Rebalancing

Often, investors want to allocate a certain percentage of their investment to each stock within their portfolio. As individual stocks rise and fall over time, the actual allocation will diverge from the initial desired allocation. *Rebalancing* is the process of selling or buying shares to return to the desired weighting in each stock.

For example, consider a portfolio made up of only two stocks, A and B. Suppose you begin the year 2014 with $100 each in A and B, so that each one makes up 50% of your portfolio, and you want to keep it that way.

After one year, stock A had a gain of 20%, whereas stock B had a loss of 10%. Your portfolio would have $120 worth of stock A, and only $90 worth of stock B, which means the weighting would be 57% in stock A and 43% in stock B. To rebalance, you would sell $15 worth of A, and buy an additional $15 worth of B. After rebalancing, the portfolio returns to the 50/50 weighting, with $105 invested in each stock as you begin the second year.

Measuring Portfolio Risk

Portfolio risk is measured by the standard deviation of the portfolio's returns. The standard deviation of a portfolio depends on the standard deviation of returns for each asset, the weightings allocated to each asset, and the correlations between those assets.[62]

When two or more stocks that are not perfectly correlated are combined into a portfolio, the portfolio will have a lower standard deviation than the average of its component's standard deviations. In a given period, when one stock loses value another might gain value, offsetting the loss. As a result, the rate of return on the portfolio will not have as many extreme values as any individual stock, and thus will have a lower standard deviation of returns.

The standard deviation of a portfolio describes the variability of the portfolio's historic rate of return around its mean annual return. Calculating the portfolio standard deviation is beyond the scope of this book, but the fundamental insight is simple and important. When the stocks in a portfolio are not perfectly correlated (i.e., the stocks do not all move in lock-step with each other), the standard deviation of the portfolio will be less than the weighted-average of the standard deviations of the individual stocks in that portfolio. This is because as some stocks go up in a given year, other stocks go down and thus some of the extremes cancel each other out.

Example: The 5-stock Portfolio

Let's return to the 5 individual stocks discussed above. By rebalancing each year, the portfolio's annual rate of return would be the arithmetic average of the 5 stocks' returns each year. Figure 11.2 shows the wealth indices for investing in the 5 individual stocks as well as the portfolio of 5 stocks, for the period 2005-2014.

[62] This calculation is a bit complex, and is beyond the scope of this text. Students interested in a detailed explanation of the portfolio variance should read
http://faculty.washington.edu/ezivot/econ424/portfolioTheoryMatrix.pdf

Figure 11.2 Annual Wealth Indices for Investment in the 5-Stock Portfolio, 2005-2014.

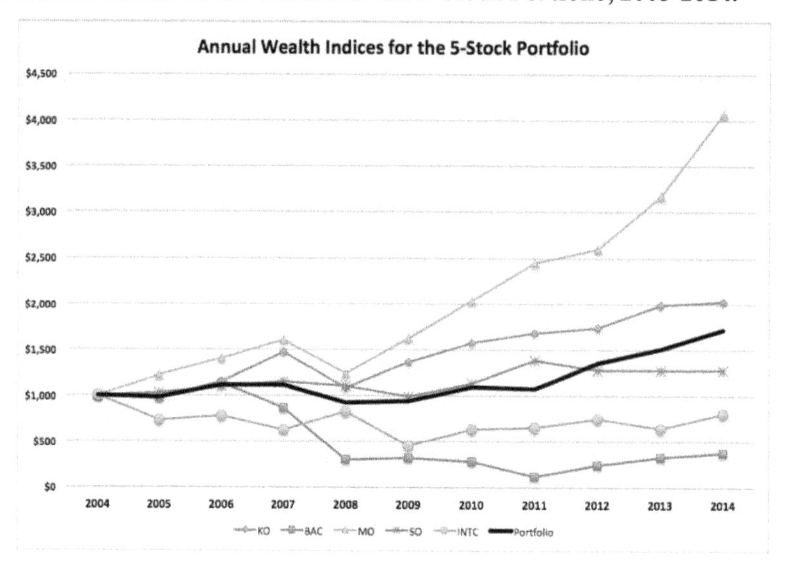

Any time you create a portfolio from stocks that are not perfectly correlated, it will have a lower standard deviation than the weighted-average of the component stocks' standard deviation. For example, the average of the standard deviations of the 5 stocks in the portfolio shown above (Figure 10.10) was about 22.6%. By contrast, the standard deviation of the *portfolio's* returns was about 11.8%. Investors benefit from diversification because some of the movements of individual stocks cancel each other out, resulting in a lower standard deviation for the annual returns of the portfolio as a whole.

Diversification and Risk

Let's explain why the portfolio of stocks had a lower standard deviation of returns, as compared to the standard deviations of the individual stocks.

Idiosyncratic risk (also known as *diversifiable risk*) is risk that is highly specific to one company (or one industry). For example, Coca-Cola's fortune is exposed to risks specific to its manufacturing and distribution operations, such as an increase in the cost of its raw materials or an increase in the cost of fuel for it's trucks. Apple Computer's business is exposed to risks about its manufacturing operations overseas as well as competition from new products that might unseat its market dominance. However, an increase in the cost of high-fructose corn syrup (one of the main ingredients in Coca-Cola) is not likely to have an effect on the fortunes of Apple Computer. Idiosyncratic risk can be eliminated by diversification.

Market risk (also known as *non-diversifiable risk*) is risk that is always present in the stock market and cannot be diversified away. All stocks, regardless of their specific business or industry, are subject to both macroeconomic conditions and the preferences of individual investors. Macroeconomic conditions, such as interest rates, inflation, unemployment, government regulation and international trade, affect all companies in the economy, but might have different effects on different companies' operations. The preferences of investors are influenced by fear and greed and are not systematically predictable.

As the number of stocks in the portfolio increases, the standard deviation of the portfolio decreases. Figure 11.3 illustrates this effect.

Figure 11.3 Portfolio Risk is Comprised of Idiosyncratic Risk and Market Risk

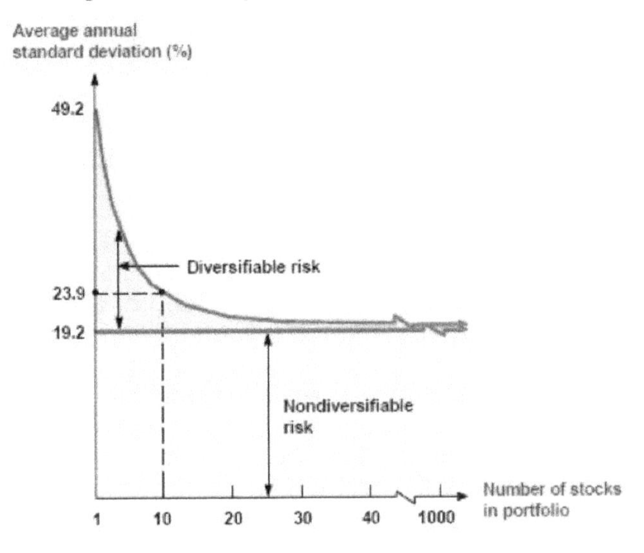

As the graph in Figure 11.3 indicates, a portfolio of as few as 30 stocks offers substantial diversification effects, i.e., it reduces the overall variability of returns as compared to a single stock. There is a decreasing marginal benefit from diversification. As the number of stocks continues to increase, the standard deviation of the portfolio becomes asymptotic toward the standard deviation of the stock market as a whole (i.e., systematic or market risk). At the limit, the most diversified portfolio would hold all of the stocks in the entire stock market in proportion to their market capitalization.

Great Minds in Applied Mathematics/Personal Finance: Harry Markowitz

Harry Markowitz (1927–) is a professor of Economics at the Baruch College of the City University of New York.

In the early 1950s, Markowitz developed portfolio theory, which describes the relationship among different assets held together in an investment portfolio. His work about diversification formalized the colloquial saying, "don't put all your eggs in one basket."

By measuring the riskiness (variability of returns) of individual securities, Markowitz created the methodology to combine securities that tend to move in opposite directions in a diversified portfolio. Modern portfolio theory allows construction of an optimal portfolio to achieve the maximum expected return for a given level of risk.[63]

Harry Markowitz was awarded the Nobel Memorial Prize in Economics in 1990 for *"pioneering work in the theory of financial economics"*.[64]

[63] Source: The Concise Encyclopedia of Economics, *Harry Markowitz*http://www.econlib.org/library/Enc/bios/Markowitz.html
[64] "The Sveriges Riksbank Prize in Economic Sciences in Memory of Alfred Nobel 1990". Nobelprize.org. http://www.nobelprize.org/nobel_prizes/economics/laureates/1990/

11.2 Modern Portfolio Theory and the Efficient Frontier

At the beginning of Module 8, we described the trade-off between investment risk and return. As an investor hopes to achieve a higher rate of return, she or he must also accept an increased amount of risk, i.e., uncertainty of returns.. Investors should not simply take on additional risk in the hopes of achieving a higher return, but rather should take only as little risk as is required for their expected level of return. *Modern Portfolio Theory* describes a process to maximize the expected investment return for a given level of risk (i.e., variability of returns).

Investors are Generally Risk Averse

The standard deviation is a proxy for the riskiness of the investor's rate of return. A *low standard deviation* indicates that the investor's returns will generally be close to the mean return (i.e., a smooth ride). A *higher standard deviation* would indicate that the investor's returns would include large swings in the rate of return (i.e., a bumpy ride). Psychologists have noted that people are generally risk-averse, i.e., they prefer to have a sure thing rather than a risky outcome. Thus, investors expect a greater rate of return to compensate for taking on additional risk.

Using this information, the next question becomes, "how should I allocate my investment among these stocks?" That is, what portfolio will have the lowest possible variability for a certain amount of expected return?

Creating Portfolios

Portfolios can be constructed with an infinite number of different weights, so there are also infinite possible portfolios. Given a set of investment choices, our objective is to determine the weights that *minimize portfolio risk* for a certain return expectation.

For example: if you have only 4 stocks from which to choose, you would want to know how to allocate your investment across those 4 stocks to achieve a specific expected return. Even with only 4 stocks, there are many possible portfolios that will achieve an expected return of 15%.

Given a set of investment choices, the *efficient portfolio* is the one that will achieve the expected rate of return with the lowest standard deviation.

The problem of finding the best possible portfolio weights is an example of constrained optimization. *Linear programming* is a method to achieve a best-possible outcome (e.g., highest expected rate of return given the standard deviation of returns) subject to some linear relationships (e.g., the weights of each individual asset in the portfolio).

The Efficient Frontier

The *efficient frontier* is the set of optimal portfolios that achieve the highest expected return at the minimal standard deviation. To construct the efficient frontier, we assume a set of assets in which to invest, the expected return and standard deviation for each asset, and the correlation coefficients among the assets. For each level of expected annual return (e.g., 10%) there is a portfolio (i.e., a series of weightings in each asset) that will achieve this expected return with the smallest possible standard deviation.

Figure 11.4 illustrates the plot of all the possible portfolios that we could achieve.[65]

[65] Image from http://toolsformoney.com/portfolio_optimization.htm.

Figure 11.4 The Efficient Frontier

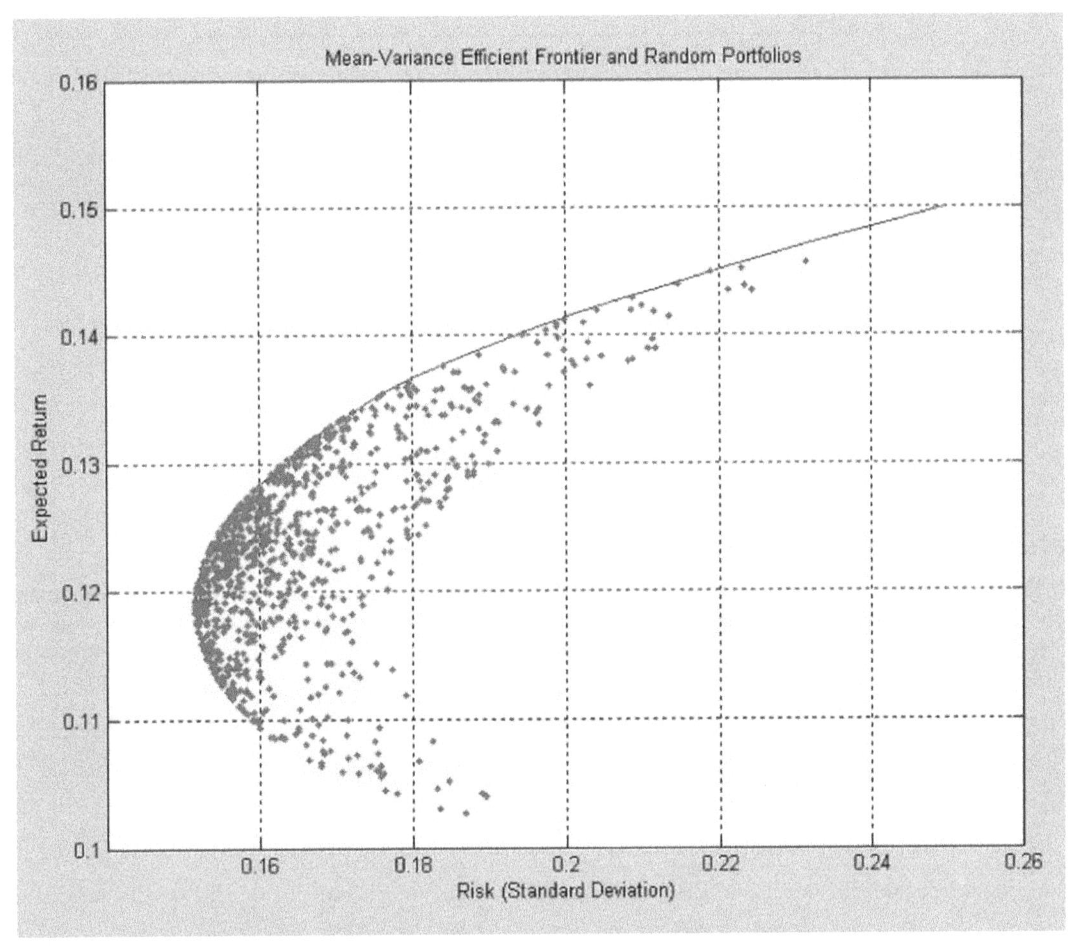

Note that using only stocks, there are no portfolios to the left of the efficient frontier. An investor cannot achieve a lower standard deviation while investing in stocks, because all stocks have market risk, which cannot be eliminated by diversification. Investors must choose stock portfolios from the frontier or to its right. However, choosing anything to the right of the frontier is not optimal (i.e., too much risk for the expected return).

Risk Free Assets and the Tangency Portfolio

The efficient frontier shows the optimal portfolios that involve taking on investment risk. Government bonds and certificates of deposit are generally considered to be risk free, at least with respect to the stock market.[66]By definition, risk free assets have a standard deviation of 0%.

An investor who wants to earn a rate of return higher than the risk free rate, but cannot tolerate as much risk as investing solely in stocks, could spread their investments between risk free assets and an optimal stock portfolio on the efficient frontier. We can plot a line between the risk free rate of return, with its standard deviation of 0% and the efficient frontier.

The *best possible capital allocation line* (CAL) is the line tangent to the efficient frontier that goes through the risk-free rate (see Figure 11.5). The *tangency portfolio* is the portfolio on the efficient frontier that is tangent to the risk-free rate of return. Any asset allocation on this line combines the risk-free asset and the tangency portfolio to achieve some expected rate of return at the lowest possible standard deviation.

[66] Treasury Bills and Certificates of Deposit are free of default risk, but still have inflation-risk. TIPS and I-Bonds provide protection against both default risk and inflation risk. None of these have stock market risk.

Figure 11.5 The Efficient Frontier and Tangency Portfolio

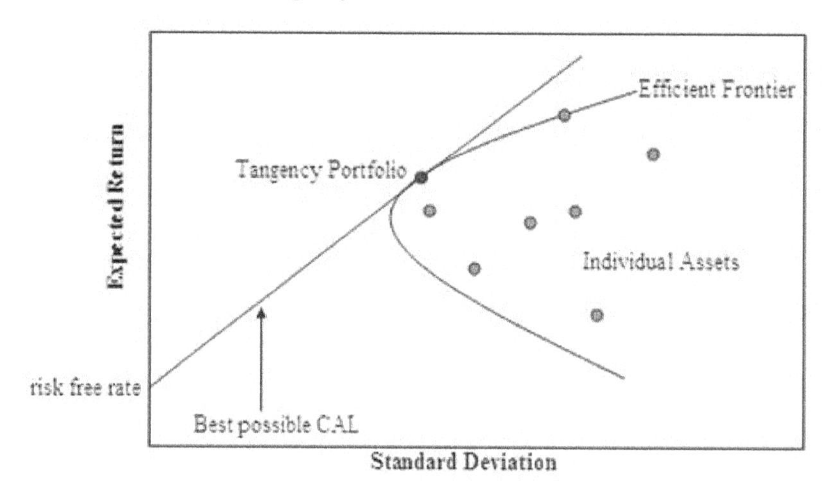

Implications for Investing

To invest along this best possible CAL, an investor would allocate their investment between the risk-free asset and the tangency portfolio. For example, suppose that the risk-free asset provides a return of 2% per year with 0% standard deviation, and that the tangency portfolio provides an expected return of 10%, with a 20% standard deviation. The table in Figure 11.6 illustrates some possible asset allocations along this best-possible CAL.

Figure 11.6 Possible Asset Allocations along the Best Possible CAL

w(risk-free)	w(stocks)	Expected Return	Standard Deviation
100%	0%	2.0%	0%
80%	20%	3.6%	4%
60%	40%	5.2%	8%
40%	60%	6.8%	12%
20%	80%	8.4%	16%
0%	100%	10.0%	20%

Some possible investors' asset allocations:

- A totally risk-averse investor could earn a safe, risk-free return of 2%, by allocating all of his or her wealth to the risk-free asset and none to stocks.
- A moderately risk-averse investor might allocate 60% to risk-free assets and 40% to stock. He or she would expect to earn a 5.2% annual rate of return with a standard deviation of 8%.
- A non-risk-averse investor might allocate only 20% to risk-free assets, and 80% to stocks. He or she would expect an 8.4% annual rate of return with a 16% standard deviation.
- It's even possible for a risk-seeking investor to invest more than 100% in stocks. Such an investor could borrow at the risk-free rate of return, and the amount borrowed in stocks. For example suppose an investor with $10,000 decides to borrow $5,000 (using a "margin" loan from his brokerage firm[67]), and invests $15,000 in stocks. This investor would expect to earn a 15% annual return, with a standard deviation of 30%.

Criticism of Modern Portfolio Theory

An important line of criticism of modern portfolio theory is that the correlation among the returns of different stocks (i.e., the extent to which stocks move together over time) is not constant. During financial crises, it has been the case

[67] Borrowing via a margin account is especially risky. If the stock portfolio loses value, the investor is still responsible to repay the margin loan in full. For example, if the portfolio suffers a 20% decline (from $15,000 to $12,000), the investor must still repay the $5,000 borrowed. This would leave him with only $5,000 ($10,000 - $5,000). While the margin loan amplifies the gains, it also amplifies the losses.

that virtually all stocks declined in lockstep together, leaving no safe havens for stock market investors. Since all stocks have market risk, the only way to reduce or eliminate market risk is by diversifying your investments out of stocks, i.e., putting a portion of your investment wealth into bonds or other investment vehicles.

11.3 The Historical Distribution of Stock Returns

In the previous section, we saw the benefits of broad diversification as a way to reduce the idiosyncratic risks of investing in stocks. The broadest possible diversification would be to own shares of every publicly traded stock in proportion to each stock's market capitalization.

The Standard and Poor's 500 Index (which comprises about 75% of the market capitalization of all publicly traded stocks in the United States) is a broad unmanaged portfolio of stocks. We can use the S&P 500 Index as a proxy for the entire stock market in the United States. The distribution of the real (i.e., net of inflation) annual rate of return on S&P500 index for the years 1926-2012 is shown in Figure 11.7.

Figure 11.7 Distribution of Historical Stock Real Returns for the S&P500 Index, 1926-2012[68]

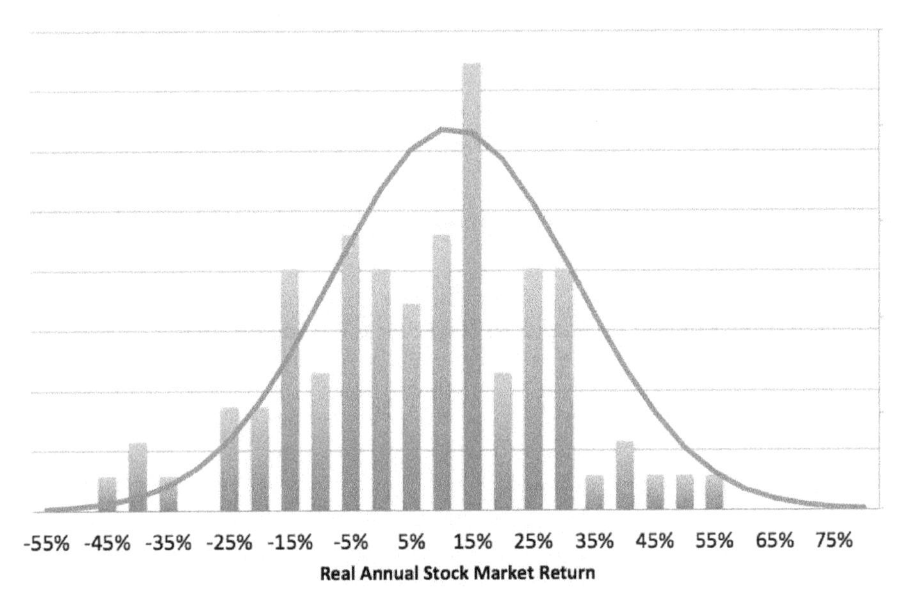

During this 87-year period, the real annual rate of return on the S&P500 ranged from a minimum of -41% in 2008 to a maximum of 59% in 1933. The distribution of stock returns follows an approximately normal distribution with a mean real return of about 8.7% and a standard deviation of 20.5%. During this period, the average real rate of return on U.S. Treasury Bonds was almost 3%, meaning that stock market investors earned a risk premium of about 6% annually.

The strongest prediction we can make about *future* stock returns is that we expect the distribution of future returns to be similar to the distribution of historical returns, i.e., having approximately the same mean and standard deviation of returns. We assume that the stock market will have approximately the same distribution of good years and bad years as it did in the past, which is by no means guaranteed. By accepting this assumption, we can make predictions about the range in which the annual rate of return might fall in any given year.

Confidence Intervals for Stock Returns

The historical rate of return on stocks follows an approximately normal distribution.[69] We can express a confidence interval about the real rate of return on stocks based on the historical mean rate of return of 8.7% and the standard deviation of 20.5%.

[68] Source: http://www.bogleheads.org/wiki/S%26P_500_index.

[69] The distribution of annual stock market returns exhibits too many extreme events to be considered normally distributed. Nonetheless, it is approximately normal and close enough for our assumptions.

About 68% of the time, we expect the real annual rate of return on stocks to be within the range of 8.7% ± 20.5%, or in the range of -11% to +29% (Figure 11.8).

Figure 11.8 A 68% Confidence Interval for Expected Real Stock Market Returns

Probability
-55% -45% -35% -25% -15% -5% 5% 15% 25% 35% 45% 55% 65% 75%
Distriubtion of Expected Stock Returns

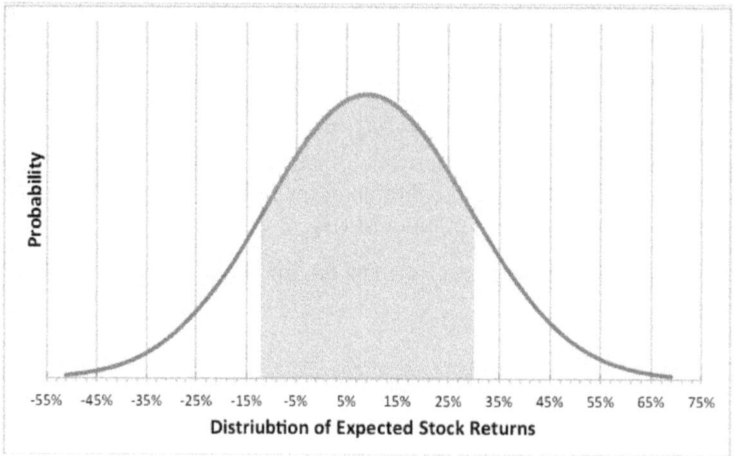

About 95% of the time, we expect the real annual rate of return on stocks to fall in the range of 8.7% ± 2(2.50%), or in the range of -32% to +49% (see Figure 11.9).

Figure 11.9 A 95% Confidence Interval for Expected Real Stock Market Returns

Probability
-55% -45% -35% -25% -15% -5% 5% 15% 25% 35% 45% 55% 65% 75%
Distriubtion of Expected Stock Returns

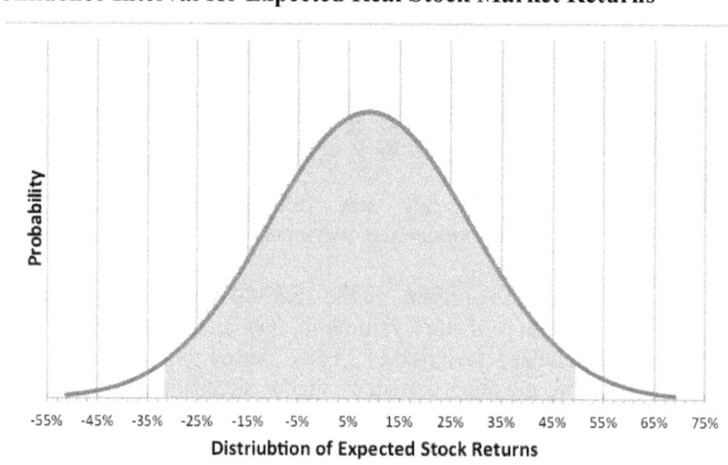

The extreme stock market returns of 2008 (-41%) and 1933 (59%) fall outside of this 95% confidence interval. Over the period 1926-2012, the real return on the S&P 500 index fell within the 95% confidence interval in 81 out of 87 years, i.e., 93% of the time. This is approximately in line with what we would expect, if stocks were in fact normally distributed.

We do not know for certain that stock returns will have a mean of 9% and a 20% standard deviation. Rather, we are assuming that the future will be similar to the historical record. In Module 10, we will use this historical record to develop a better understanding of the risk-return trade off for investors as we make estimates about the level of future wealth.

11.4 Summary

Diversification is a risk-management strategy that involves trading a large exposure to one risk for smaller exposures to many risks. A diversified portfolio of stocks helps to reduce the standard deviation of returns. In constructing a diversified portfolio, the correlation between stocks' returns helps us to identify assets that are not perfectly correlated with each other (and thus create better diversified portfolios).

By investing in several stocks that are not perfectly correlation with one another, it is possible to eliminate the idiosyncratic risk of investing in individual stocks. However, the systematic risk of the stock market cannot be eliminated through diversification.

Given a set of possible assets in which to invest and the correlation coefficients among those assets, an efficient portfolio is one that minimizes the expected standard deviation of the portfolio for the desired level of expected return. The efficient frontier is the set of all efficient portfolios. To maximize the risk-return tradeoff, investors should choose to invest along the top-half of the efficient frontier. The best possible capital allocation line runs between the risk-free asset and the efficient frontier. To obtain the best risk-return trade off, investors should select the level of stock-market risk they are willing to accept, invest in a combination of the risk-free asset and the tangency.

At the limit, the most diversified portfolio would hold all of the stocks in the entire stock market in proportion to their market capitalization (i.e., the way an index mutual fund would).

11.5 Review Questions

6. What is diversification? How does it work? What kind of risk does diversification eliminate?
7. Differentiate between idiosyncratic risk and market risk. Give examples of each kind of risk, and explain how one can manage (or reduce) each.
8. From a risk-management perspective, what is an efficient portfolio?
9. What is the efficient frontier?
10. Why should an investor select a portfolio along the best possible capital allocation line?
11. How can investors reduce their exposure to stock market risk?

12 The Truth About Risk in the Stock Market

Jessep: You want answers?
Kaffee: I want the truth!
Jessep: You can't handle the truth!
- A Few Good Men, written by Aaron Sorkin

Learning Objectives

- Discuss the conventional wisdom about why you should invest in stocks and the relationship between investment time horizon and risk of investing in the stock market.
- Develop a scientific/statistical understanding of the effect of your investment time horizon on the risk of investing in the stock market.
- Understand the idea of asset allocation as a risk-reduction strategy and discuss the shortcomings of this strategy.
- Consider the effects of a bad sequence of returns on your stock market wealth.
- Integrate investment in the stock market, Monte Carlo simulation, and the Life-Cycle model.

12.1 Conventional Wisdom: Stocks are not Risky in the Long Run

The conventional wisdom about investing in the stock market is that your choice of which asset(s) to invest in depends on your investment time horizon. If you are saving for a short-term goal (e.g., to purchase a car or save up a down payment for a house), you might not be able to weather the ups and downs of the stock market. For a short investment horizon, you should have your assets in bank accounts, CDs, or short-term Government bonds. However, if you have a long investment time horizon (i.e., retirement) you should invest in common stocks.

The conventional wisdom makes the case that, over the long run, investors who hold stocks will be rewarded by a risk premium, i.e., by taking on risk they will earn a higher rate of return than they would earn with risk-free investments. Countless personal financial advisors, investment trade books and textbooks, and even the Securities and Exchange Commissions' www.investor.gov website repeat this conventional wisdom.[70] The conventional wisdom is false. Despite the historical record, there is no law of nature that guarantees that this will hold true in the future.

The actual rate of return on stocks in any given year is effectively a random variable; it is unknown in advance. But more importantly, how much wealth you will have two years from now depends on how much wealth you have next year, and so forth. On account of randomness, the predictions we make about the distant future depend on the predictions we make in the near future. Wouldn't it be great to have a confidence interval around our long-run predictions?

A Weather Analogy: Predicting the Future one Step at a Time

Randomness is a factor in various fields of prediction. Weather forecasting provides a good example. Consider meteorologists predicting the path of a hurricane. The meteorologists at the National Hurricane Center have discovered that the path of a hurricane depends on a host of factors, such as its wind speed, direction, ocean and air temperatures. However, randomness is also a factor in the direction of the storm and as the time horizon increases, so does the uncertainty about the storm's path. Predictions about the storm's path have greater uncertainty as the time horizon increases. Meteorologists use a cone of uncertainty to illustrate the effect of randomness on their predictions (Figure 12.1).

[70] See for example: the discussion of risk and return at http://www.investor.gov/classroom/teachers/classroom-resources/risk-return.

Figure 12.1 The Path of a Weather System is Uncertain, and the Uncertainty Grows with the Time Horizon[71]

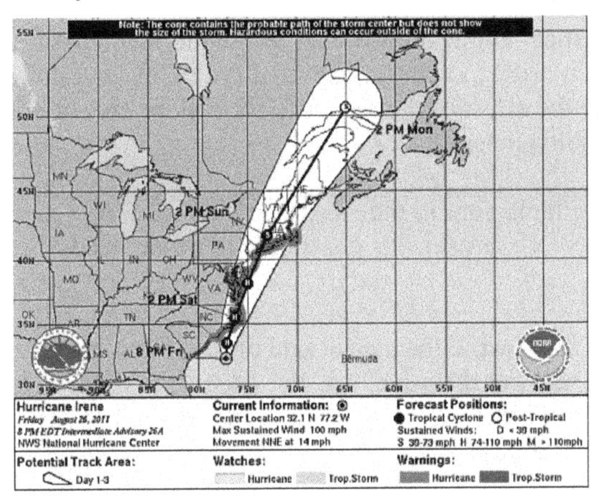

The graph in Figure 12.1 provides a *cone of uncertainty*, which expresses the NHC's 68% confidence interval about the storm's path (i.e., a 1-standard deviation confidence interval). As the time horizon increases, the cone of uncertainty widens due to the sequence of predictions. Mathematically, the *standard error* of the prediction grows in proportion to the square root of the time horizon. As the time horizon increases, so does the uncertainty about the ending location. As a result of randomness, it's difficult to make accurate predictions beyond several hours in advance; the path of the storm tomorrow depends on where it goes today.

Predicting Future Stock Market Wealth

The future wealth achieved by an investor in the stock market is similar to the future location of a hurricane. Over the short-run, predictions about the future path of the storm are likely to be reasonably accurate. Over a longer time horizon, the amount of final wealth depends on the sequence of multiple random events, which is discussed in the next section.

The best we can do is to think about the historical pattern of the distribution of annual stock returns. We can describe how annual stock returns were distributed in the past, and guess that the distribution of stock returns in the future will be similar.

[71] http://www.nhc.noaa.gov

12.2 Monte Carlo Simulation

Monte Carlo simulation is a statistical and computational technique involving a series of random events to make predictions. The technique is widely used in many disciplines, including investments. In particular, a Monte Carlo simulation can be used to illustrate the effect of the sequence of stock returns, which determines the amount of final wealth at the end of an investor's holding period.

Consider an investor who invests a lump sum in the stock market with a 30-year time horizon. What would his or her wealth be at the end of 30 years? To answer this question, we would need to know the annual rate of return for each year.

If we were doing this by hand, we could write the annual rate of return (on stocks) for each of the last 80 years on a little piece of paper, fold them up, and put them in a hat. For each year in our investment simulation, we draw one piece of paper, and use that rate of return for the year. We put the papers back into the hat to allow re-use for other years (i.e., a random draw with replacement).

Implementation of the Monte Carlo Process

The Monte Carlo simulation presented here will draw values (in this case, of annual stock returns) randomly. Instead of drawing little pieces of paper with stock returns on them out of a hat, we use a random-number generator to generate simulated stock returns. The simulated returns follow the historic distribution of real stock returns, which means they will be clustered around the mean historical return, and will be distributed randomly around that value based on the historical standard deviation of annual returns.

Recall that the *standard normal distribution* is a normal distribution with a mean of 0 and a standard deviation of 1 (see Figure 10.4). In the Monte Carlo process, we will use a random number generator to generate values according to the standard normal distribution, and then use these values to create simulated stock returns. Figure 12.2 provides an overview of the steps involved.

Figure 12.2 Overview of the Monte Carlo Process to Generate Simulated Stock Returns

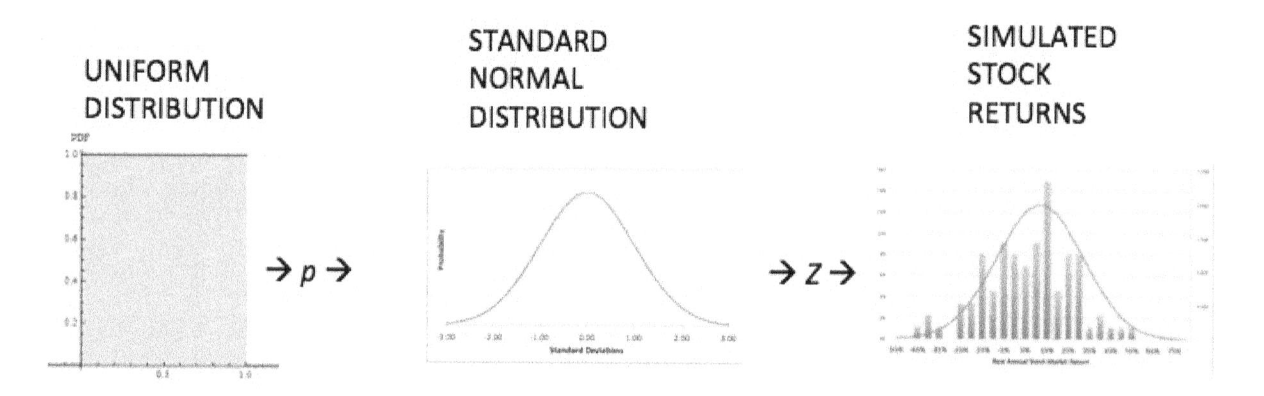

Each randomly drawn rate of return is calculated as:

$$return = \mu + Z\sigma$$

Where μ is the historical mean return, σ is the historical standard deviation, and Z is a random number that follows the probability density function of the standard normal distribution. Note that this calculation is very similar to the method to calculate the bounds of a confidence interval. In this case, we use a Z-score to determine the number of standard deviations above or below the mean to use for our random stock return.

The probability density function for the standard normal distribution allows us to select a probability level (between 0% and 100%) and find the corresponding Z-score. For example, there is a 50% probability of achieving a value below the mean, and this corresponds to a Z score of 0. A 10% probability level corresponds to achieving a value that is 1.28 standard deviations below the mean, i.e., a Z score of -1.28, and a 90% probability level corresponds to a value that is 1.28 standard deviations above the mean, i.e., a Z score of +1.28.

Excel Functions to Work with the Normal Distribution

In Excel, we can calculate the Z value using the NORM.S.INV() function. NORM.S.INV() finds the value of Z based a probability value in the range of (0,1). Figure 12.3 shows an example of using the NORM.S.INV() function.

Figure 12.3 Using the Excel NORM.S.INV Function

	A	B	C
1	mean	9%	
2	stdev	20%	
3			
4		UNIFORM	Z-SCORES
5	YEAR	RAND()	NORM.S.INV(p)
6	1	0.99	=NORM.S.INV(B6)

The Excel function RAND() draws a random number that is uniformly distributed over the range (0,1) and serves as the probability seed to the NORM.S.INV() function. Together, the expression NORM.S.INV(RAND()) will generate a random value from the standard normal distribution. After obtaining a randomly-generated Z-score, we can use this to generate a simulated stock return (Figure 12.4).

Figure 12.4 Generating a Simulated Stock Return in Excel

	A	B	C	D
1	mean	9%		
2	stdev	20%		
3				
4		UNIFORM	Z-SCORES	mu + Z*sigma
5	YEAR	RAND()	NORM.S.INV(p)	RETURN
6	1	0.99	2.46	=B1+C6*B2

We calculate a separate rate of return for each year in the hypothetical investment using the above formula. For a 30-year simulation, we would generate 30 separate rates of return, one for each year. Finally, we use these randomly generated rates of return to find the evolution of an investor's wealth over time.

Sample Results

A stock market investor should think about his or her investment horizon like the score in a basketball game. It's fun to score some points early in the game or to be leading at halftime, but what really matters is the score at the end of the game. Similarly, the results of an investment simulation can be compared at any point along the time horizon, but the ending wealth is what is most important to support future consumption.

The simulation calculates the ending wealth for an investor who invests $100 at the start of the investment period and holds that investment for all 30 years. The most important question we want to address is how the investment in stocks compares to alternatives. In particular, we need to compare the simulated future wealth achieved by investing in stocks against a risk-free portfolio invested in Treasury Inflation-Protected Securities (TIPS). Historically, TIPS have earned a real return of about 3% above in the inflation rate.[72] An investor who earns a 3% real return for 30 years will see $100 invested grow to a future value $242:

$$FV = PV(1 + r)^n = \$100(1.03)^{30} = \$242.72$$

Does an investment in the stock market result in higher ending wealth than investing in TIPS?

[72] As of this writing in 2013, the real yield on long-term TIPS was about 1%. The long run average real rate of return on Treasury bonds has been closer to 3%. Source: http://www.federalreserve.gov/releases/h15/data.htm.

Figure 12.5 shows a sample run of the Monte Carlo simulation. In this example, the simulation run was a lucky draw, and achieved ending wealth after 30 years of $646. However, it failed to match actual investment returns from the S&P 500 index during the period 1976-2005.

Figure 12.5 Sample Run of the Monte Carlo Simulation

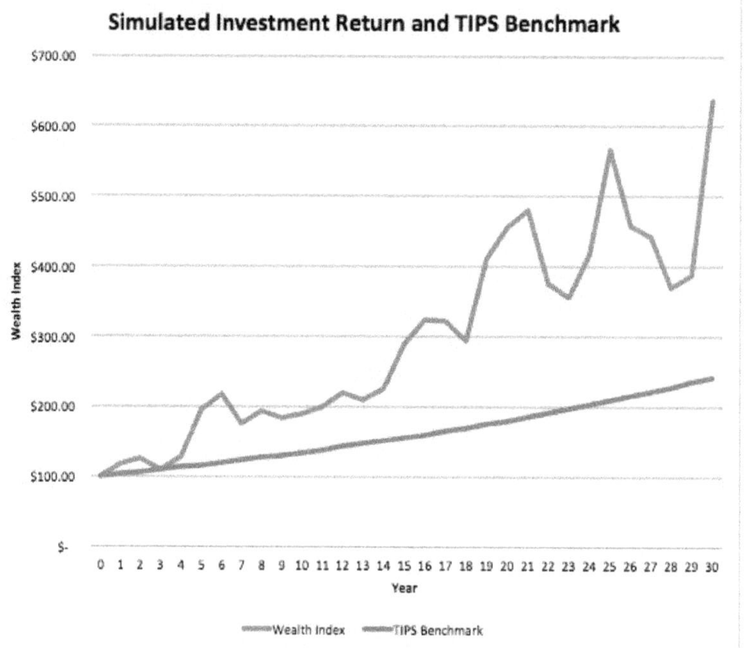

However, the graph in Figure 12.6 shows the result of a less fortunate simulation run, with an ending real wealth of about $141 – missing the $242 benchmark of the all-TIPS portfolio.

Figure 12.6 Sample Run of the Monte Carlo Simulation

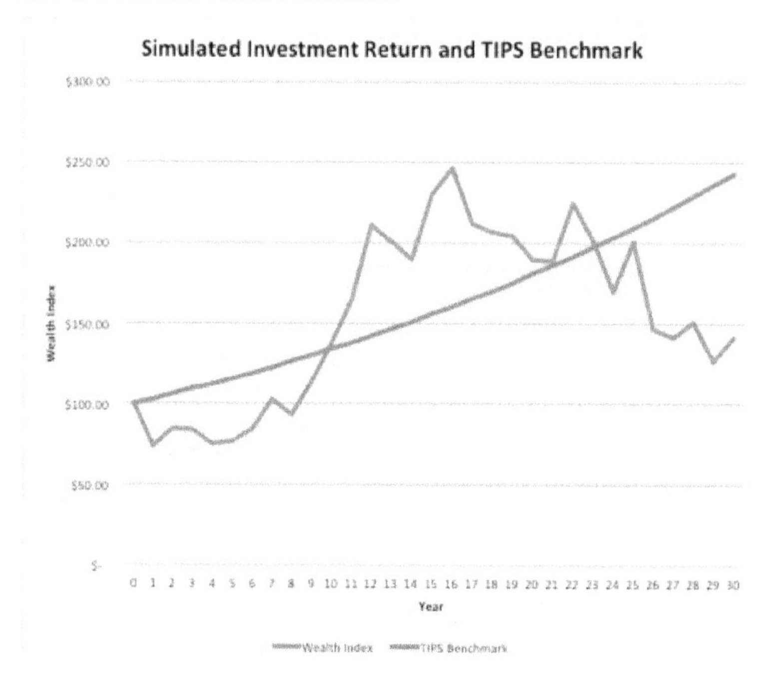

Many Random Trials

One random trial is hardly indicative of what an individual investor should expect for future wealth. The power of the Monte Carlo simulation technique is being able to run a large number of trials and make observations about the distribution of the outcomes. Figure 12.7 illustrates 10 Monte Carlo trials, along with the TIPS portfolio as a benchmark.

Figure 12.7 Monte Carlo Simulation with 10 Random Trials

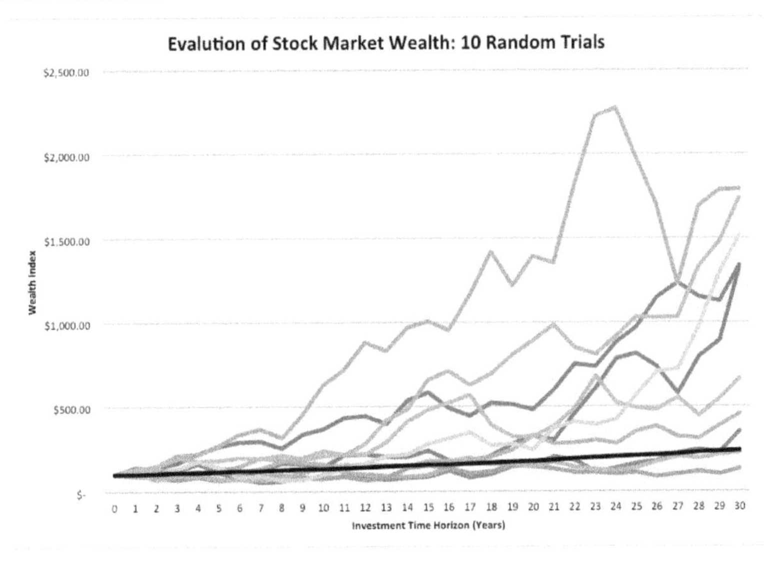

Even with only ten trials, we can see the wide distribution of the final wealth at the end of the investment time horizon. The ending wealth achieved in each of these ten trials range from less than $50 to more than $1,700. The results presented in Figure 12.8 are the same annual returns as in Figure 12.7 but are shown with a logarithmic scale.

Figure 12.8 Monte Carlo Simulation with 10 Random Trials (Logarithmic Scale)

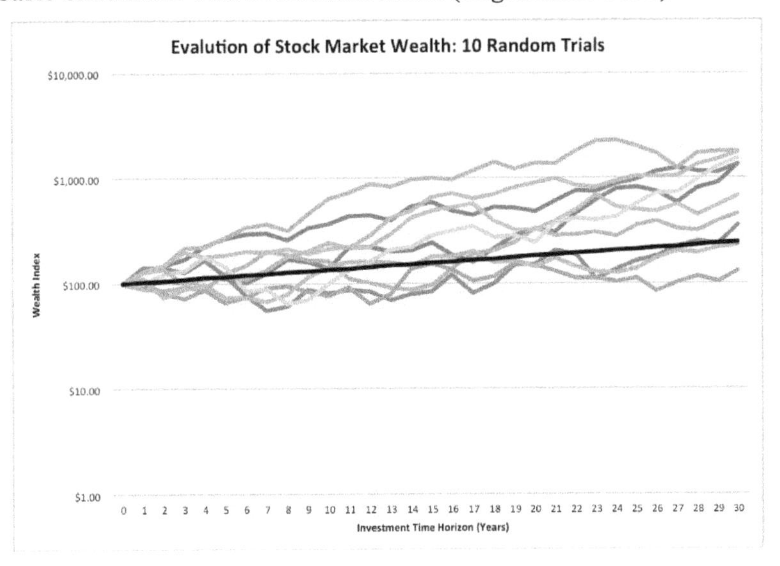

In Figure 12.8, notice how the vertical axis has reference marks in increasing orders of magnitude (i.e., $1, $10, $100). The logarithmic scale makes it easier to visualize the distribution of returns without the scale being biased by outliers. As well, notice what happens to the benchmark portfolio of all TIPS (e.g., a 3% real rate of return): on a logarithmic scale, the exponential growth of compounding at a constant rate of interest plots as a straight line.

Why is there so much variation among the trails?

The annual wealth achieved by investing in stocks depends on the rate of return achieved in any particular year, but also on the previous years' rates of return. Several good years in a row result in a tremendous amount of compounded growth. Several bad years in a row can decimate an investor's wealth. Moreover, the amount of ending wealth (after 30 years) depends on the returns earned in the earlier years as well as the sequence of returns.

Risk and Time

With 50 trials, we begin to see a more pronounced pattern of possible outcomes (Figure 12.9).

Figure 12.9 Monte Carlo Simulation with 50 Random Trials (Logarithmic Scale)

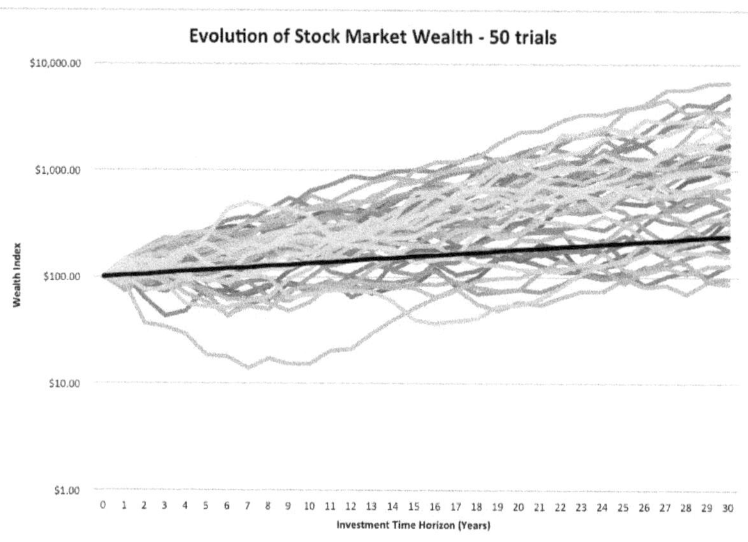

It is important to note that none of these random trials are predictive of a *specific* outcome that will actually be achieved in the future. Nonetheless, the distribution of possible outcomes can help us understand how a stock market investor would do compared to a risk-free investment in TIPS.

The majority of the trials show the ending stock market wealth that exceeds the TIPS benchmark, sometimes by orders of magnitude. In this set of 50 trials, the ending wealth achieved by investing in stocks fell short of the TIPS benchmark 8 times in 50. In these cases, the investors would be worse off after 30 years by investing in stocks – possibly much worse off.

Increasing the number of random trials (i.e., years in the simulation) widens the range of possible outcomes. This is similar to the cone of uncertainty that we saw with the hurricane example (Figure 12.1).

Using Confidence Intervals to Estimate Future Stock Market Wealth

We can describe the distribution of possible future wealth using confidence intervals. Consider an initial investment of $100 and stock returns drawn randomly from a distribution with a mean real rate of return of 9% and standard deviation of 20% per year. Figure 12.10 provides some confidence interval bounds to describe the likely amount of future wealth over a 30-year time horizon.

Figure 12.10 Confidence Intervals for Future Investment Returns

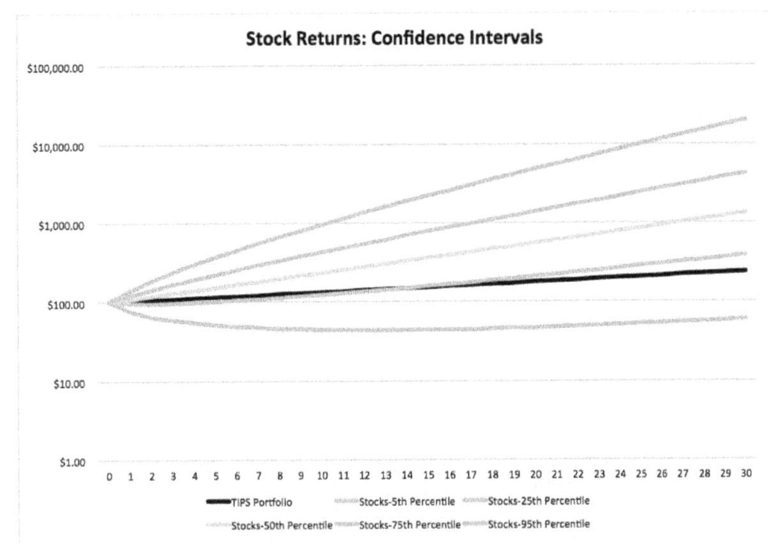

As the time horizon increases, the standard deviation of the compounded rate of return grows with the square root of time. Whereas the distribution of outcomes in one year is described by the mean (μ) and standard deviation (σ), the distribution of wealth in n years is described by the mean μ and standard deviation $\sigma\sqrt{n}$.

The implication of these confidence intervals (Figure 12.10) is that the risk of stock market investment growth with the time horizon. The conventional wisdom, that a long time horizon makes it safe to invest in stocks, is false. Time does not diversify stock market risk. Rather, a long time horizon amplifies stock market risk.

Distribution of Final Wealth

What is the chance that you will miss your investment objective, and not have enough wealth to support your future consumption (i.e., in old age)? The traditional way of measuring investment risk[73] fails to measure the true economic risk to an individual investor. Figure 12.11 addresses this question by plotting a frequency distribution of 1,000 sample runs from the Monte Carlo simulation.

[73] The variability of returns (i.e., the standard deviation) is often used as a proxy for riskiness, such that a higher standard deviation indicates greater risk.

Figure 12.11 Results of 1,000 Sample Runs of the Monte Carlo Simulation

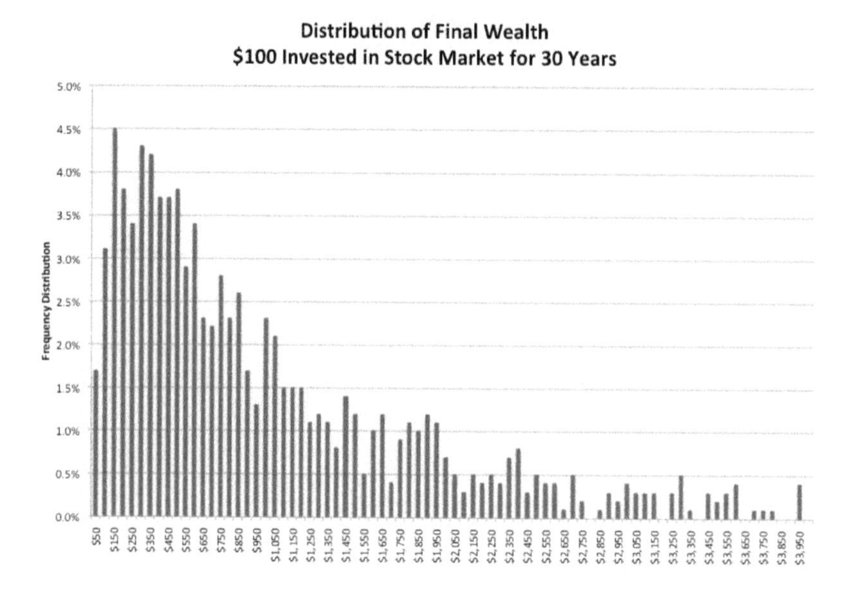

After 1,000 trials, each consisting of 30 years of investment returns drawn randomly, we see that about 17% of the time investing in stocks fails to beat the final wealth from the 3% TIPS benchmark. That is, even if you have a 30-year time horizon, there is about a 17% chance that investing in stocks will leave you worse off than investing in TIPS.

What should investors do about the risk of investing in stocks? Asset allocation strategies (presented below) can help to mitigate the risk of stock market investment by systematically reducing an investor's exposure to stocks.

12.3 Asset Allocation

Even without detailed statistical knowledge, investors intuitively know that stocks provide a wild ride through periods of gains and losses. *Asset allocation* is the practice of spreading your assets across multiple asset classes, such as stocks, bonds, cash, real estate, or precious metals.

An investor's *weighted-average rate of return* is the rate of return for each asset class, times the percentage weight of the portfolio held in that asset class. For a portfolio consisting of only stocks and bonds, the portfolio return (i.e., the weighted average rate of return) is:

$$portfolio\ return = w_s \times r_s + w_b \times r_b$$

where w_s is the weight (i.e., the percentage of the portfolio) held in stocks and r_s is the rate of return on stocks, and w_b is the weight held in bonds and r_b is the rate of return on bonds.

There are some diversification benefits from having your assets spread across multiple asset classes, so long as the annual returns from these different assets classes are not perfectly correlated. For example, if you allocate your wealth equally between stocks and bonds, losses in your stock investment might be offset in whole or in part by gains in your bond investment, or vice-versa.

A portfolio's risk and return characteristics are dependent on asset allocation. An investor who wishes to increase his or her expected rate of return should increase the allocation to risky assets (i.e., stocks) and simultaneously reduce allocation to safe assets (i.e. TIPS). On the other hand, an investor who wishes to reduce risk exposure should reduce his or her allocation to risky assets (i.e., stocks) and simultaneously increase the allocation to risk-free assets (i.e., TIPS).

Many investors end up buying and selling stocks as an emotional reaction to events beyond their control (i.e., greed leads to buying and fear leads to selling). To remove the emotional component from investing, investors decide on an asset allocation plan and stick to it irrespective of the ups and downs in the market.

Periodically, as the value of each asset in the portfolio rises or falls, the asset allocation will diverge from the target allocation. Rebalancing within a portfolio is the process of selling or buying within specific asset classes to return to your target asset allocation. Annual rebalancing is common, and quarterly rebalancing has additional benefits for reducing risk due to changes in asset allocation away from the target allocation, so long as the transaction costs of rebalancing are sufficiently small.

Rebalancing Example

Suppose that an investor has a 50/50 (stocks/bonds) asset allocation as of January 1, 2012, with $250,000 invested in stocks and $250,000 invested in bonds. As of January 1, 2013, the portfolio consists of $285,000 invested in stocks and $257,500 invested in bonds, for a total of $542,500. The annual rate of return on the stocks in the portfolio was:

$$r_s = \frac{\$285{,}000}{\$250{,}000} - 1 = 14\%$$

and the annual rate of return on the bonds in the portfolio was:

$$r_b = \frac{\$257{,}250}{\$250{,}000} - 1 = 3\%$$

The investor's weighted average rate of return was:

$$portfolio\ return = w_s \times r_s + w_b \times r_b$$
$$portfolio\ return = 0.5 \times 14\% + 0.5 \times 3\%$$
$$portfolio\ return = 0.07 + 0.015 = 0.085 = 8.5\%$$

As of January 1, 2013, the investor's asset allocation would be $\$285{,}000/\$542{,}500 = 52.5\%$ in stocks and $\$257{,}500/\$542{,}500 = 47.5\%$ in bonds. As a result of the investment returns, the portfolio has drifted away from its target asset allocation.

To return to the target 50/50 asset allocation, the investor would want to have $0.5(\$542{,}500) = \$271{,}250$ invested in stocks and $\$271{,}250$ invested in bonds. To rebalance the portfolio, the investor would sell $\$285{,}000 - \$271{,}500 = \$13{,}500$ worth of stocks, and buy an additional $\$13{,}500$ worth of bonds.

Asset Allocation Strategies

Many different asset allocation strategies exist, to suit investor's individual needs. An infinite number of gradations of asset allocation are possible. Figure 12.12 shows some examples.

Figure 12.12 Sample Asset Allocation Strategies

Strategy	Objective	Sample Weighting	Riskiness
Aggressive Growth	Grow the investor's capital as quickly as possible	100% in stocks	Very Risky
Growth and Income	Grow the investor's capital while generating some current income.	50% in stocks 50% in TIPS	Moderately Risky
Capital Preservation and Income	Preserve the investor's capital while generating current income.	80% in TIPS 20% in money market (i.e., near cash)	Risk Averse

Rule of Thumb: 100 Minus Your Age

Given the volatility of returns in the stock market, many financial advisors often recommend an age-based asset allocation. The idea is simple enough: as investors become more risk averse with age, they should systematically reduce risk exposure by gradually changing their asset allocation to have a smaller proportion in stocks. As you age, this strategy will reduce your weighted-average expected returns and the expected standard deviation of the portfolio's returns.

A common rule of thumb is to invest a percentage of your assets in stocks equal to 100 minus your age. In recent years, some sources have suggested 110 or 120 minus your age.[74] The rule of thumb is based on the idea of reducing risk exposure as you age, but the selection of 100 or 110 or 120 minus age is arbitrary. While the 100 minus your age asset allocation is easy to calculate, it creates an administrative burden that many small investors might not want to undertake on their own. The financial services industry has responded to this perceived need by creating ***target-date mutual funds***.[75]

Evidence from Target-Date Mutual Funds

How has the asset-allocation glide path worked for investors nearing retirement? In the recent 2008 stock market decline, losses in the 2010 target-date funds were in the range of 9%-41%.[76] Investors who were only a couple of years away from retirement experienced substantial declines in their retirement wealth. Since the funds continued to re-allocate assets away from stocks, this had the effect of selling out positions at a loss. Many investors responded by deciding to delay retirement, or by accepting they would have to consume less in retirement as a result of investment losses. While the glide path describes the asset allocation, it fails to describe the risk of loss.

Asset Allocation and Value at Risk

It is possible to reduce the standard deviation of returns for an overall portfolio by reducing the exposure to risky assets and increasing the exposure to safe assets (i.e., converting assets from stocks to TIPS). A portfolio of all TIPS will have very low annual standard deviation of returns, as compared to the stock market portfolio. However, investors are not merely concerned with the standard deviation of returns. Simply put: investors do not want to lose wealth (or at least not much), and especially do not want to reduce their standard of living due to investment losses.

Due to the accumulation of assets as you age, you will have a lot more wealth when old than when young. A young investor might begin their investing career at age 22 with assets of $0, and accumulate assets by saving several thousand dollars each year. An investor nearing retirement might have accumulated assets of $1,000,000 or more by age 65.

A financial metric called the ***value at risk*** is the amount of wealth you might lose in the case of a bad event in the stock market. Let's define a bad event as one that causes an annual rate of return in the stock market that is 2 standard deviations below the mean return. We would expect such a bad event to occur about once in 40 years.[77] Recall that the historical mean return on stocks during the years 1926-2012 was about 9% above inflation, and that the annual standard deviation was about 20% per year. Thus, in the case of a bad event in the stock market, we would expect an annual return of:

$$r_s = \mu + Z\sigma$$
$$r_s = 0.09 + (-2)(0.2) = -31\%$$

[74] See for example: http://money.cnn.com/retirement/guide/investing_basics.moneymag/index7.htm

[75] Target-date mutual funds promise to reduce the allocation to stocks in a systematic way so as to reduce the investor's stock market exposure by the targeted retirement date. In recent years, target-date mutual funds have become very popular in 401(k) retirement accounts, with over 40% of workers selecting a target-date fund if one was available in their plan. In many cases a target-date fund is the default investment option.

[76] https://www.sec.gov/spotlight/investor-advisory-committee-2012/iac-recommendation-target-date-fund.pdf

[77] If stock returns are normally distributed, about 19 out of 20 years the annual stock returns will be within 2 standard deviations above or below the mean. About 1 time out of twenty, the annual stock return will be outside this 2 standard deviation range. We can thus assume that about 1 time in 40 years it will be more than 2 standard deviations above the mean, and about 1 time in 40 years it will be more than 2 standard deviations below the mean.

Put differently, a bad event in the stock market will result in losing about 31% of your stock market assets. Based on the distribution of historic stock market returns, most investors will experience a bad event at least once in their investing career. We cannot predict if or when such a bad event will happen – it might occur when you are young and have very little financial wealth, or might occur when you are nearing retirement and have a lot of financial wealth. You might experience multiple bad events in a row, or none at all.

The severity of the bad event on your overall portfolio will depend on your asset allocation, and on your total amount of invested wealth. The value at risk in dollars is defined by:

$$value\ at\ risk = A \times w_s \times r_s$$

where A is the amount of invested assets, w_s is the weighting or proportion of assets held in stocks, and r_s is the rate of return on stocks from a bad event.

Value at Risk Examples

Consider a 25-year-old investor with $10,000 of financial assets, who has a 75% allocation to stocks, and 25% to TIPS. What would be his value at risk, in the case that a 2 standard deviation "bad event" occurs in the stock market? We expect that stocks would lose 31% of their value, and with a 75% exposure to stocks, his loss would be about:

$$value\ at\ risk = \$10{,}000 \times 0.75 \times -0.31 = \$2{,}325$$

For a 25-year-old with $10,000 of retirement assets, we can estimate that his value at risk is about $2,325.

Now consider the case of a 65-year-old investor with $1,000,000 of financial assets in a portfolio allocated 35% in stocks and 65% in TIPS. Again, we would expect that a 2 standard deviation bad event would result in a 39% loss in her stock portfolio. With a 35% exposure to stocks and a 31% loss, we would expect her portfolio to lose about:

$$value\ at\ risk = \$1{,}000{,}000 \times 0.35 \times -0.31 = \$108{,}500$$

The investor with a lower allocation to stocks would expect to have a smaller loss in percentage terms. However, the value at risk approach makes it clear that the magnitude of the loss is much greater to the investor with more financial assets. The risk to one's living standard is small for the 25-year-old, who has many years in which to earn an income and accumulate more assets. On the other hand, the severity of the loss is further increased for the 65-year-old, who has little (or no) human capital, and *cannot* recover from such a loss.

There are many unknowns about future stock market returns due to the effects of randomness. We do not know if a bad event will occur; what the severity of such a bad event might be; when it might occur; and whether there might be multiple bad events in successive years. However, expressing the potential loss to a portfolio in dollars helps to quantify the magnitude of the risk.

12.4 Living Standard Risk

From our Monte Carlo simulations, we found evidence that investing in the stock market is risky, especially in the long run. Asset allocation strategies intended to reduce the risk of stock market investment are conceptually appealing, but leave investors with too much value at risk. A bad event of a large enough magnitude that occurs near retirement could result in a drastic reduction of wealth just when your human capital is used up.

The Sequence of Returns

The *sequence of returns* refers to the order in which the rates of return are achieved, such as whether you have gains followed by losses, or losses followed by gains. For example in a 3-year investment horizon you could have any of these possible sequences of returns: [11%, -9%, 7%], or [-9%, 7%, 11%], or [7%, 11%, -9%]. Although the same rates of returns are achieved in different sequence, the average return over the 3-year period remains 3% per year and the compounded annual rate of return remains 2.624%.

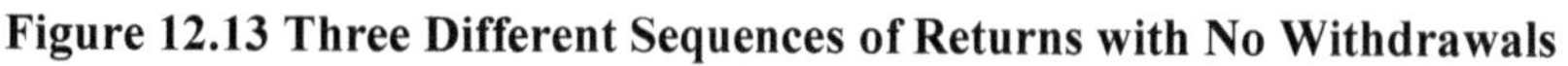

When the Sequence of Returns Does Not Matter

Suppose that an investor held a portfolio of stocks for a period of ten years, during which time he or she did not contribute additional funds, or make any withdrawals. For the investor who remains *fully invested* for the entire period, the sequence of returns does not change the end result (Figure 12.13).

Figure 12.13 Three Different Sequences of Returns with No Withdrawals

Put differently: if you will remaining invested and do not need to withdraw the funds, it does not matter whether you have good years followed by bad years, or bad years followed by good years.

When the Sequence of Returns Matters

When an investor is making contributions or withdrawals, the sequence of returns will have a significant impact on the amount of final wealth. For example, consider an investor who is making withdrawals from the portfolio during retirement. Poor returns during the first few years, combined with withdrawals, could lead to premature depletion of the assets.A bad sequence of returns can have a severe impact on his or her standard of living.

For example, suppose an investor reaches retirement age with $1,000,000 of financial assets. The investor wants to withdraw $60,000 per year to spend on consumption. The remainder of these assets will stay invested in stocks. Will this plan be sustainable?

Figure 12.14 Spending Down $1,000,000 with Good Sequence of Returns

1	MONTE CARLO SIMULATION				
2	Seeds for simulated stock return				
3	mean	9%			
4	stdev	20%			
5					
6	Initial Wealth		$1,000,000		
7	Annual Withdrawal		$60,000		
8					
9	Year	r(stocks)	Change in Wealth	Withdrawal	Wealth
10					$1,000,000
11	1	21%	$205,528.39	($60,000)	$ 1,145,528
12	2	5%	$60,149.35	($60,000)	$ 1,145,678
13	3	26%	$294,506.97	($60,000)	$ 1,380,185
14	4	7%	$100,846.70	($60,000)	$ 1,421,031
15	5	19%	$264,199.54	($60,000)	$ 1,625,231
16	6	-5%	($87,407.53)	($60,000)	$ 1,477,823
17	7	4%	$62,688.62	($60,000)	$ 1,480,512
18	8	11%	$157,309.48	($60,000)	$ 1,577,822
19	9	39%	$616,677.47	($60,000)	$ 2,134,499
20	10	21%	$447,827.77	($60,000)	$ 2,522,327
21	11	16%	$392,460.17	($60,000)	$ 2,854,787
22	12	-1%	($15,060.87)	($60,000)	$ 2,779,726
23	13	17%	$464,081.16	($60,000)	$ 3,183,807
24	14	-19%	($611,503.00)	($60,000)	$ 2,512,304
25	15	-10%	($260,389.88)	($60,000)	$ 2,191,914
26	16	-27%	($592,512.20)	($60,000)	$ 1,539,402
27	17	-11%	($165,354.27)	($60,000)	$ 1,314,048
28	18	-6%	($83,250.20)	($60,000)	$ 1,170,798
29	19	40%	$468,468.28	($60,000)	$ 1,579,266
30	20	9%	$137,444.52	($60,000)	$ 1,656,710
31	21	4%	$65,057.06	($60,000)	$ 1,661,768
32	22	-39%	($643,979.44)	($60,000)	$ 957,788
33	23	5%	$48,243.65	($60,000)	$ 946,032
34	24	-13%	($119,618.96)	($60,000)	$ 766,413
35	25	37%	$281,055.17	($60,000)	$ 987,468
36	26	22%	$216,815.87	($60,000)	$ 1,144,284
37	27	-1%	($12,947.09)	($60,000)	$ 1,071,337
38	28	1%	$10,906.43	($60,000)	$ 1,022,243
39	29	9%	$92,073.30	($60,000)	$ 1,054,316
40	30	-20%	($206,071.16)	($60,000)	$ 788,245

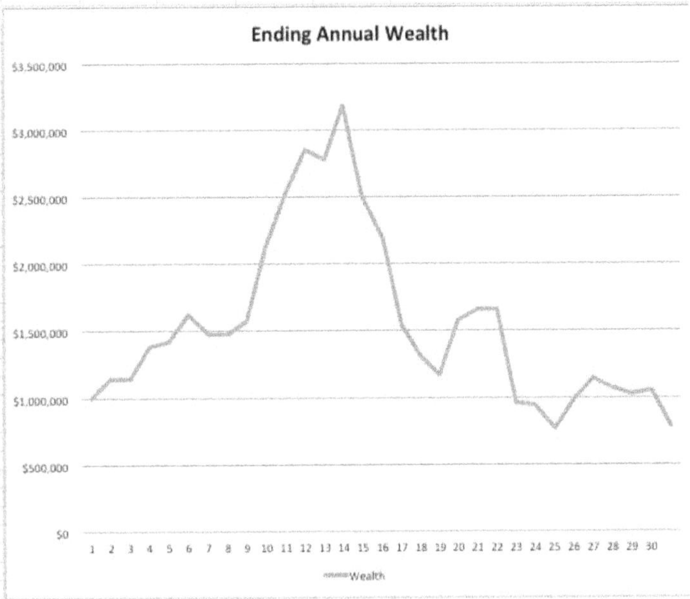

The graph in Figure 12.14 illustrates the result of a favorable sequence of returns. In this case, spending $60,000 per year (6% of initial assets) appears to be sustainable – in fact, this leaves the investor with over $1,000,000 of assets after 30-years of retirement.

Now consider the graph in Figure 12.15, in which the investor suffers a from poor stock market returns in the initial years of retirement.

Figure 12.15 Spending Down $1,000,000 with Poor Sequence of Returns

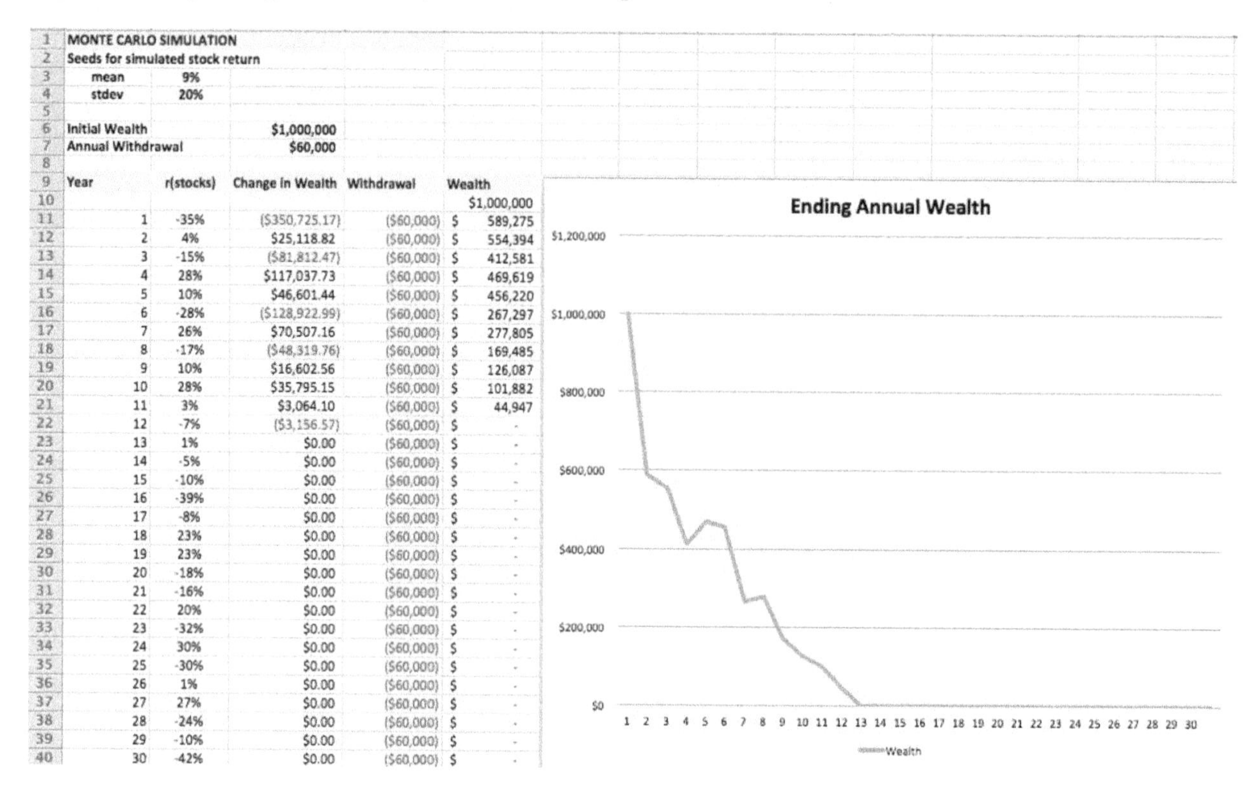

MONTE CARLO SIMULATION
Seeds for simulated stock return
mean 9%
stdev 20%

Initial Wealth $1,000,000
Annual Withdrawal $60,000

Year	r(stocks)	Change in Wealth	Withdrawal	Wealth
				$1,000,000
1	-35%	($350,725.17)	($60,000)	$ 589,275
2	4%	$25,118.82	($60,000)	$ 554,394
3	-15%	($81,812.47)	($60,000)	$ 412,581
4	28%	$117,037.73	($60,000)	$ 469,619
5	10%	$46,601.44	($60,000)	$ 456,220
6	-28%	($128,922.99)	($60,000)	$ 267,297
7	26%	$70,507.16	($60,000)	$ 277,805
8	-17%	($48,319.76)	($60,000)	$ 169,485
9	10%	$16,602.56	($60,000)	$ 126,087
10	28%	$35,795.15	($60,000)	$ 101,882
11	3%	$3,064.10	($60,000)	$ 44,947
12	-7%	($3,156.57)	($60,000)	$ -
13	1%	$0.00	($60,000)	$ -
14	-5%	$0.00	($60,000)	$ -
15	-10%	$0.00	($60,000)	$ -
16	-39%	$0.00	($60,000)	$ -
17	-8%	$0.00	($60,000)	$ -
18	23%	$0.00	($60,000)	$ -
19	23%	$0.00	($60,000)	$ -
20	-18%	$0.00	($60,000)	$ -
21	-16%	$0.00	($60,000)	$ -
22	20%	$0.00	($60,000)	$ -
23	-32%	$0.00	($60,000)	$ -
24	30%	$0.00	($60,000)	$ -
25	-30%	$0.00	($60,000)	$ -
26	1%	$0.00	($60,000)	$ -
27	27%	$0.00	($60,000)	$ -
28	-24%	$0.00	($60,000)	$ -
29	-10%	$0.00	($60,000)	$ -
30	-42%	$0.00	($60,000)	$ -

In this example, the investor has substantial losses in the first several years of retirement. As a result, their retirement wealth is quickly lost and used up, and they run out of assets within 11 years of retirement. For someone who retires at age 65, this would mean running out of assets at age 76. They're not dead yet!

The 4% Withdrawal Rule

The conventional wisdom recommends that retirees plan to spend down their retirement assets at a constant rate. One common method is to set a spending rate at an amount equal to 4% of initial wealth on the date of retirement, and to adjust this starting amount annually for inflation. For example, if you had $1,000,000 of retirement wealth at your retirement date, you would spend $40,000 in the first year. If there was 3% inflation during the first year, then in the second year you would adjust your spending to account for inflation by withdrawing $40,000(1.03) = $41,200.

For a portfolio invested in risk-free assets, such as Treasury Inflation Protected Securities, the 4% withdrawal rate might be appropriate, or might even be too low. However, for a portfolio that is even partially invested in risky assets (e.g., 40% stocks and 60% bonds), the 4% withdrawal rate might be unsustainable depending on the sequence of investment returns. A bad sequence of returns early in retirement combined with regular withdrawals will cause the value of your portfolio to decline quickly. When you are making withdrawals from a portfolio of risky assets, you have the risk of exhausting and outliving your financial assets.

Living Standard Risk

Using the life-cycle model, the correct metric for the risk of investment returns is the sustainable living standard you can afford, given your income and assets. Figure 12.16 shows the relationship between investment returns and future living standard.

Figure 12.16 Future Living Standard with Investment only in Risky Assets

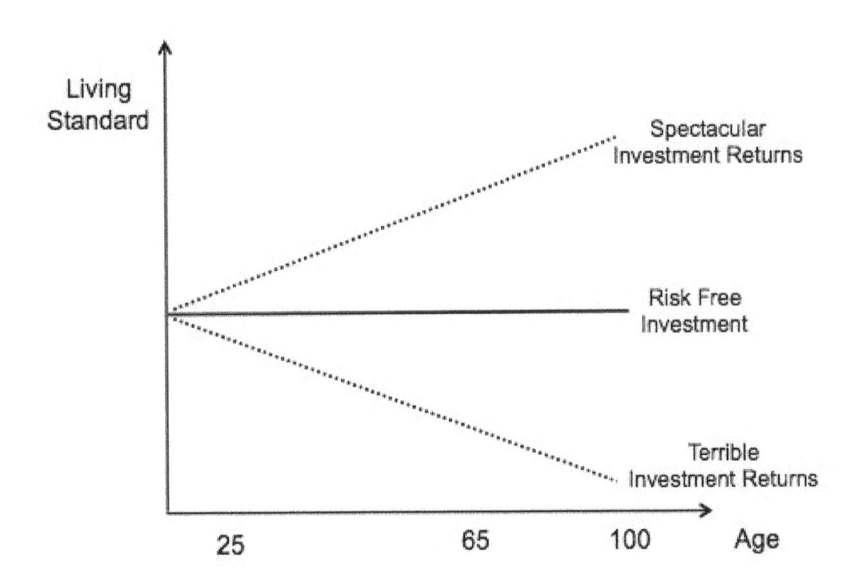

Your future standard of living depends directly on your financial wealth. If your investments do well as you age, you will have more wealth available for consumption, and your future living standard will increase due to the good investment returns. On the other hand, if your investment does poorly, you will arrive at retirement age with less wealth than anticipated. The result of poor investment returns is that your living standard will fall below that of the risk-free alternative.

By investing in risky assets, you give up the certainty about how much wealth you will have in the future (e.g., at retirement), but instead you get the chance of having more future wealth as compared to investing only in risk free assets. To the extent that you invest in risky assets, the level of your future living standard depends on the success of your investments.

12.5 Upside Investing

"If you're gonna play the game boy, you gotta learn to play it right...
You never count your money, when you're sitting at the table.
There'll be time enough for countin', when the dealin's done."
–Kenny Rogers, country singer

In the previous sections, we've seen two opposing ideas about investing in stocks. On the one hand, the historic rates of return generated by stocks have been spectacular. Investors want higher returns and are willing to invest in stocks to earn the risk premium. On the other hand, stocks are risky, even in the long run. Time does not diversify against stock market risk, and the chance of missing your goal (i.e., sustainable consumption in retirement) increases with the length of the investing time horizon. A rational investor would like to get the upside potential from investing in stocks, without the risk to his or her living standard.

Upside investing is a strategy to help you achieve some of the upside gains of investing in risky assets, without the risk of a decline in your living standard due to poor investment results.[78] This strategy is consistent with the life-cycle model and smoothing your sustainable standard of living. By planning your future consumption based only on safe assets, you can set a minimum level below which your future living standard cannot decline (see Figure 12.17).

Figure 12.17 Future Living Standard with Upside Investing

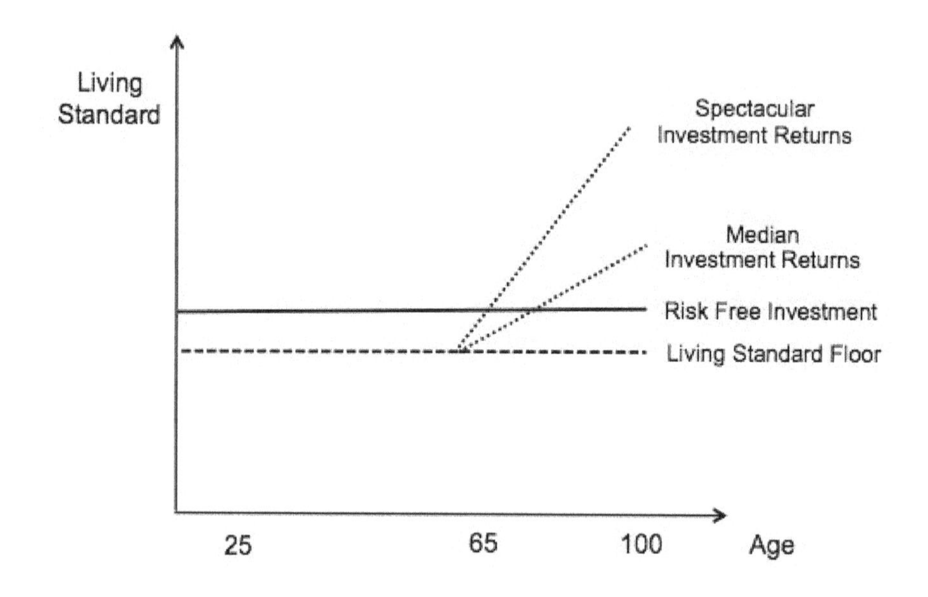

[78] http://www.mint.com/blog/goals/upside-investing-02082011/

Asset Allocation: Separating Needs from Wants

A sound approach to dealing with living standard risk is to separate retirement needs (e.g., housing, food, medical care/insurance, etc.) from retirement wants (e.g., travel, fine foods and wines, etc.). Your asset allocation should use risk-free sources of income and investments (i.e., Social Security and TIPS) to ensure that you can fully fund your minimum required needs (i.e., your living standard floor) until your maximum age of life.

We can use the Life-Cycle Model and the minimum level of consumption (i.e., needs), to find out how much saving would be required to fund those needs using only risk-free sources. After you have funded your needs in a way that does not include taking on stock market risk, you can allocate your remaining funds to investments in the stock market in the hope of achieving a higher standard of living.

Gambling Safely

Casino patrons (who are not addicted, compulsive gamblers) follow two simple rules about how to play at the casino:

- Gamble only with money you can truly afford to lose.
- Don't start spending money until you leave the casino.

You can use the same rules about investing in the stock market. The investment equivalent of these rules is as follows:

- Only invest money in the stock market that you can afford to lose. Don't invest in the stock market with assets that you will rely on to fund your future consumption.
- Begin by assuming the worst-case scenario. If you increase your wealth by investing in stock, that's great, but your future living standard should not be dependent on your success in your stock market investments.

Once you have established a floor for your future living standard, you can invest in risky assets as a gamble: good stock market returns would result in additional wealth at retirement, which might raise your future living standard. Over a period of many years, you will reduce your exposure to stock market risk by systematically converting your assets from risky to risk-free assets.

Some Numeric Examples

Consider a 25-year old, James, who earns $50,000 per year from age 25 to age 65. For simplicity, we assume no longevity risk (i.e., James will live to age 100) and no taxes and no benefits (i.e., Social Security income when old). From age 66 to 100, James must consume by spending down his assets.

In the examples below, we assume that the "safe assets" are risk-free TIPS, with a 1% real rate of return. We also assume that the "risky assets" are stocks, with a mean real return on 9% per year, and a standard deviation of 20% per year.

Baseline Scenario

To illustrate the effect of the upside investing strategy on your sustainable standard of living, we begin with a risk-free baseline scenario against which to compare the results of investing in risky assets. If James were to invest only in safe assets, his sustainable annual living standard (i.e., consumption) would be $31,410 from age 25 to age 100. This is the "benchmark" against which to compare other investment choices.

Example: 100 Minus Your Age in Stocks

Now suppose that James decides to invest using the "100% minus you age" in stocks strategy. In this case, his consumption depends on future stock market returns. To ensure he cannot run out of assets, James must consume from safe assets only. As he ages and converts more assets from risky to safe, he will be able to increase his living standard. Nonetheless, his initial living standard is low because it is based on safe-assets only. Figure 12.18 illustrates this strategy as compared to the baseline of safe assets only.

Figure 12.18 Upside Investing: "100 Minus Your Age" in Stocks Compared to All-TIPS

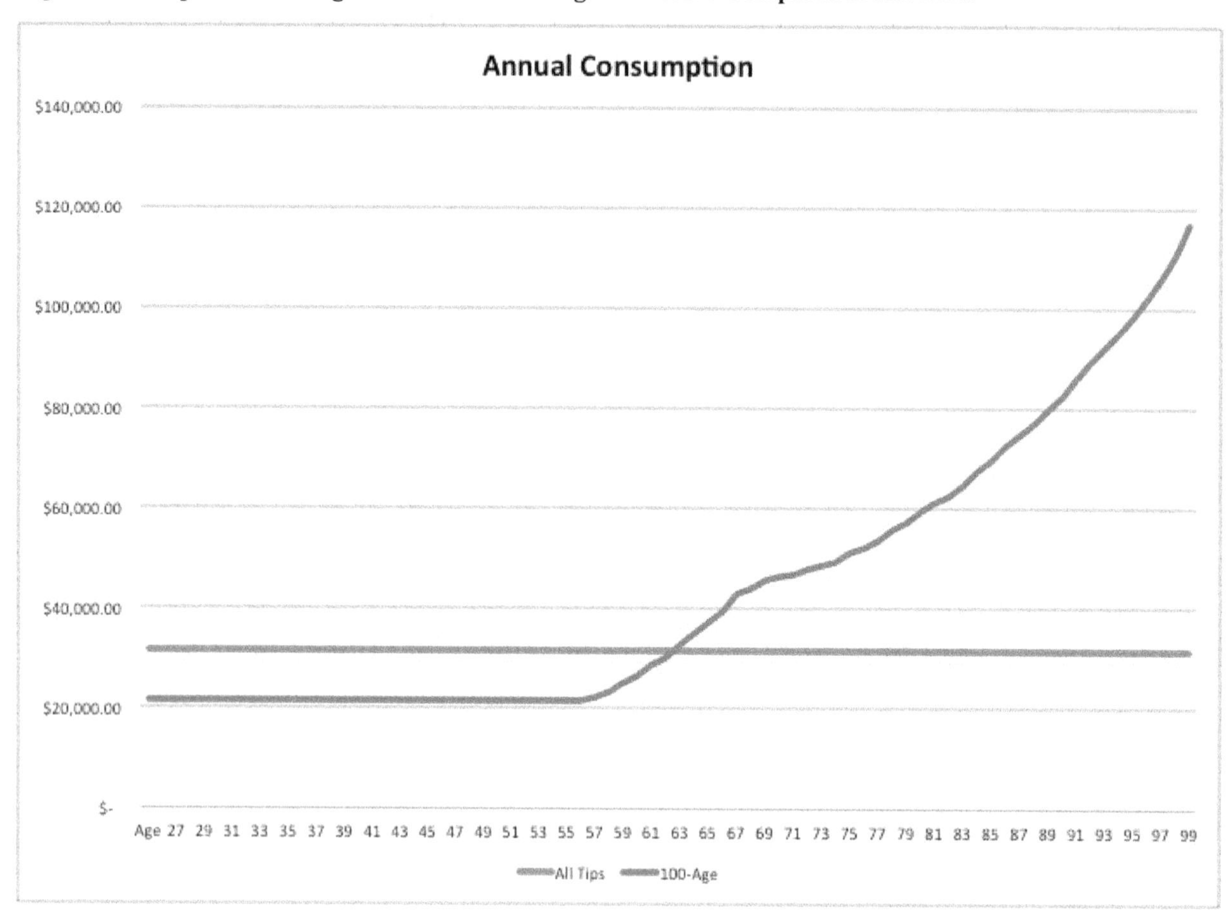

During the years before retirement, James' living standard is set at the living standard floor, i.e., the guaranteed minimum level he could sustain using safe assets only. The living standard floor assumes that he suffered the worst-case scenario and lost all of his risky assets. His living standard floor is $21,463 per year. In other words, to use the "100 minus your age" investment strategy, and eliminate the risk of running out of assets due to bad investment returns, James would need to begin by consuming 32% less than in the baseline scenario:

$$\frac{\$21,462}{\$31,410} - 1 = -32\%$$

In other words, James needs to spend less when young, so that he can be sure to weather the risk of investing in the stock market. When old, as he converts assets from risky to safe, James sees his sustainable level of consumption increase each year.

Example: Invest 50% in Stocks, Convert to all TIPS at Retirement

As a second example, suppose that James invested only 50% of his retirement account in risky assets when young, and planned to convert to safe assets immediate at retirement at age 66. Figure 12.19 illustrates this strategy compared with the baseline scenario of all TIPS.

Figure 12.19 Upside Investing: "50-50" until age 65, then 100% TIPS

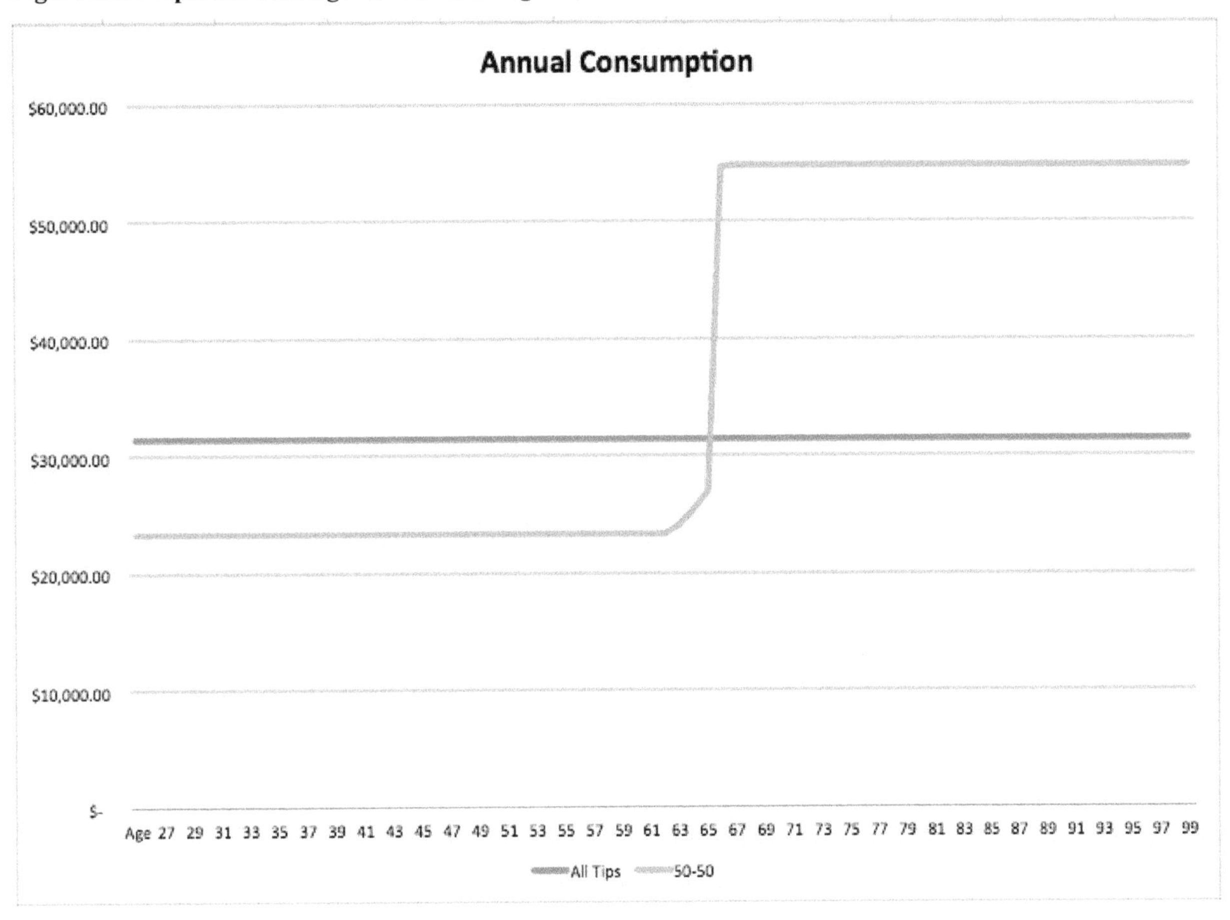

In Figure 12.19, we, we see the result of investing more conservatively when young, and converting to safe assets entirely at retirement. In this example, the living standard floor is $23,220 per year. It's still much lower than the baseline scenario, but it is $1,800 higher than in the "100 minus your age" strategy. Moreover, the standard of living increases rapidly at retirement. As soon as the assets are converted to safe assets they are available for consumption.

After retirement, James' consumption increases substantially, but the amount depends on the rate of return earned on stock investments during his working years. Each trial will be different, but even in a worse-than median scenario, he would still see an increase in his standard of living compared to the baseline scenario.

Example: Invest 50% in Stocks and Increase Percentage in Bonds Each Year

To look at final example, suppose that James invested only 50% of his retirement account in risky assets at age 25, and held that "50-50" allocation until age 38.

Beginning at age 39, James will gradually change his asset allocation toward safe assets by 1% each year. That is, at age 39 his asset allocation would be 51% TIPS and 49% stocks, and at age 40 it would be 52% TIPS and 48% stocks. At retirement (age 66), James would have 78% in TIPS and only 22% in stocks. Figure 12.20 illustrates this strategy compared to the baseline scenario.

Figure 12.20 Upside Investing: Begin with "50-50"and Increase Percentage in Bonds Each Year

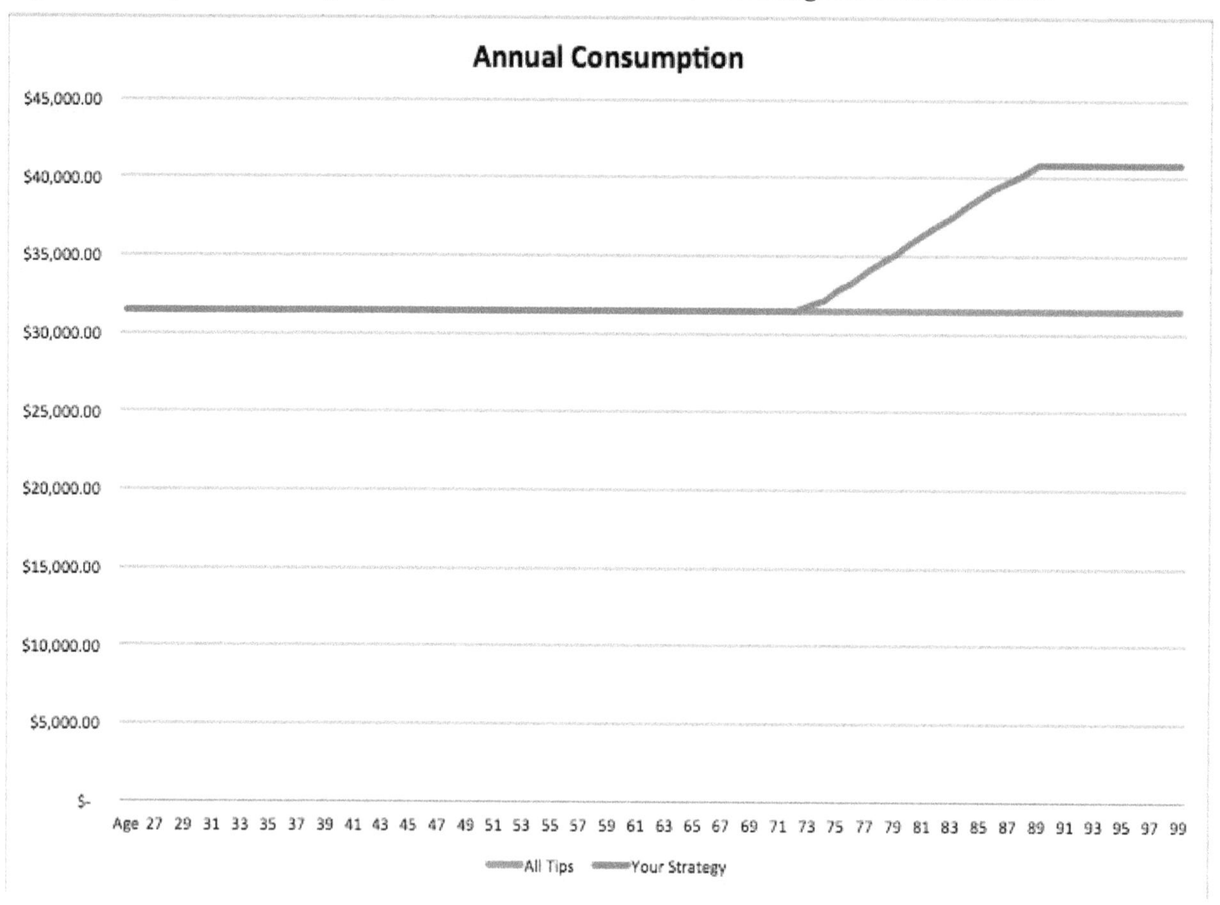

With this strategy, James is able to achieve a living standard floor that virtually matches the baseline scenario. In other words, he doesn't need to give up much living standard during the early years to still invest in stocks and hope to achieve a higher living standard later on. As he ages, he will likely achieve an increased standard of living due to his investments in the stock market.

Recap of Upside Investing

When you invest a portion of your wealth in the stock market, you are giving up some of your risk-free standard of living for a chance at a higher standard of living during retirement. The decision about what fraction of your wealth to invest in risky assets should be based on how much you need to allocate to safe assets to secure your minimum standard of living.

12.6 Summary

The conventional wisdom is that stocks are a safe investment for the long run (i.e., having a long time horizon will reduce the riskiness of investing in stocks). In this module, we used Monte Carlo simulation to illustrate the distribution of future stock returns. As we have seen, the distribution of possible outcomes becomes more spread out with time. Moreover, while the probability of loss might decrease with a longer time horizon, the magnitude (in dollars) of the potential losses increases with time.

Asset allocation is the practice of spreading your assets across multiple asset classes, such as stocks, bonds, cash, real estate, and precious metals. Conventional wisdom tells investors to change their asset allocation to be more conservative as they age, for example investing 100-minus-your-age percent of your wealth in risky assets. While this asset allocation reduces your exposure to stocks as you age, it does not reduce the risk of losing money in the stock market, nor does it guarantee that you will have more wealth at retirement than investing in TIPS.

The value at risk provides a single number to describe how many dollars your investment portfolio might lose due to a large-magnitude bad event in the stock market. Due to the sequence of returns risk, it is still possible that several years of bad investment returns right near retirement could unravel a lifetime of good saving habits and investment results.

Upside Investing is an investment strategy that treats stock market investment like casino gambling. Consumption is planned out of safe assets only, and not out of risky assets. A living standard floor below which your future consumption cannot fall is calculated based on labor income, Social Security benefits, and investments in safe assets. As you age, you convert risky assets to safe assets, thus allowing these assets to be used for consumption. If you experience gains in stock market investments, this will result in a rising standard of living.

12.7 Review Questions

1. Explain the conventional wisdom about investing in stocks for the long run. What is the evidence that supports this conventional wisdom, and what is incorrect about the conventional interpretation of this evidence?
2. Explain the general idea behind a Monte Carlo simulation and how it models the randomness in a series of events. Discuss how we developed simulated stock market returns using this technique.
3. Explain why stocks are risky even in the long run. Discuss the results of the Monte Carlo simulation presented in the module and in class.
4. Explain asset allocation and discuss the conventional wisdom about asset allocation over an investor's lifetime.
5. What is a target-date mutual fund? How does it attempt to help investors change their asset allocation over time? In what ways are target-date mutual funds more or less risky than investors expect?
6. Define the concept of value at risk, and give a numerical example to explain what it measures.
7. Explain, giving an example, how the sequence of returns could adversely affect your standard of living.
8. What are the main ideas and benefits of upside investing? How does it help us evaluate the risk-reward tradeoff of investing in risky assets?

13 Uncle Sam and You

"If you don't drink, smoke, or drive a car, you're a tax evader."
- Thomas Foley, Former Speaker of the US House of Representatives

Learning Objectives

- Extend the life-cycle model by incorporating a real return on investment.
- Define the fundamental concepts and terminology of taxation: what can be taxed, how taxes are raised, and the rate of taxation.
- Identify and explain the different types of taxes you are likely to face, both in the United States and abroad.
- Provide a broad overview of personal taxation in the United States.
- Demonstrate by example how to calculate the Federal income tax on labor income and investment income.
- Introduce Social Security benefits as the primary old-age income security program in the United States.
- Extend the life-cycle model to incorporate taxation of labor and asset income, as well as Social Security benefits.

13.1 Introduction

Governments provide goods and services for the common good. *Taxes* are the revenue collected to pay for government goods and services.

According to Nobel Laureate Economist Paul Krugman, "The basic picture of the federal government you should have in mind is that it's essentially a huge insurance company with an army." The "insurance company" describes the main programs of the government: Medicare, which provides medical care for individuals over age 65; Medicaid, which does the same for the poor; and Social Security, which provides old-age income insurance to retirees. The "insurance" and "army" functions of government make up the vast majority of the federal budget with the remaining part going to a host of programs such as NASA, the Food and Drug Administration, National Science Foundation, and home-heating assistance to the poor. The cost of these services is paid for via taxes.

13.2 Fundamental Concepts of Taxation

Tax Bases and Taxes

How should taxes be raised? A ***tax base*** is an area of the economy on which taxes can be assessed. The four main tax bases are income, labor, wealth, and consumption.

- *Income* includes flows of income from employment, business profits and capital (e.g., interest, dividends, and capital gains), and transfer payments like Social Security and unemployment insurance. An *income tax* is a tax assessed on income. For example, in the United States there are both Federal and state taxes on income.

- *Labor income* only describes income from employment. A *payroll tax* is a tax on labor income. For example, in the United States the *Federal Insurance Contributions Act* (*FICA*)assesses taxes on payroll income alone.

- *Wealth* includes fixed assets like real estate as well as financial assets. A *wealth tax* is a tax assessed on the value of a household's assets. For example, many municipalities collect property taxes assessed on the value of real property such as land and houses.

- *Consumption* is money spent on goods and services. A *consumption tax* is a tax assessed on the value of a purchase of goods or services. For example, many states collect sales tax on retail sales of goods. An ***excise tax*** is also a consumption tax on a specific good or service. For example, a tax on the consumption of gasoline or cigarettes is an excise tax.

Excise Taxes, Externalities, and Incentives

An *externality* is a benefit or cost that is not included in the retail price of a good or service. For example, one positive externality of bike riding is reduced use of fossil fuels and smog. The "benefit" of these externalities is not included in the price of a bicycle. If it were, perhaps bicycles would be free or even come with a subsidy (a negative price). Cigarettes produce a negative externality, which is the health care cost associated with smoking-related diseases. The price of cigarettes does not include the cost of the future health care for sick smokers.

Economists and marketers have long understood that individuals are motivated by incentives. In marketing, incentives often take the form of promotional pricing or discount coupons, which have a noted effect in motivating consumer behavior. Excise taxes often work as disincentives to discourage "bad" behavior.

As one example, excise taxes on cigarettes could be set at the level that would be sufficient to pay the future health care costs of sick smokers. The excise taxes on cigarettes provide a disincentive to smoking. As another example, the retail price of gasoline could be set at a level that would pay for the environmental and national security costs associated with the use of fossil fuels. An additional benefit of an excise tax is that it could be set at a level that would pay for the external costs associated with the action.

Value Added Taxes

Many countries around the world use a *value added tax* (*VAT*) as a primary source of government revenue. From the buyer's perspective it is a consumption tax on goods or services. In general the VAT is a tax collected at each phase of production or sale of goods or services based on the incremental value added by that producer or merchant. The seller remits to the government the tax on the difference between the cost of inputs and the sale price of outputs. For example, in Canada the Goods and Services Tax (GST) is a national VAT on the purchase of all goods and services. The United States does not have any national VAT, but many states have a sales tax that functions in this way.

Types of Personal Income Taxes

Personal income taxes can be raised in various ways, including per-capita taxation, flat taxation, or progressive taxation. In this section, we introduce the types of personal income taxes and provide examples of each in a general way that is not specific to the United States. In the following section, we will turn our attention specifically to income taxes in the United States.

Per-Capita Income Tax

A *per-capita tax* is administratively simple: each household is taxed the same amount per person (or per household) regardless of income. Per-capita taxes put a much higher tax burden on low-income households as they take a higher percentage of income from those households as compared to high-income households.

For example, with a per-capita income tax of $5,000 per person, a family of 4 would pay $20,000 in taxes. For a family with an income of $25,000, this would be an 80% tax rate; for a family with an income of $100,000 per year, this would be a 20% tax rate; and for a family with an income of $400,000 per year, this would be only a 5% tax rate.

Flat Income Tax

A *flat tax* imposes the same burden on all income. For example, a 10-percent flat tax would be assessed at $5,000 for a household with income of $50,000, and $7,500 for a household with income of $75,000. While a flat tax appeals to the idea of fairness with each household paying the same "fair share," the burden of a flat tax is greater on low-income households. A flat tax does not mean that every household pays the same amount of tax, but rather that every household pays the same *percentage* of their income in taxes.

For example, in 2014 the poverty-level income for a family of 4 in the United States was $23,850[79]. A 10% tax on this household would be $2,385, which would leave less than $21,500 for consumption. A 10% tax on a household

---

[79]http://aspe.hhs.gov/poverty/14poverty.cfm

earning $95,400 (4 times the poverty level) would be $9,540, leaving almost $86,000 for consumption. Given the minimum level of consumption needed for survival, the flat tax would impose a much harder burden on low-income households.

Progressive Income Tax

A *progressive tax* increases the burden of payment on those most able to pay. In a progressive tax system, income is classified into different tax brackets with each bracket subject to a different marginal tax rate.

For example, consider this simple progressive tax scheme for the Land of Make Believe:
- 0% tax on income up to $20,000;
- 10% tax on income above $20,000 and below $40,000; and
- 20% tax on income above $40,000.

Incomes are taxed at the margin, meaning that each additional dollar earned is taxed at the highest applicable marginal taxrate. A progressive income tax is a step function: as income increases, the marginal tax rate increases in steps according to the lower and upper bounds of the income brackets. You can think of a progressive tax being like a staircase with income on the run (horizontal axis) and marginal tax rate on the rise (vertical axis). As income increases, the marginal tax rate goes up in steps.

A *regressive tax* is the opposite of a progressive tax: the burden of payment decreases as income rises. For example, in the United States, the Social Security tax on payroll income operates as a regressive income tax. Examples of the Social Security tax will follow later in this module.

Taxing Authorities in the United States

Residents of the United States are subject to the taxing authorities of the Federal government, the state governments in which they live, work, or shop, and the local or municipal governments in which they live, work or shop. These taxing authorities use various combinations of taxes on the four main tax bases discussed above.

The Federal government collects many different taxes, primarily on income. Individual households are subject to the *Federal Income Tax* and the *Alternative Minimum Tax* (AMT), and corporations are subject to the *Federal Corporate Income Tax*. In addition to the Federal income tax and state income taxes, labor income is subject to *payroll taxes* that include the FICA taxes to fund Social Security and Medicare. The Federal government also collects excise taxes on goods including tobacco and gasoline.

State governments collect taxes as well, both on income and consumption. About 40 states assess income taxes on households and corporations. There are many state-level consumption taxes including sales taxes assessed by retail vendors (in about 45 states) and various state excise taxes including taxes on the sale of alcohol, tobacco, and gasoline.

At the municipal government level, the most common taxes are real estate taxes and motor vehicle excise taxes. Many municipalities collect taxes on restaurant meals, hotel stays, and car rentals.In some areas like New York City, there are municipal payroll taxes as well.

While households in the United States face all of these (and many more) taxes, the primary concern with respect to life-cycle economics is about taxes on household income. The remainder of this module focuses on the concepts and application of Federal Income Tax and FICA payroll taxes on household income.

13.3 Federal Income Tax in the United States

In the United States, the Federal income tax is generally a progressive taxation system that applies to income from labor and capital, with some aspects of per-capita taxation mixed in. Federal income taxes are due on income earned from work, unemployment insurance, rents, royalties, self-employment, business income, alimony received, interest on deposits and bonds, and in some cases Social Security income. Capital income, including capital gains and dividends[80], are also reported along with earned income, but these are taxed at a flat tax rate of 15% as of 2012. Taxes on personal income are calculated for each household (as opposed to for individual workers). Singles and unmarried couples pay taxes individually. Married couples, whether one or both spouses work, pay taxes on their combined household income – likely for reasons dating back to the agricultural focus of the economy in the early 20th century.

Federal income taxes are calculated at the margin. The U.S. tax code has 4 different tax rate schedules, depending on the filing status of the taxpayer (single, head of household, married filing jointly, and married filing separately). The United States Congress has the authority to set tax rates and marginal tax brackets. In general, the marginal tax rates have remained consistent from year to year. The marginal income brackets are adjusted for inflation by increasing the lower and upper bounds for each tax bracket slightly each year.

Federal Income Tax Reporting and Collection

The Internal Revenue Services (IRS) is the tax collection and enforcement authority in the United States, ensuring accurate tax reporting and payments by workers and corporations. Workers are responsible for reporting their income to the IRS and paying all of the taxes due. To facilitate tax collection, most employers withhold estimated taxes from workers' paychecks and remit these amounts directly to the IRS. Self-employed workers must remit their own estimated tax payments quarterly. In January of each year, employers send to workers an annual statement (called a form W2) of the previous year's wages and taxes withheld, and a duplicate copy is sent to the IRS for control purposes.

Since the amounts withheld by employers (and estimated tax payments by the self-employed) are only estimates, an annual tax calculation is made to determine the actual amount of tax due. For example, if your tax withholding or estimated tax payments are $2,500 and the tax due is $2,000, you are entitled to a refund of $500. If your tax withholding was too low (or you didn't pay enough estimated taxes), you must pay the remainder due by the tax-filing deadline (see below).

Income Tax Returns

An *income tax return*is an annual report of a household's tax obligations. The tax return reports all income and certain allowable adjustments to income, deductions and exemptions that legally reduce taxable income and any credits that reduce the tax due. Also reported are the amount of taxes withheld from paychecks (or estimated taxes paid) and finally the amount overpaid (or underpaid). Every household must file a tax return (IRS Form 1040, 1040A, or 1040EZ) by April 15, the tax-filing deadline for the previous calendar year.

Failure to file your tax return (or request an extension) by the deadline will lead to penalty fees. Moreover, even if you do not file your tax return you are responsible to pay all taxes due by the deadline, or else interest charges will be added to any unpaid balance. *Tax evasion* or non-payment of taxes legally due is a criminal offense in the United States.[81]

After filing a tax return, households will receive a refund for any over-payment or must make a payment to cover any underpayment.

---

[80] Capital gains are gains that occur from the sale of appreciated property, such as an investment. The capital gain is the difference between the purchase price called the basis and the sale price. Capital gains are realized only upon the sale of appreciated property, and unrealized gains are not taxable. Dividends are periodic cash payments made from a corporation to its shareowners.

[81] It should be noted that *tax avoidance* (i.e., structuring one's income and expenses to minimize the amount of tax due) is perfectly legal.

Deductions, Exemptions, and Credits

There are many legal ways to reduce one's tax liability, i.e., how much you must pay in taxes in a given year. Taxpayers use tax deductions and exemptions to legally reduce the amount of income on which taxes are due.

Every tax filer is allowed to choose either the standard deduction or to itemize deductions to reduce his or her tax liability. The *standard deduction* is a fixed amount of income one can earn free of taxes. The standard deduction for 2015 is $6,300 for single tax filers and $12,600 for married couples filing a joint tax return.[82] *Itemized deductions* are allowable expenses that can be deducted on a pre-tax basis. Some of the most popular tax deductions are deductions for home mortgage interest expenses, moving expenses (if the move was required for a new job), home office expenses, and health-care expenses. Deductions are reported on a separate form called Schedule A. Taxpayers choose the greater of either the standard deduction or itemized deductions.

The personal *exemption* is a per-capita reduction in taxable income. In 2015, the personal exemption amount is $4,000 per person. A single person can claim only a $4,000 exemption. A family of five may take five exemptions, worth a total of $5 \times \$4{,}000 = \$20{,}000$ of reduction in taxable income.

Finally, a *tax credit* is a dollar-for-dollar reduction in the tax due. The most common tax credits are the child tax credit and the earned-income tax credit (EITC). For example, the child tax credit provides a $1,000 reduction in taxes for each child in the household. The EITC provides a credit to low-income workers, in part to offset the payroll taxes they pay (see the section on payroll taxes below). The value of tax credits is gradually phased out as the adjusted gross income increases.

IRS Form 1040EZ

Single and married tax filers without children who earn all their income from employment or interest and do not claim itemized deductions may use a simplified tax form called Form 1040EZ. Figure 13.1 shows a sample tax return Form 1040EZ for tax year 2014.

While the 1040EZ Income Tax Return form may seem daunting at first, most tax filers will find that only a few lines apply to them. The following line numbers refer to the line numbers in Form 1040EZ (refer to Figure 13.1).

- Lines 1, 2, and 3 show the filer's income for the year. Line 4 is the total of all income, referred to as the *adjusted gross income* or AGI.

- Line 5 reports whether or not the filer is entitled to a personal exemption. Most people are able to claim a personal exemption of $4,000 and a standard deduction of $6,300, for a total of $10,300 (in 2015). The $10,300 is deducted from adjusted gross income, so that the first $10,300 of income each year is tax-free. A married couple can claim 2 personal exemptions of $4,000 each, and a standard deduction of $12,600 for a total exemption of $2 \times \$4{,}000 + \$12{,}600 = \$20{,}600$.

- Line 6 is the taxable income: AGI less the amount of the exemption. This is the amount of income on which the filer will be taxed.

- Line 7 is the amount of tax that was withheld by the filer's employer as reported on form W2.

- Lines 8a and 8b are for tax credits, which lower the amount of tax due, but only apply to some workers. Line 9 is the total of lines 7 and 8.

- Line 10 is the amount of tax owed as calculated by either the tax tables or the marginal tax rates (see the section "Marginal Tax Calculations" for a complete explanation).

---

[82] Note that the amounts for the personal exemption, standard deduction, and tax brackets are adjusted each year to roughly keep pace with the consumer price index. See http://www.irs.gov/uac/Newsroom/In-2015,-Various-Tax-Benefits-Increase-Due-to-Inflation-Adjustments

- Line 11 is the net refund (if taxes withheld were more than the tax owed). This is the amount that will be returned by the government if you overpaid
- Line 12 is the net tax payment owed (if taxes withheld were less than the tax owed). This is the amount you still owe to the government if you underpaid.

Figure 13.1 The 1040EZ Federal Income Tax Form for the Year 2014

Form **1040EZ**

Department of the Treasury—Internal Revenue Service

Income Tax Return for Single and Joint Filers With No Dependents (99)

2014

OMB No. 1545-0074

Your first name and initial | Last name | **Your social security number**

If a joint return, spouse's first name and initial | Last name | **Spouse's social security number**

Home address (number and street). If you have a P.O. box, see instructions. | Apt. no. | ▲ Make sure the SSN(s) above are correct.

City, town or post office, state, and ZIP code. If you have a foreign address, also complete spaces below (see instructions).

Presidential Election Campaign Check here if you, or your spouse if filing jointly, want $3 to go to this fund. Checking a box below will not change your tax or refund. ☐ You ☐ Spouse

Foreign country name | Foreign province/state/county | Foreign postal code

Income

Attach Form(s) W-2 here.

Enclose, but do not attach, any payment.

1 Wages, salaries, and tips. This should be shown in box 1 of your Form(s) W-2. Attach your Form(s) W-2. **1**

2 Taxable interest. If the total is over $1,500, you cannot use Form 1040EZ. **2**

3 Unemployment compensation and Alaska Permanent Fund dividends (see instructions). **3**

4 Add lines 1, 2, and 3. This is your **adjusted gross income.** **4**

5 If someone can claim you (or your spouse if a joint return) as a dependent, check the applicable box(es) below and enter the amount from the worksheet on back.
☐ **You** ☐ **Spouse**
If no one can claim you (or your spouse if a joint return), enter $10,150 if **single;** $20,300 if **married filing jointly.** See back for explanation. **5**

6 Subtract line 5 from line 4. If line 5 is larger than line 4, enter -0-. This is your **taxable income.** ▶ **6**

Payments, Credits, and Tax

7 Federal income tax withheld from Form(s) W-2 and 1099. **7**

8a **Earned income credit (EIC)** (see instructions) **8a**

b Nontaxable combat pay election. 8b

9 Add lines 7 and 8a. These are your **total payments and credits.** ▶ **9**

10 **Tax.** Use the amount on **line 6 above** to find your tax in the tax table in the instructions. Then, enter the tax from the table on this line. **10**

11 Health care: individual responsibility (see instructions) Full-year coverage ☐ **11**

12 Add lines 10 and 11. This is your **total tax.** **12**

Refund

Have it directly deposited! See instructions and fill in 13b, 13c, and 13d, or Form 8888.

13a If line 9 is larger than line 12, subtract line 12 from line 9. This is your **refund.** If Form 8888 is attached, check here ▶ ☐ **13a**

▶ b Routing number ▶ c Type: ☐ Checking ☐ Savings

▶ d Account number

Amount You Owe

14 If line 12 is larger than line 9, subtract line 9 from line 12. This is the **amount you owe.** For details on how to pay, see instructions. ▶ **14**

Third Party Designee

Do you want to allow another person to discuss this return with the IRS (see instructions)? ☐ **Yes.** Complete below. ☐ **No**

Designee's name ▶ Phone no. ▶ Personal identification number (PIN) ▶

Sign Here

Joint return? See instructions.

Keep a copy for your records.

Under penalties of perjury, I declare that I have examined this return and, to the best of my knowledge and belief, it is true, correct, and accurately lists all amounts and sources of income I received during the tax year. Declaration of preparer (other than the taxpayer) is based on all information of which the preparer has any knowledge.

Your signature | Date | Your occupation | Daytime phone number

Spouse's signature. If a joint return, **both** must sign. | Date | Spouse's occupation | If the IRS sent you an Identity Protection PIN, enter it here (see inst.)

Lines which do not apply to your personal situation may be left blank or marked 0. For example, if you did not collect unemployment insurance you may mark the amount in line 3 as 0.

IRS Form 1040

In general, the specific types of income earned and the deductions or credits claimed would determine which form must be used. The 1040EZ form (Figure 13.1) is the "short form," which is applicable to single or married workers without itemized deductions and without dependent children. Tax filers whose income includes self-employment, business income, rent, royalties, capital gains, or dividends, and those wishing to itemize deductions are required to file Form 1040 (the "long form"). In addition, households with dependent children must file Form 1040 to be able to claim the exemption for dependents. Depending on specific situations, additional forms or schedules might be required (for example, *Schedule A* is used to report itemized deductions).

Income Taxes for Dependent Children

Children whose parents can claim them as dependents are still required to file tax returns separate from their parents, provided they earn enough income to pay taxes. The income threshold depends on the type of income. As of 2014, dependent children with more than $1,000 of unearned income (e.g., interest income) or more than $6,100 of earned income. In many cases, the tax filing will require only a 1040EZ and can be completed on paper in only a few minutes. Refer to IRS Publication 929 for a full explanation.[83]

Federal Income Tax is Calculated at the Margin

The amount of Federal Income Tax due is calculated based on a schedule of marginal tax rates. The marginal tax rates that apply to your income depend on your tax filing status. The two most common tax filing statuses are *single* and*married filing jointly*, and the two uncommon are *married filing separately* and *head of household.*

Marginal Tax Rates for Single Tax Filers

For single tax filers, the 2015 Federal tax rates are:

- **10%** on taxable income from $0 to $9,225, plus
- **15%** on taxable income from $9,225 to $37,450, plus
- **25%** on taxable income from $37,450 to $90,750, plus
- **28%** on taxable income from $90,750 to $189,300, plus
- **33%** on taxable income from $189,300 to $411,500, plus
- **35%** on taxable income from $411,500 to $413,200, plus
- **39.6%** on taxable income over $413,200.

For visual types, the tax rates are plotted on a graph in Figure 13.2.

Marginal Tax Rates for Married Tax Filers

For married couples filing jointly, the 2015 Federal tax ratesare:

- **10%** on taxable income from $0 to $18,450, plus
- **15%** on taxable income from $18,450 to $74,900, plus
- **25%** on taxable income from $74,900 to $151,200, plus
- **28%** on taxable income from $151,200 to $230,450, plus
- **33%** on taxable income from $230,450 to $411,500, plus
- **35%** on taxable income from $411,500 to $464,850, plus
- **39.6%** on taxable income over $464,850.

Notice that the tax rates for married couples are not symmetrical with the tax rates for singles. At the lower end of the income spectrum, the tax rates for married couples are double those for singles. At the highest end of the income spectrum, the tax rate for married couples is the same as for singles. A *marriage subsidy* exists when a high-income earner is married to a low-incomer earner, and the result is paying less in taxes as a couple then they would pay as singles. A *marriage penalty* occurs when two high-income people are married, and the result is that as a couple they pay more in taxes than they would pay as two singles.

[83]http://www.irs.gov/publications/p929/index.html

Marginal tax rates create a *step function*, where the rate of taxation that applies to income depends on the level of income. As your income rises, so does your marginal tax rate. Each dollar of income is taxed in only one tax bracket. The total amount of tax is the sum of the tax due in each bracket:

$$tax\ due = \sum TI_b \times \tau_b = \sum (upper\ bound - lower\ bound) \times \tau_b$$

where TI_b is the amount of taxable income that falls within the bracket, and τ_b is the marginal tax rate that applies to that bracket.

Marginal Tax Calculation Algorithm

An *algorithm* is a series of steps that describe a procedure to accomplish a particular task. The algorithm for calculating marginal taxes can be described as follows.

For each tax bracket, find out how much income is taxable in that bracket.
  If the taxable income exceeds the upper bound of the tax bracket:
    The amount taxable in this bracket is: upper bound – lower bound
    Tax in this bracket is: (upper bound – lower bound) * marginal rate
  Else (when the taxable income does not exceed the upper bound of the tax bracket):
    The amount taxable in this bracket is: taxable income – lower bound
    Tax in this bracket is: (taxable income – lower bound) * marginal rate
    Stop
Add together the tax in each tax bracket to arrive at the total tax due.

Example A: Dennis

Consider Dennis, a single taxpayer with an adjusted gross income of $65,000 (i.e., line 4 of Form 1040). As a single, Dennis can deduct $10,300 for the standard deduction and personal exemption, which leaves taxable income of $54,700.

Go through each tax bracket and find the taxable income in that bracket:
10% bracket: is taxable income over $9,225? Yes.
  Taxable income in this bracket is $\$9{,}225 - \$0 = \$9{,}225$.
  Tax in this bracket is $10\% \times \$9{,}225 = \$922.50$.
15% bracket: is taxable income over $37,450? Yes.
  Taxable income in this bracket is $\$37{,}450 - \$9{,}225 = \$28{,}225$.
  Tax in this bracket is $15\% \times \$28{,}225 = \$4{,}233.75$.
25% bracket: is taxable income over $90,750? No.
  Taxable income in this bracket is $\$54{,}700 - \$37{,}450? = \$17{,}250$.
  Tax in this bracket is $25\% \times \$17{,}250 = \$4{,}312.50$.
  Stop.
Now, we add all these amounts to find the total tax due:
$Tax\ due = \$922.50 + \$4{,}233.75 + \$4{,}312.50 = \$9{,}468.75$.

Dennis is in the 25% marginal tax bracket, because his taxable income exceeded $37,450. His marginal tax bracket is the highest bracket in which he has taxable income. However, Dennis' marginal tax rate is not a fair indication of his overall tax burden.

The *effective tax rate*shows how much of an individual's or household's income is contributed toward taxes. This rate is what we really care about when we talk about fairness and taxes. The effective tax rate is:

$$Effective\ Tax\ Rate = \frac{Total\ Taxes}{Total\ Income}$$

Dennis did not pay 25% of his income in taxes: his effective tax rate is:

$$Effective\ Tax\ Rate = \frac{\$9.468.75}{\$65{,}000} = 14.57\%$$

Figure 13.2 shows the marginal tax rate (the step function) and effective tax rate (the curved line) for a single tax filer as a function of adjusted gross income.

Figure 13.2 Federal Marginal and Effective Tax Rates as a Function of Adjusted Gross Income, for a Single Tax Filer in 2015.

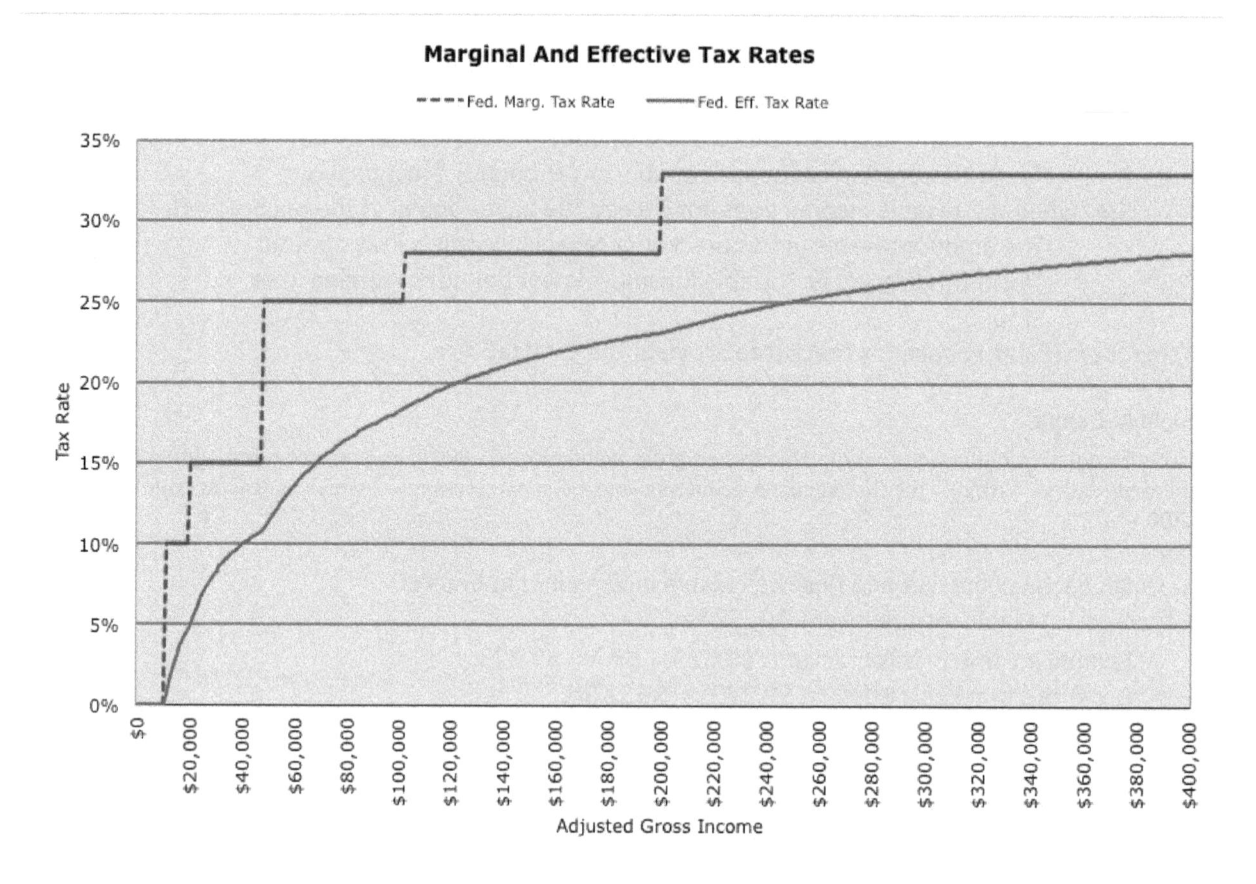

Note that effective tax rates will always be lower than marginal tax rates, because no matter your total income, at least part of it is being taxed in the lowest tax bracket(s). For example, if you have $20,000 of adjusted gross income, your taxable income would be $9,700. The first $9,225 of taxable income is taxed at the 10% marginal tax rate, and the remaining $475 is taxed at the 15% rate. Total taxes due are: $\$922.50 + 0.15 \times \$475 = \$993.75$. The effective tax rate is:

$$Effective\ Tax\ Rate = \frac{\$993.75}{\$20{,}000} = 4.97\%$$

Example B: Al and Mary Ann
Consider Al and Marry Ann, a married couple with an adjusted gross income of $103,630 (line 4). Al and Mary Ann deduct $20,600 for the standard deduction and personal exemption, which leaves taxable income of $83,030.

Go through each tax bracket and find the taxable income in that bracket.
10% bracket: is taxable income over $18,450? Yes.

Taxable income in this bracket is $18,450 – $0 = $18,450.
Tax in this bracket is 10% × $18,450 = $1,845.

15% bracket: is taxable income over $74,900? Yes.

Taxable income in this bracket is $74,900 – $18,450 = $56,450.
Tax in this bracket is 15% × $56,450 = $8,467.50.

25% bracket: is taxable income over $151,200? No.

Taxable income in this bracket is $83,030 – $74,900? = $8,130.
Tax in this bracket is 25% × $8,130 = $2,032.50.
Stop.

Now, we add all these amounts to find the total tax due:
$Tax\ due = \$1{,}845 + 8{,}467.50 + 2{,}032.50 = \$12{,}345.$

Al and Mary Ann earned enough to put them into the 25% marginal tax bracket. However, their effective tax rate is: about 12%:

$$Effective\ Tax\ Rate = \frac{\$12{,}345}{\$103{,}630} = 11.91\%$$

13.4 Other Income Taxes

In addition to the Federal income taxes discussed above, all labor income is subject to the Federal Insurance Contributions Act (FICA), commonly referred to as *payroll taxes*. FICA taxes are used to fund the largest Federal insurance programs, Social Security[84] and Medicare. Employers and employees each pay half of the FICA taxes. The Social Security tax rate is 12.4% (6.2% paid by employers and 6.2% paid by employees). The maximum income on which Social Security tax is collected is $118,500 in 2015 and income above the Social Security wage limit is not subject to the Social Security tax.[85] This type of tax is an example of a regressive tax, since it only affects income at the lower end of the income spectrum. The Medicare tax rate is 1.45% on all payroll income (i.e., the amount reported by employers on Form W2) with no upper limit. Taking employer and employee contributions together, payroll taxes add up to 15.3% of payroll income.

Many states also assess a state income tax. The details vary by state, but in many cases state taxes are based on your Federal adjusted gross income or taxable income. Some state taxes are progressive, and some are flat taxes. For example, in Massachusetts state income taxes are assessed at a flat rate of 5.3% on your Federal adjusted gross income with some additional (different) deductions. Figure 13.3 graphically depicts the marginal and effective tax rates including state and payroll taxes.

Figure 13.3 Marginal and Effective Tax Rates, including Federal, FICA and State Taxes, as a Function of Adjusted Gross Income, for a Single Tax Filer in 2015

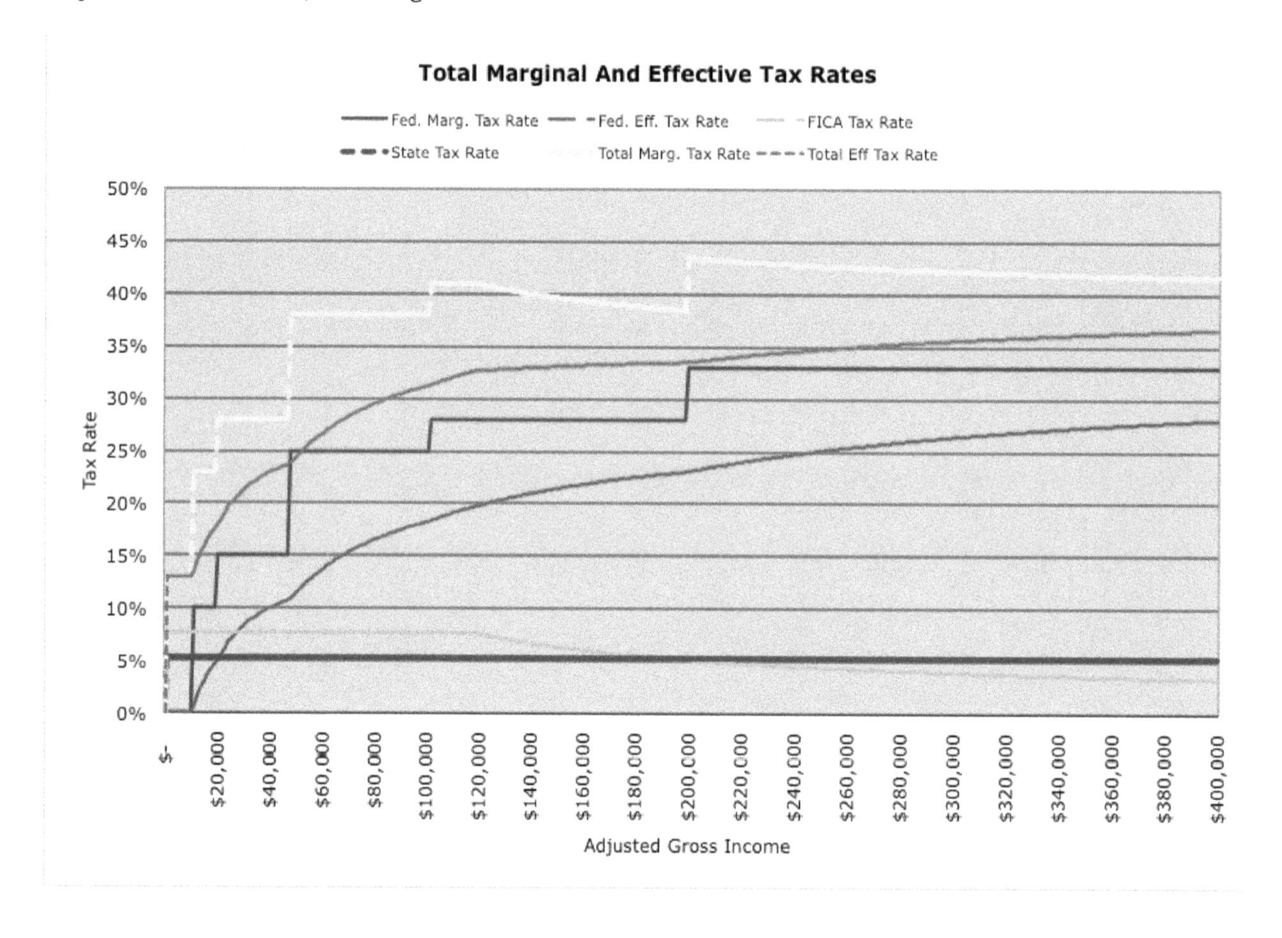

[84] Social Security is further discussed in the next section of this module.

[85] The maximum wage limit for the Social Security tax is adjusted every year to keep pace with inflation. It was $117,000 in 2014, and is $118,500 for 2015.

Taxes on Asset Income

Asset income is the income generated by one's assets. Asset income takes many different forms, including interest, rents, royalties, capital gains and dividends. Under the US Tax code, different tax rates apply to different types of income.

Asset income can be further subdivided into two categories based on their tax treatment. Rents, royalties, and interest are subject to the progressive Federal income taxes and state income taxes, but are not subject to FICA payroll taxes. Capital gains and dividends received from owning shares in a corporation are subject to a Federal flat-rate tax of 15% and might also be subject to state taxes. Figure 13.4 depicts the income tax bases and the taxes to which each type of income is subjected.

Figure 13.4Types of Income and the Taxes to which the Income is Subjected

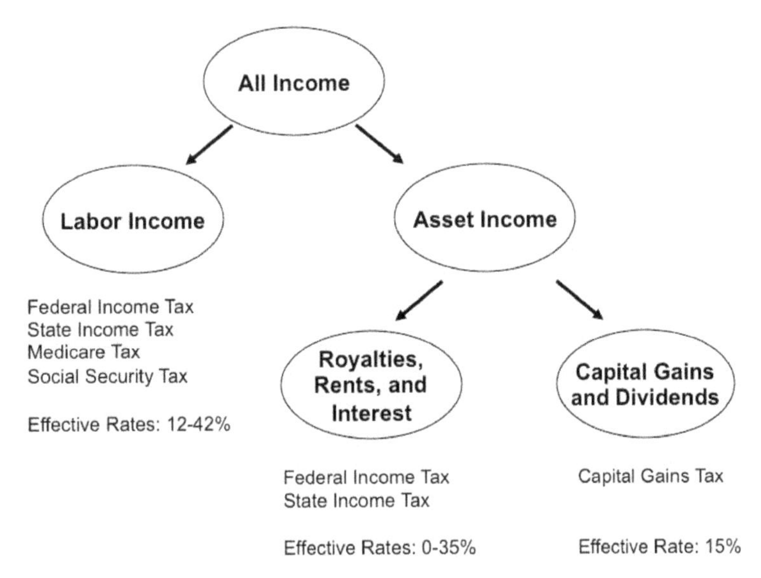

It is important to note that the taxes on asset income apply to *income generated by one's assets*, but not on the assets themselves.

13.5 Social Security Benefits

Franklin Delano Roosevelt's New Deal legislation in the 1930s created the *Social Security* system to combat old-age poverty by providing old-age income insurance. *Social Security benefits* are monthly cash payments received by qualified elderly citizens who have earned enough work credits (or are entitled to receive spousal benefits on the record of a qualified worker). Benefits are indexed for inflation (so they maintain their real purchasing power) and provide longevity insurance (i.e., one cannot outlive the benefits). Social security benefits are sufficient to prevent extreme poverty, but generally not sufficient to afford a comfortable middle-class lifestyle. Benefits under Social Security are effectively an inflation-indexed lifetime annuity (e.g., a permanent income, which is adjusted annually to maintain the recipient's purchasing power). For most workers, Social Security benefits represent the most important part of retirement income.

When Should Workers Begin Receiving Benefits?

Workers choose when to sign up for benefits and can sign up online or at a Social Security field office. The age at which you start to receive benefits determines the level of real annual benefits you will receive *for the rest of your life*. Covered workers are eligible to receive benefits at their normal retirement age (age 66 for workers born before 1960, or age 67 for workers born after 1960). It is possible to receive benefits beginning as early as age 62 and 1 month, in which case your benefit amount reduced by about 8 percent per year for each year before your normal retirement age. It is also possible to take benefits as late as age 70 (delayed retirement benefits), in which case your benefit amount is increased by about 8 percent for each year you wait after your full retirement age (until age 70).

Benefits Amount

The level of Social Security benefits are determined by a worker's pre-retirement income in a highly *regressive* manner: lower-wage workers will receive benefits at a higher ratio of their pre-retirement income, and higher-wage workers will receive benefits at a lower ratio of their pre-retirement income. For most workers, earning more income will result in receiving greater retirement benefits, albeit at a decreasing marginal rate (look ahead to Figure 13.5). There is a maximum level of benefits, and very-high wage earners (earning more than $118,500 in 2015) will receive only the maximum amount no matter how much wage income they received. The graph in Figure 13.5 shows estimated Social Security benefits as a function of income earned by a sample worker born in 1960.

Figure 13.5 Estimated Social Security Benefits as a Function of Income

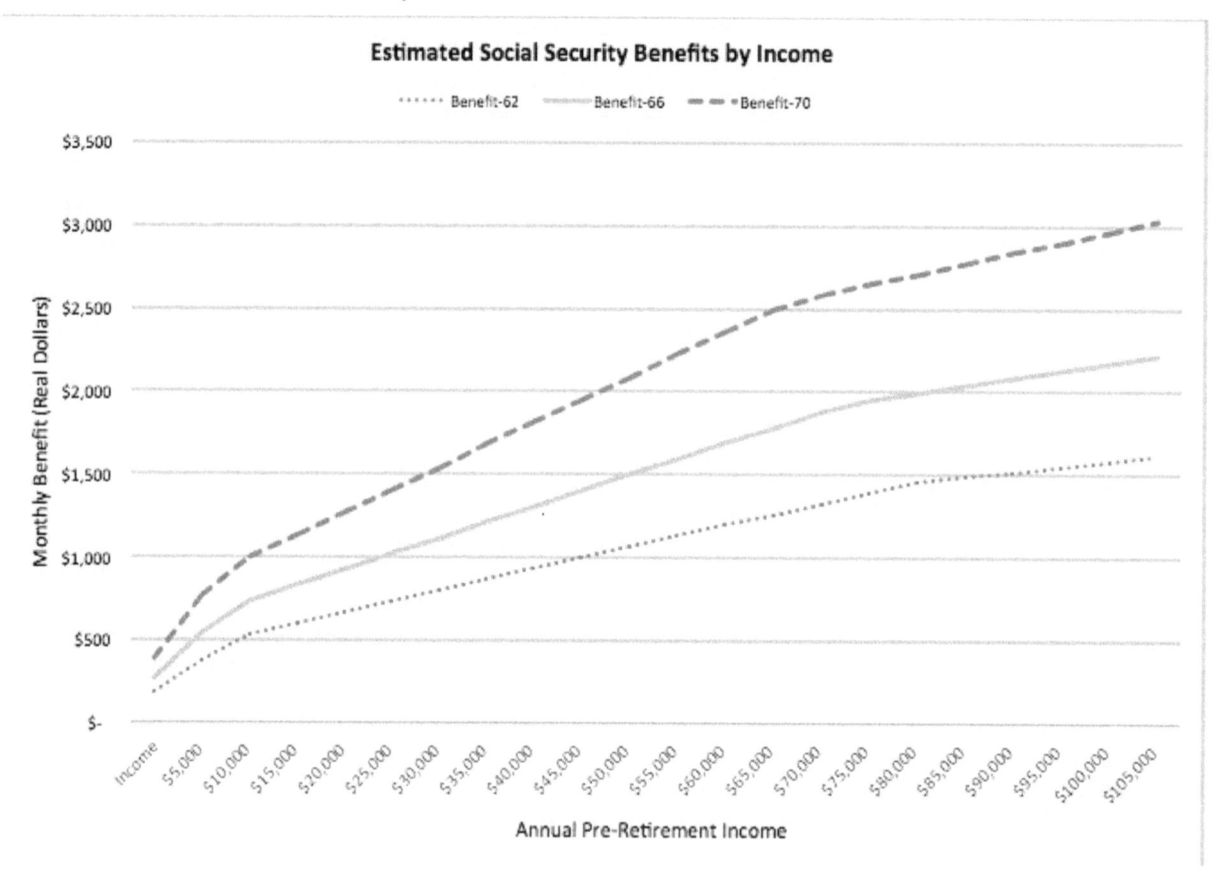

In the graph in Figure 13.5, the benefits amounts are in real dollars, so we can compare these amounts directly to real wages and prices. The three curves in Figure 13.5, correspond to 3 possible ages at which workers may elect to begin receiving benefits. For a worker born in 1960, it is possible to receive full retirement benefits at age 66 ("Benefit-66), early benefits beginning 62 and 1 month ("Benefit-62"), and delayed benefits as late as age 70 ("Benefit-70").

Workers often choose to receive benefits based on the date they stop working, but this is not required. For example, you could stop working at age 62 and wait until age 70 to begin receiving benefits, or even begin receiving benefits while still working. How should workers choose the date at which to begin receiving benefits? Based on the life-cycle model, you should chose to begin receiving benefits such that you maximize and smooth your lifetime standard of living.

The Social Security Administration's website provides a quick calculator, which can be used to estimate benefit amounts based on age and pre-retirement income. Figure 13.6 shows an estimate of social security benefits for a young worker earning $50,000 per year.[86]

Figure 13.6 Social Security Benefits Estimate

Information you submitted
Date of birth: **6/15/1990**
Current earnings: **$50,000.00**
Benefit in **year-2015** dollars

Retirement Benefit Estimates

Retirement age	Monthly benefit amount [1]
62 and 1 month in 2052	$1,276.00
67 in 2057	$1,812.00
70 in 2060	$2,247.00

[1] Assumes no future increases in prices or earnings.

Notice that the full retirement benefit amount for a worker born in 1990 and earning $50,000 per year is $1,812 per month, beginning at age 67, or $21,744 per year. In most cases, workers will benefit most (in actuarial present value) by waiting until ago 70 before beginning to receive benefits.

[86]Social Security Quick Calculator, http://www.ssa.gov/oact/quickcalc/. Benefit estimates depend on your date of birth and on your earnings history. For actual benefits, refer to your annual Social Security Statement.

13.6 Incorporating Taxes and Benefits into the Lifetime Budget Constraint

Taxes on ordinary (labor) income are straightforward enough: the more you earn, the more taxes you pay. But the effect of taxes on the lifetime budget constraint is subtly more complicated. Let's begin by reviewing the lifetime budget constraint for a two-period model with income earned only when young without taxes or benefits.

We can write the budget constraint when young as:

$$Cy = W - S$$

And the budget constraint when old as:

$$Co = S(1 + r)$$

By writing both equations in present value, we can combine to create a lifetime budget constraint:

$$W = Cy + \frac{Co}{1 + r}$$

In words: consumption when young and the present value of consumption when old are bounded by lifetime income.

Taxes and Benefits

Along with the burden of taxation come the benefits of transfer payments (from the government to households). The most common transfer payments are received in old age in the form of Social Security and Medicare benefits, but other common benefits include Temporary Aid to Families with Dependent Children ("welfare"), Food Stamps, assistance with home heating costs, and tax credits for work, children, and education. Here we describe the lifetime budget constraint in a world with taxes and benefits:

Revised budget constraint when young: $Cy = W - S - Ty + By$
Revised budget constraint when old: $Co = S(1 + r) - To + Bo$

The new terms *Ty* and *To* refer to taxes when young and taxes when old, and the new terms *By* and *Bo* refer to benefits received when young and benefits received when old.

The revised lifetime budget constraint (in present value terms) is given by:

$$Cy + \frac{Co}{1 + r} = W - Ty - \frac{To}{1 + r} + By + \frac{Bo}{1 + r}$$

In words, we can say that the present value of lifetime consumption, $Cy + \frac{Co}{1+r}$, is limited by the present value of all income (*W*) minus the present value of all taxes, $Ty - \frac{To}{1+r}$, plus the present value of all benefits, $By + \frac{Bo}{1+r}$.

Uncle Sam's Lifetime Net Tax Take

You have, or your household has, a lifetime business partner in Uncle Sam (a colloquial name for the US Government). Of everything you earn, you give part to Sam in taxes. When you are old (or maybe when you are young), Sam gives some to you in benefits. In the previous section, we described the lifetime budget constraint including taxes *Ty* and *To* and benefits *By* and *Bo*:

$$Cy + \frac{Co}{1 + r} = W - Ty - \frac{To}{1 + r} + By + \frac{Bo}{1 + r}$$

Let's simplify by describing the net value (in the present) of all of your interactions with Uncle Sam by the symbol *H,* which is Uncle Sam's lifetime net tax take. *H* describes how much you will pay in lifetime taxes less what you will receive in lifetime benefits.[87]

$$H = Ty + \frac{To}{1+r} - By - \frac{Bo}{1+r}$$

Simplifying the lifetime budget constraint, we can write:

$$Cy + \frac{Co}{1+r} = W - H$$

Taxes on Labor Income Alone

If taxes were only assessed on labor income, figuring out how much you owe would be simple. Figure 13.7 depicts the lifetime budget constraint for a two-period model with taxes on labor income alone, as compared to the income you would have with no government at all (i.e., Uncle Sam is Dead).

Figure 13.7 Two-Period Lifetime Budget Constraint if Taxes Depended only on Labor Income

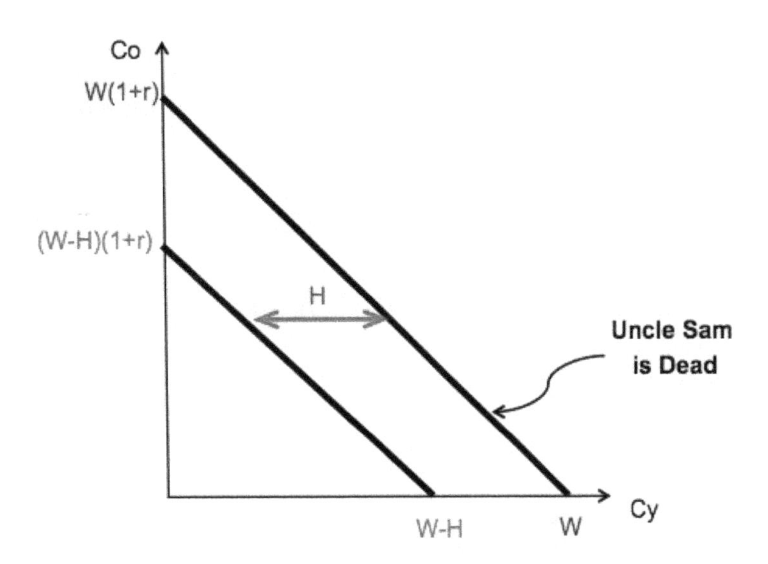

In Figure 13.7, the level of earned income directly determines the level of *H*, Uncle Sam's net lifetime tax take. Some fraction of labor income is taxed in the year in which the income is earned, and the remainder *W-H* is available to consume or save. *W-H* depends directly on the level of income taxed when the income is earned.

Taxes on Both Labor Income and Asset Income

In addition to the taxes assessed on labor income, taxes in the United States are assessed on asset income (e.g., income from interest, dividends, capital gains, etc.). Taxes on asset income are due in the year in which that asset income is *realized* (or earned from a tax perspective). Income from interest and dividends is realized in the year in which it is received. However, income from capital gains is taxable in the year in which the asset is sold. By choosing when to sell an asset, you have some control over the year in which the gains will be taxed.

Consider Pete, who earns \$50,000 per year who saves 20% of his income each year. Pete's savings S = \$10,000 per year is invested at r = 5% and earns \$500 of interest income in year 2 (the first full year for which it was invested).

[87] Why are we using letter H? In many economics textbooks, the letter G is used to indicate the government-spending component of the economy. To avoid confusion about G, we are using the next available letter H.

Figure 13.8 Savings, Asset Income, and Gross Income

	A	B	C	D	E	F	G	H	I
1	S=	$ 10,000							
2	r =	5%							
3	Year	Assets	Asset Income	Savings	Ending Assets		Wage Income	Asset Income	Gross Income
4	1	$ -	$ -	$ 10,000	$ 10,000		$ 50,000	$ -	$ 50,000
5	2	$ 10,000	$ 500	$ 10,000	$ 20,500		$ 50,000	$ 500	$ 50,500
6	3	$ 20,500	$ 1,025	$ 10,000	$ 31,525		$ 50,000	$ 1,025	$ 51,025
43									
44	40	$ 1,140,950	$ 57,048	$ 10,000	$ 1,207,998		$ 50,000	$ 57,048	$ 107,048

As we see in Figure 13.8, each year Pete earns wage income of $50,000, and also earns asset income on his accumulated savings from previous years. As his wealth grows, so does his asset income, and as his asset income grows, so does his taxable gross income.

Taxes are a Disincentive to Saving

Recall that savings is what's left over from income after consumption ($S = W - C$). The more you save, the more assets you will accumulate. Having more assets lead to more asset income, and asset income is taxable. The more asset *income* you have, the more you will pay in future taxes.

Uncle Sam's net tax burden depends upon (is a function of) the level of consumption when young. If you consume less when young, you will save more, accumulate more assets, earn more asset income in the future, and thus owe more future taxes. If you consume more when young, you will save, accumulate fewer assets, earn less asset income, and owe fewer future taxes. After reviewing the effect of taxation on the lifetime budget constraint, we arrive at a most unhappy conclusion: *taxes are a disincentive to saving.*

Figure 13.9 shows the revised lifetime budget constraint for a two-period model in a world with taxation on both earned income *and* asset income.

Figure 13.9 Two-Period Lifetime Budget Constraint for a World with Taxation of Both Labor Income and Asset Income

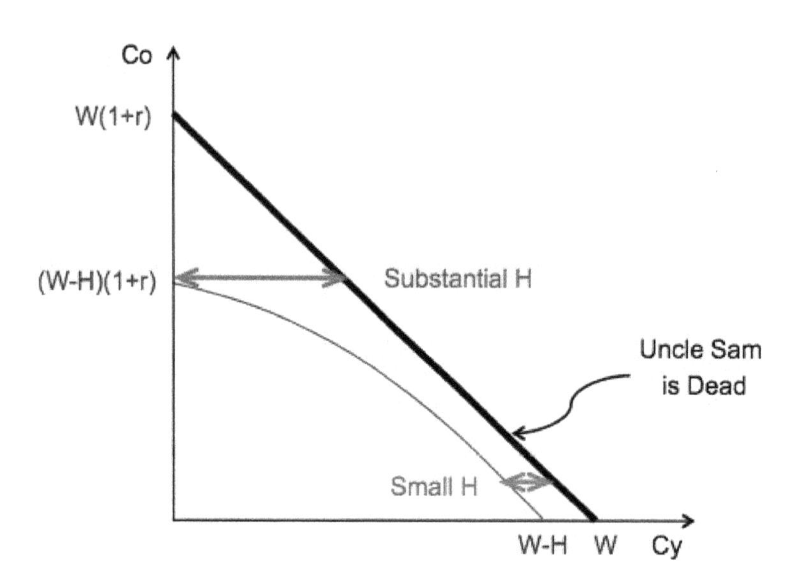

Given that taxes when old depend on choices about consuming (and thus saving) when young, we write H(Cy). In words, taxes are a function of consumption when young. Incorporating this notation, we can rewrite the two period lifetime budget constraint as:

$$Cy + \frac{Co}{1+r} = W - H(Cy)$$

The choice of *how much to consume* when young affects *how much you pay in taxes*. We now turn to the problem of smoothing consumption when taxes depend on consumption choices.

Smoothing Consumption in a World with Taxes

As we have just seen, the amount of one's total lifetime net tax burden depends on household choices about how much to consume and save when young. This creates a messy situation for determining the level of smooth consumption: The amount you can consume when young depends in part on the taxes you will pay when old, and the amount of taxes you pay when you're old depends on your saving (and consumption) when young. This is a case of the classic "chicken and egg" problem: which comes first?

We can solve this kind of problem by re-writing the lifetime budget constraint to set the level of $\bar{C}$ for smooth consumption, i.e., $\bar{C} = Cy = Co$. The equation now simplifies as:

$$Cy + \frac{Co}{1+r} = W - H(\bar{C})$$

Finally, all we need to do is guess values of $\bar{C}$ until we discover the *single* value of $\bar{C}$ for which the equation will be equal. Given the tax rate structure and benefits structure, there is only one value of $\bar{C}$ for which this equation will be solved, and that is the level of smooth consumption that can be achieved.

13.7 The Life-Cycle Model In a World With Taxes: A Spreadsheet Solution

The two period life-cycle model is useful to illustrate conceptual ideas about consumption smoothing in a simple way and to graphically show the tradeoff between consuming when young and when old. In this section, we return to a multi-period LC model to show how this actually works in practice.

Previously, we developed a multi-period LC model. In Module 2 (*The Life-Cycle Model*) we developed the basic framework including using lifetime income, smoothing consumption. In Module 4 (*The Life-Cycle Model with Interest*), we included a positive real rate of return on investment, and accounted for asset income, which added to lifetime consumption.

In this iteration of the LC model we will include income from employment (W) as well as Social Security benefits income (B) and taxes (T). Our revised annual budget constraint is given by the equation:

$$C = W + B + AI - T - S$$

Recall that AI is asset income, C is the level of consumption in a given period, and S is the amount of savings required to satisfy the annual budget constraint.

General Assumptions

The general assumptions made by this calculator are as follows:

- All dollar amounts are in real dollars.
- The real rate of return is constant for all years.
- The tax calculations use the marginal tax rates for 2015, using a single filing status on the 1040EZ return.[88]

Description of the Spreadsheet Solution

The spreadsheet has 3 main sections that produce calculations for each year of the life-cycle:

- Section A: The Life-Cycle Model
- Section B: The Tax Calculator
- Section C: Federal Marginal Income Tax Calculations

Section A: The Life-Cycle Model

Figure 13.10 shows the set up of the life-cycle model in the spreadsheet.

Figure 13.10 Section A, The Life-Cycle Model

	A	B	C	D	E	F	G	H	I
1	Real Rate	1.0%	* assumes all assets invested in TIPS						
2	Initial Assets	$ -	* negative for initial debts						
3	Final Assets	$ -	SECTION A: LIFE-CYCLE MODEL						
4	Consumption	$ -							
5									
6			W	B	A	AI	T	S	C
7	Age	Year	Labor Income	Soc Sec Benefits	Assets	Asset Income	Taxes	Savings	Consumption
8	23	2015	$ -	$ -	$ -	$ -	$ -	$ -	$ -
9	24	2016	$ -	$ -	$ -	$ -	$ -	$ -	$ -
10	25	2017	$ -	$ -	$ -	$ -	$ -	$ -	$ -

The inputs required for the life-cycle model are: the amounts of annual income, whether from wages (W) or Social Security Benefits (B), and the level of initial assets (A), which could be 0. The columns marked asset income (AI), savings (S), consumption (C), and taxes (T) are calculated by sections B and C of the spreadsheet (see below).

Section B: The Tax Calculator

Figure 13.11 shows the general tax calculator, which implements several calculations from Form 1040EZ, as well as calculating state and payroll taxes.

[88] The 1040EZ is used here for demonstration purposes because it is a reasonable approximation of the tax situation facing a single tax payer with a straightforward tax situation. Many taxpayers will not be able to use form 1040EZ, because of dependents, self-employment income, or itemized deductions.

Figure 13.11 Section B, the Tax Calculator

	A	B	K	L	M	N	O	P	Q	R
1	Real Rate	1.0%								
2	Initial Assets	$ -						Social Sec. Limit		
3	Final Assets	$ -	SECTION B: TAX CALCULATOR					$ 118,500		
4	Consumption	$ -								
5										
6			W	AI	AGI		5%	6.20%	1.45%	T
7	Age	Year	Labor Income	Asset Income	Gross Income	Fed Inc. Tax	State Tax	Soc. Sec Tax	Medicare Tax	Total Taxes
8	23	2015	$ -	$ -	$ -	$ -	$ -	$ -	$ -	$ -
9	24	2016	$ -	$ -	$ -	$ -	$ -	$ -	$ -	$ -
10	25	2017	$ -	$ -	$ -	$ -	$ -	$ -	$ -	$ -

The calculations made in Section B include:

- Adjusted Gross Income (AGI), which is the sum of labor income (W) plus asset income (AI).
- The amount of Federal income tax show in column N is calculated in section C (described below), and referenced here for convenience in determining total taxes.
- The state income tax is calculated assuming a flat tax rate on adjusted income, a scheme followed by many states. For example, in Massachusetts the state income tax is 5% of AGI, but obviously this input can be changed for other states.
- Social Security taxes due. Social Security taxes are collected at 6.2% on all labor income (but not on asset income) up to $118,500, the limit in 2015.
- Medicare Tax due. Medicare taxes are collected at the rate of 1.45% on all labor income (but not on asset income).

Section C: Federal Marginal Income Tax Calculations

Figure 13.12 shows the Federal Marginal Income Tax calculator, which implements the step-function for marginal tax rate calculations.[89]

Figure 13.12 Section C, the Federal Marginal Income Tax Calculator

	A	B	U	V	W	X	Y	Z	AA	AB
1	Real Rate	1.0%								
2	Initial Assets	$ -								
3	Final Assets	$ -	SECTION C: MARGINAL TAX RATE CALCULATIONS							
4	Consumption	$ -								
5				2015 Marginal Tax Rates						
6			Cutoffs	$ 9,225.00	$ 37,450.00	$ 90,750.00	$ 189,300.00	$ 411,500.00	$ 413,200.00	no limit
7	Age	Year	Rates	10%	15%	25%	28%	33%	35%	39.60%
8	23	2015		$ -	$ -	$ -	$ -	$ -	$ -	$ -
9	24	2016		$ -	$ -	$ -	$ -	$ -	$ -	$ -
10	25	2017		$ -	$ -	$ -	$ -	$ -	$ -	$ -

The column labeled "Taxable Income" takes the total adjusted gross income from Section C, and subtracts the Federal Exemption amount for a single filer.

The columns V through AB calculate the amount of taxable income in each marginal tax bracket, using the rates in row 6. For each cell in the tax calculator, an IF statement implements the logic to determine the amount of income that falls within the tax bracket. Since the amount of taxable income in a bracket cannot be less than 0, the MAX function is used to take the greater amount of the taxable income in this bracket, or 0. Finally, this amount of taxable income is multiplied by the applicable marginal tax rate.

For example, the cell V8 contains this formula: =MAX(IF($T8<V$6,$T8,V$6),0)*V$7. The same calculation is made for each tax bracket and for each year.

Using Inputs for Wages and Benefits

We now begin to fill in the inputs to the LC model and apply the calculations. First, we need to make assumptions about the level of labor income and social security benefits using a conservative human capital estimate for all wages and using the Social Security Administration's (SSA) quick benefits calculator to determine Social Security benefits.

---

[89] This calculator is "tuned" to work for the single tax filing status using the marginal tax rates from 2015. To implement this spreadsheet for a married couple filing jointly, it would be necessary to use the appropriate marginal tax brackets for that filing status. Similar, if marginal tax rates or brackets change, the spreadsheet will need to be adjusted.

Figure 13.13 shows the inputs for labor income and Social Security benefits. In this example, we will assume the worker will work until age 66 and then receive Social Security benefits from age 67 until age 100, the maximum age of life. To keep things simple and conservative, we assume that wages are flat in real dollar terms for all working years.

a. We enter $50,000 as the amount of wages in cell B3 and set each year's labor income to reference cell B3 (for all years until retirement at age 66, cells C7 through C52).
b. Based on the estimate provided by the SSA's quick calculator (Figure 13.6), we will enter real social security benefits of $21,492 in cell B4.
c. Notice also how the field labeled Taxes has been filled in for each year for which there is labor income. In general, Social Security benefits are not taxable, but in some cases benefits are taxable. For now we will assume that benefits are not taxable for the sake of simplicity (notice that we set AGI = W + AI without including B).

Figure 13.13 Filling in Inputs to the Life-cycle model

	A	B	C	D	E	F	G	H	I
1	Real Rate	1.0%	* assumes all assets invested in TIPS						
2	Initial Assets	$ -	* negative for initial debts						
3	Final Assets	$ -	SECTION A: LIFE-CYCLE MODEL						
4	Consumption	$ -							
5									
6			W	B	A	AI	T	S	C
7	Age	Year	Labor Income	Soc Sec Benefits	Assets	Asset Income	Taxes	Savings	Consumption
8	23	2015	$ 50,000	$ -	$ -	$ -	$ 12,044	$ -	$ -
9	24	2016	$ 50,000	$ -	$ -	$ -	$ 12,044	$ -	$ -
10									
51	65	2057	$ 50,000	$ -	$ -	$ -	$ 12,044	$ -	$ -
52	66	2058	$ 50,000	$ -	$ -	$ -	$ 12,044	$ -	$ -
53	67	2059	$ -	$ 21,744	$ -	$ -	$ -	$ -	$ -
54	68	2060	$ -	$ 21,744	$ -	$ -	$ -	$ -	$ -
55									
86	99	2091	$ -	$ 21,744	$ -	$ -	$ -	$ -	$ -
87	100	2092	$ -	$ 21,744	$ -	$ -	$ -	$ -	$ -
88			FINAL ASSETS AFTER AGE 100		$ -				

Tax Calculations

The calculations of taxes for each year are performed in sections 2 and 3 of the spreadsheet. Figure 13.14 shows the Federal income tax calculation. Notice how the $50,000 of Adjusted Gross Income was reduced to $39,700 of Taxable Income. This is because of the Federal exemption + standard deduction amount of $10,300.

Figure 13.14 Section C, showing the Federal Marginal Income Tax Calculation

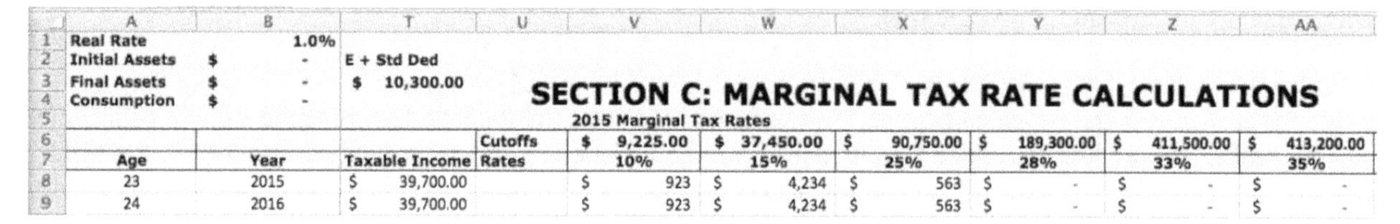

	A	B	T	U	V	W	X	Y	Z	AA
1	Real Rate	1.0%								
2	Initial Assets	$ -	E + Std Ded							
3	Final Assets	$ -	$ 10,300.00		SECTION C: MARGINAL TAX RATE CALCULATIONS					
4	Consumption	$ -								
5					2015 Marginal Tax Rates					
6				Cutoffs	$ 9,225.00	$ 37,450.00	$ 90,750.00	$ 189,300.00	$ 411,500.00	$ 413,200.00
7	Age	Year	Taxable Income	Rates	10%	15%	25%	28%	33%	35%
8	23	2015	$ 39,700.00		$ 923	$ 4,234	$ 563	$ -	$ -	$ -
9	24	2016	$ 39,700.00		$ 923	$ 4,234	$ 563	$ -	$ -	$ -

The sum of Federal income taxes for all applicable brackets is $5,718.75, which is calculated in column N. Figure 13.15 shows the calculation of taxes for state and payroll taxes). The column labeled "Total Taxes" is the sum of Federal income tax, state income tax, Social Security tax and Medicate tax.

Figure 13.15State and Payroll Tax Calculations, and Calculation of Total Taxes

	A	B	K	L	M	N	O	P	Q	R
1	Real Rate	1.0%								
2	Initial Assets	$ -						Social Sec. Limit		
3	Final Assets	$ -	SECTION B: TAX CALCULATOR					$ 118,500		
4	Consumption	$ -								
5										
6			W	AI	AGI		5%	6.20%	1.45%	T
7	Age	Year	Labor Income	Asset Income	Gross Income	Fed Inc. Tax	State Tax	Soc. Sec Tax	Medicare Tax	Total Taxes
8	23	2015	$ 50,000	$ -	$ 50,000	$ 5,718.75	$ 2,500.00	$ 3,100.00	$ 725.00	$ 12,043.75
9	24	2016	$ 50,000	$ -	$ 50,000	$ 5,718.75	$ 2,500.00	$ 3,100.00	$ 725.00	$ 12,043.75

The amount of Federal and State income tax is a function of Adjusted Gross Income, but the amount of payroll taxes for Social Security and Medicare is a function of labor income. The total taxes, based on the adjusted gross income of \$50,000 are \$12,043.75. It is recommended to play around with some different inputs for labor income (W) to see how the tax calculations adjust to the inputs.

Implementing the Annual Budget Constraint

Now we move onto implementing the life-cycle model. Recall the annual budget constraint for the life-cycle model with taxes is $C = W + B + AI - T - S$. To smooth consumption, a single value of C̲ should solve the annual budget constraint for all periods and achieve final assets of \$0 at the maximum age of life. We want to find that value of C̲, but we need to begin by guessing a value for consumption and solving the annual budget constraint to find the value of savings.

We begin with a "guess" for consumption of \$30,000 and input this value in cell B4. We will refer to this amount of consumption for each year's consumption, i.e., in column I. We can now solve for annual savings by re-arranging the annual budget constraint to $S = W + B + AI - T - C$. Figure 13.16 shows the updated spreadsheet with the annual budget constraint implemented in cell H7.

Figure 13.16 The Annual Budget Constraint, Shown for Age 22

	A	B	C	D	E	F	G	H	I
1	Real Rate	1.0%	* assumes all assets invested in TIPS						
2	Initial Assets	$ -	* negative for initial debts						
3	Final Assets	$ 254,377.92	SECTION A: LIFE-CYCLE MODEL						
4	Consumption	$ 30,000.00							
5									
6			W	B	A	AI	T	S	C
7	Age	Year	Labor Income	Soc Sec Benefits	Assets	Asset Income	Taxes	Savings	Consumption
8	23	2015	$ 50,000	$ -	$ -	$ -	$ 12,044	$ 7,956	$ 30,000
9	24	2016	$ 50,000	$ -	$ 7,956	$ 80	$ 12,068	$ 8,012	$ 30,000

In Figure 13.16, we started with a "guess" for the level of consumption of \$30,000. Income was assumed, and taxes were calculated in the manner previously described. We can re-arrange the annual budget constraint to solve for the level of savings, $S = W + B + AI - T - C$. This formula is entered into cell H7 using the formula `=C8+D8+F8-G8-I8`. Figure 13.17 shows the calculation for the amount of savings at age 23.

Figure 13.17 Formula to Implement the Annual Budget Constraint

	A	B	C	D	E	F	G	H	I
1	Real Rate	1.0%	* assumes all assets invested in TIPS						
2	Initial Assets	$ -	* negative for initial debts						
3	Final Assets	$ 254,377.92	SECTION A: LIFE-CYCLE MODEL						
4	Consumption	$ 30,000.00							
5									
6			W	B	A	AI	T	S	C
7	Age	Year	Labor Income	Soc Sec Benefits	Assets	Asset Income	Taxes	Savings	Consumption
8	23	2015	$ 50,000	$ -	$ -	$ -	$ 12,044	=C8+D8+F8-G8-I8	$ 30,000
9	24	2016	$ 50,000	$ -	$ 7,956	$ 80	$ 12,068	$ 8,012	$ 30,000

We interpret this as follows: given a labor income of \$50,000, benefits of \$0, asset income of \$0, total taxes of \$12,044.75, and consumption of \$30,000, we arrive at a level of savings at age 23 of \$7,956. The amount saved at age 23 is added to assets at age 24. Hence the amount of \$7,956 appears as assets at age 24 (i.e., in cell E9).

It is important to note that savings is a "plug" which we use to solve the annual budget constraint, given the level of consumption. We will return to the amount of savings shortly. Notice also that the \$7,946 of regular assets owned at the beginning of the year for age 24 earned asset income of approximately \$80. The asset income of \$80 is added to adjusted gross income in year 2016, thereby increasing taxes in year 2016 to \$12,068. This is the effect that was described earlier: as we convert income to regular assets by the process of saving, we thus increase the annual tax burden.

Now we can complete the spreadsheet for all years, by making one more adjustment: regular assets in the "running total" of all years of savings. For all years after age 23, regular assets is calculated as:

$$A_n = A_{n-1} + S_{n-1}$$

In words: the level of regular assets in year n is the sum of the regular assets in year *n-1*, plus the savings in year *n-1*. (The savings in year *n-1* includes the asset income earned in year *n-1*.) The equations for the lifetime budget constraint (column H), which calculates the amount of savings each year and the amount of regular assets (column E), are extended down for all years until age 100. Figure 13.18 shows the completed spreadsheet with several rows omitted for brevity.

Figure 13.18 The Completed LC Model, with Consumption of $30,000 per Year

	A	B	C	D	E	F	G	H	I
1	**Real Rate**	**1.0%**	*** assumes all assets invested in TIPS**						
2	**Initial Assets**	**$ -**	*** negative for initial debts**						
3	**Final Assets**	**$ 233,623.62**	**SECTION A: LIFE-CYCLE MODEL**						
4	**Consumption**	**$ 30,000.00**							
5									
6			**W**	**B**	**A**	**AI**	**T**	**S**	**C**
7	**Age**	**Year**	**Labor Income**	**Soc Sec Benefits**	**Assets**	**Asset Income**	**Taxes**	**Savings**	**Consumption**
8	23	2015	$ 50,000	$ -	$ -	$ -	$ 12,044	$ 7,956	$ 30,000
9	24	2016	$ 50,000	$ -	$ 7,956	$ 80	$ 12,068	$ 8,012	$ 30,000
10									
51	65	2057	$ 50,000	$ -	$ 386,913	$ 3,869	$ 13,204	$ 10,665	$ 30,000
52	66	2058	$ 50,000	$ -	$ 397,577	$ 3,976	$ 13,236	$ 10,739	$ 30,000
53	67	2059	$ -	$ 21,744	$ 408,317	$ 4,083	$ 204	$ (4,377)	$ 30,000
54	68	2060	$ -	$ 21,744	$ 403,940	$ 4,039	$ 202	$ (4,419)	$ 30,000
55									
86	99	2091	$ -	$ 21,744	$ 245,527	$ 2,455	$ 123	$ (5,923)	$ 30,000
87	100	2092	$ -	$ 21,744	$ 239,603	$ 2,396	$ 120	$ (5,980)	$ 30,000
88				**FINAL ASSETS AFTER AGE 100**	$ 233,624				

Now that we have completed the spreadsheet, we discuss a few important elements. First, notice how the level of consumption is smooth at $30,000 for all years. We began the process by choosing a level of consumption, and this selection led to the level of savings. The amount of savings determined the amount of assets, asset income, and thus taxes on asset income. With income known (assumed), everything else is a *function of consumption*.

Also notice the amount of assets left over after year 100, which in this case is $233,624. This is evidence of under-consumption, i.e., over-saving. By setting the level of consumption at $30,000, the result is that at age 100 the subject has $208,702 of assets left over at the end of life.

If one has the goal to leave a bequest at the end of their life, it would be possible to plan ahead to have any arbitrary amount of assets left over at age 100, for example $100,000. For now, we will assume no bequest motive and treat the remaining assets at age 100 as under consumption.

Since the level of consumption of $30,000 was just a guess that was too low, we can try another guess. Figure 13.19shows what happens when the level of consumption is set at $31,000 per year.

Figure 13.19 The Completed LC Model, with Consumption of $31,000 per Year

	A	B	C	D	E	F	G	H	I
1	**Real Rate**	**1.0%**	*** assumes all assets invested in TIPS**						
2	**Initial Assets**	**$ -**	*** negative for initial debts**						
3	**Final Assets**	**$ 122,933.03**	**SECTION A: LIFE-CYCLE MODEL**						
4	**Consumption**	**$ 31,000.00**							
5									
6			**W**	**B**	**A**	**AI**	**T**	**S**	**C**
7	**Age**	**Year**	**Labor Income**	**Soc Sec Benefits**	**Assets**	**Asset Income**	**Taxes**	**Savings**	**Consumption**
8	23	2015	$ 50,000	$ -	$ -	$ -	$ 12,044	$ 6,956	$ 31,000
9	24	2016	$ 50,000	$ -	$ 6,956	$ 70	$ 12,065	$ 7,005	$ 31,000
10									
51	65	2057	$ 50,000	$ -	$ 338,283	$ 3,383	$ 13,059	$ 9,324	$ 31,000
52	66	2058	$ 50,000	$ -	$ 347,607	$ 3,476	$ 13,087	$ 9,389	$ 31,000
53	67	2059	$ -	$ 21,744	$ 356,996	$ 3,570	$ 178	$ (5,865)	$ 31,000
54	68	2060	$ -	$ 21,744	$ 351,132	$ 3,511	$ 176	$ (5,920)	$ 31,000
55									
86	99	2091	$ -	$ 21,744	$ 138,882	$ 1,389	$ 69	$ (7,937)	$ 31,000
87	100	2092	$ -	$ 21,744	$ 130,945	$ 1,309	$ 65	$ (8,012)	$ 31,000
88				**FINAL ASSETS AFTER AGE 100**	$ 122,933				

Notice now that assets remaining at the end of life are $122,933. Having assets of $122,933 (versus $233,624) left over at age 100 suggests that $31,000 is a better guess than $30,000, but this number is still too low. We could keep guessing until we get it right, since there is only one value of consumption that will result in neither under-consuming nor over-consuming.

Using Goal Seek to Solve for Smooth Consumption

Excel provides a built-in tool called Goal Seek, which implements the guess and check algorithm. It works by repeating trying values until some ending condition is satisfied. In this case, we ask Goal Seek to set the assets at the end of life (shown in cell E88 and repeated in cell B3)to $0, by changing the level of consumption (cell B4). Figure 13.20 shows the Goal Seek dialog box with values selected to solve for the smooth level of consumption.

Figure 13.20 Using Goal Seek to Solve for the Level of Consumption

	A	B	C	D	E	F	G	H	I
1	Real Rate	1.0%	* assumes all assets invested in TIPS						
2	Initial Assets	$ -	* negative for initial debts						
3	Final Assets	$ 122,933.03	SECTION A: LIFE-CYCLE MODEL						
4	Consumption	$ 31,000.00							
5									
6			W	B	A	AI	T	S	C
7	Age	Year	Labor Income	Soc Sec Benefits	Assets	Asset Income	Taxes	Savings	Consumption
8	23	2015	$ 50,000	$ -	$ -	$ -	$ 12,044	$ 6,956	$ 31,000
9	24	2016	$ 50,000	$ -	$ 6,956	$ 70	$ 12,065	$ 7,005	$ 31,000
10									
51	65	2057	$ 50,000	$ -	$ 338,283			$ 9,324	$ 31,000
52	66	2058	$ 50,000	$ -	$ 347,607			$ 9,389	$ 31,000
53	67	2059	$ -	$ 21,744	$ 356,996			$ (5,865)	$ 31,000
54	68	2060	$ -	$ 21,744	$ 351,132			$ (5,920)	$ 31,000
55									
86	99	2091	$ -	$ 21,744	$ 138,882			$ (7,937)	$ 31,000
87	100	2092	$ -	$ 21,744	$ 130,945			$ (8,012)	$ 31,000
88			FINAL ASSETS AFTER AGE 100		$ 122,933				
89									

Goal Seek
Set cell: B3
To value: 0
By changing cell: B4
Cancel OK

Goal Seek works by guessing values until it finds the value that satisfies its constraint.

In Figure 13.21, we see that Goal Seek has solved the life-cycle model for the one value of consumption that satisfies the lifetime budget constraint while leaving no assets left over at age 100. Given the lifetime income and benefits that we used as inputs, and the tax scheme and rates built into the calculator, we found that smooth consumption of $32,110.60 can be sustained to age 100.

Figure 13.21 The LC Model with Smooth Consumption and $0 Final Assets

	A	B	C	D	E	F	G	H	I
1	Real Rate	1.0%	* assumes all assets invested in TIPS						
2	Initial Assets	$ -	* negative for initial debts						
3	Final Assets	$ 0.00	SECTION A: LIFE-CYCLE MODEL						
4	Consumption	$ 32,110.60							
5									
6			W	B	A	AI	T	S	C
7	Age	Year	Labor Income	Soc Sec Benefits	Assets	Asset Income	Taxes	Savings	Consumption
8	23	2015	$ 50,000	$ -	$ -	$ -	$ 12,044	$ 5,846	$ 32,111
9	24	2016	$ 50,000	$ -	$ 5,846	$ 58	$ 12,061	$ 5,887	$ 32,111
10									
51	65	2057	$ 50,000	$ -	$ 284,274	$ 2,843	$ 12,897	$ 7,836	$ 32,111
52	66	2058	$ 50,000	$ -	$ 292,110	$ 2,921	$ 12,920	$ 7,890	$ 32,111
53	67	2059	$ -	$ 21,744	$ 300,000	$ 3,000	$ 150	$ (7,517)	$ 32,111
54	68	2060	$ -	$ 21,744	$ 292,484	$ 2,925	$ 146	$ (7,588)	$ 32,111
55									
86	99	2091	$ -	$ 21,744	$ 20,441	$ 204	$ 10	$ (10,172)	$ 32,111
87	100	2092	$ -	$ 21,744	$ 10,269	$ 103	$ 5	$ (10,269)	$ 32,111
88			FINAL ASSETS AFTER AGE 100		$ 0				

Earlier we described savings as a plug: an amount chosen to make the annual budget constraint work. In the life-cycle model, the amount of savings is different every year. Conventional financial advice usually suggests constant savings targets for each year, but this necessarily will lead to shocks in the level of consumption that negatively impact utility. Since the goal is to smooth consumption (i.e., to maximize your utility), the level of savings will adjust annually to achieve this goal. During the years of labor income (age 23 to 66 in this example), savings is advised, and during years without labor income (years 67 to 100), dissaving is advised.

13.8 Summary

In this module, we introduced taxation as the way to raise money to pay for government provided goods and services. A discussion of tax bases and systems for taxation was presented with an emphasis on taxation in the United States. The Federal income tax, which applies to individual income, was discussed as a detailed example of a progressive tax system. The marginal tax rate is the rate applied to the last dollar of earnings, and the effective tax rate is the fraction of household income that goes to taxes.

Social Security benefits provide old-age income assurance to qualified individuals in the United States. The amounts of benefits are determined by several factors including pre-retirement income and the age at which the beneficiary begins to receive benefits.

After the discussion of taxation and benefits in the United States, we turned our attention to the implications of taxation on the life-cycle model. In particular, the taxation of asset income creates a tax disincentive to savings. The implication for consumption smoothing is that the level of smooth consumption depends on the level of future taxes, but the level of future taxes depends on the amount of savings when young, which in turn depends on consumption when young (i.e., it is a circular problem). This kind of problem must be solved by a guess-and-check method, which we demonstrated using a spreadsheet model in Microsoft Excel. The end result of consumption smoothing is that we have a practical financial plan: a recommendation of how much to consume and how much to save each year.

The main shortcoming of this consumption-smoothing spreadsheet is that it cannot correctly smooth consumption for a household that is borrowing constrained. This shortcoming was address in Module 3, where we discussed dynamic programming as a way to find the smoothest path without borrowing under conditions of borrowing constraints.

This shortcoming will be addressed in Module 7, when we introduce Economic Security Planner, a software program that provides a comprehensive implementation of the life-cycle model. ESPlanner is fundamentally similar to the consumption smoothing spreadsheet developed here, but includes accurate tax calculations for all filing statuses, state taxes, as well as household composition (e.g., married couples, children), housing, both regular assets and tax-favored retirement accounts, and special expenses (e.g., loan payments, future college tuition expenses, or bequests).

13.9 Review Questions

1. Identify and explain the main tax bases and which taxes apply to each in the USA.
2. Explain how the following taxes work, and give an example of each: per-capita taxes, flat taxes, progressive taxes, and regressive taxes.
3. Explain these parts of the tax calculation process: adjusted gross income, exemptions, deductions, the marginal tax calculation process, and tax refunds.
4. Explain by giving a numerical example why effective tax rates are always lower than marginal tax rates.
5. Identify the main benefits households receive from the U.S. Federal Government.
6. Discuss the relationship between Social Security benefits and income.
7. Describe the various types of personal income, and which taxes apply to each.
8. Consider Joan, a single tax filer who earned $57,000 in 2015. Assuming she will file her taxes using form 1040EZ:
    a. What is her adjusted gross income?
    b. What is her taxable income?
    c. What is her Federal income tax due?
    d. What is her marginal tax rate?
    e. What is her effective tax rate?
9. Consider Carl and Gina, a married couple, who earned $137,000 in 2015. Assuming they will file their taxes jointly using form 1040EZ:
    a. What is their adjusted gross income?
    b. What is their taxable income?
    c. What is their Federal income tax due?
    d. What is their marginal tax rate?
    e. What is their effective tax rate?
10. Describe the trade off between receiving early Social Security benefits, normal benefits, and delayed benefits.
11. Explain, by drawing a graph (or giving a numerical example), how taxes are a disincentive to savings. For simplicity, you may assume a two-period life-cycle.
12. Write the equation for a two-period lifetime budget constraint in a world with taxes. Explain the implications of taxes on asset income on trying to smooth consumption. How do we solve this kind of problem and why does it work (hint, we did this in Excel)?

True or False, and Explain Why

13. Saving money will increase your taxes.
14. All taxes collected on wage income are progressive.

CPSIA information can be obtained
at www.ICGtesting.com
Printed in the USA
LVHW012017301121
704746LV00009B/812